Sustainable Design for t Environment

Sustainable Design for the Built Environment marks the transition of sustainable design from a specialty service to the mainstream approach for creating a healthy and resilient built environment. This groundbreaking and transformative book introduces sustainable design in a clear, concise, easy-to-read format. This book takes the reader deep into the foundations of sustainable design and creates a holistic and integrative approach addressing the social, cultural, ecological, and aesthetic aspects in addition to the typical performance-driven goals.

The first section of the book is themed around the origins, principles, and frameworks of sustainable design aimed at inspiring a deeper, broader, and more inclusive view of sustainability. The second section examines strategies such as biophilia and biomimicry, adaptation and resilience, health and well-being. The third section examines the application of sustainability principles from the global, urban, district, building, and human scale, illustrating how a systems thinking approach allows sustainable design to span the context of time, space, and varied perspectives.

This textbook is intended to inspire a new vision for the future that unites human activity with natural processes to form a regenerative, coevolutionary model for sustainable design. By allowing the reader an insightful look into the history, motivations, and values of sustainable design, they begin to see sustainable design, not only as a way to deliver green buildings, but as a comprehensive and transformative meta-framework that is so needed in every sector of society. Supported by extensive online resources including videos and PowerPoints for each chapter, this book will be an essential reading for students of sustainability and sustainable design.

Rob Fleming is the Salaman Family Chair in Sustainable Design and Professor and Director of the MS in Sustainable Design Program at Philadelphia University, USA.

Saglinda H Roberts is currently an Assistant Professor in the Michael Graves College at Kean University, USA, and has over 30 years extensive design experience with a large variety of projects.

Sustainable Design for the Built Environment

Rob Fleming and Saglinda H Roberts

Routledge
Taylor & Francis Group

LONDON AND NEW YORK

First published 2019
by Routledge
2 Park Square, Milton Park, Abingdon, Oxon OX14 4RN

and by Routledge
52 Vanderbilt Avenue, New York, NY 10017

Routledge is an imprint of the Taylor & Francis Group, an informa business

British Library Cataloguing-in-Publication Data
A catalogue record for this book is available from the British Library

Library of Congress Cataloging-in-Publication Data
A catalog record has been requested for this book

ISBN: 978-1-138-06617-5 (hbk)
ISBN: 978-1-138-06618-2 (pbk)
ISBN: 978-1-315-15930-0 (ebk)

Typeset in Baskerville
by codeMantra

Visit the companion website: www.routledge.com/cw/fleming

Contents

Figures

Tables

Acknowledgments

There were so many people that played a role in getting this book published. Shane Clark was a graduate assistant who provided much needed research, patience with the authors, and words of encouragement. JP Bhullar contributed her creativity and illustrator skills to design the Table of Sustainable Design Elements. Laura Parisi was the one who kept the office functioning while the writing was taking place. Laur Hesse Fisher offered critical help at just the right time on the Global Sustainable Design Chapter. Jazmin Toledo was the hero who took on the tedious task of sourcing a large portion of the book. A special thanks is on order for the 2018 and 2019 graduating classes of the MS in Sustainable Design Program at Jefferson University who made many sacrifices so that this book could be completed. Howard Ways, a longtime friend and colleague at Jefferson University, is responsible for the use of the word "Place" as the fourth "P" in the Quadruple Bottom Line. Jeffrey Zarnoch's persistent encouragement was always welcome. Thanks to the Healthcare Studio at Kean University for vetting the Bio-inspired Design principles and to Aidan Kleckner for acting as our enthusiastic student reader.

Thanks and appreciation to all the colleagues, friends, and family that tirelessly encouraged and inspired us on this long journey.

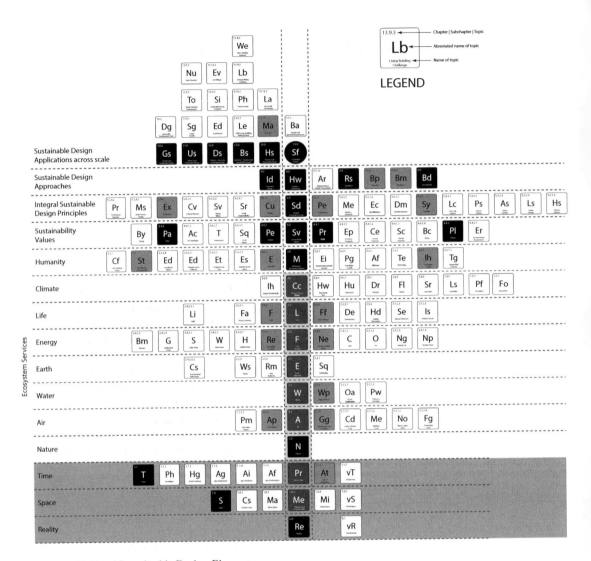

Figure 0.1a Table of Sustainable Design Elements.
Source: Created and drawn by the authors.

The table of sustainable design elements

How do you write a book about everything? After all, sustainability is an all-encompassing set of values that change the way we think about problems and help to discover new opportunities to make the world a better place – not just for today, but for future generations as well. That sounds great, but the realm of sustainable design is made richer, but also more complex, by the interdependence of each topic that needs be covered in this book. In the end, everything is connected, so we decided to start with a "Table of Sustainable Design Elements", not unlike the Periodic Table that scientists use to capture a comprehensive list of the elements. This became our *playing field* to make sure that we could include and explore all the major ideas, concepts, frameworks, and approaches for sustainable design. Later, we found ourselves drawn to the metaphor of a tree – its roots, trunk, and canopy signifying a particular aspect of sustainable design. In Figure 0.1a, the "roots" of sustainable design can be found in the ability to think across time and space in meaningful and powerful ways. The trunk of the tree represents nature itself, a sturdy and consistent force that holds up the entire top of the tree which is humanity. We are the latest and greatest species on earth capable of so much good, but also able to so much destruction. Humanity is completely reliant upon the "trunk" of nature, but once cut, the entire tree collapses, and with it, the entire human race. We are so far up in the "tree canopy," looking out over the horizon for new opportunities and new levels of happiness, that we overlook just how vulnerable our existence has become. And this is why sustainable design has arrived at just the right moment, a time of massive climatic changes, significant world view shifts, and remarkably powerful technologies. The opportunity to repair and eventually regenerate nature is within our grasp, and so the top of tree offers a wide array of strategies and concepts aimed at achieving a sustainable future.

The table also became the Table of Contents for the book and allowed us to discover a *nested ordering system* for the chapters and subchapters, thereby allowing a nonlinear organization of information that better reflects the networked reality we live in today. The number of each "tile" in the table corresponds to its location in the book and to an important topic within the realm of sustainable design. For example, the topic of the Living Building Challenge, a rating system for regenerative buildings, is covered in Section 13.9.3. The number **13** is referred to Chapter 13 which covers sustainable design at the building scale. The number **9** is referred to Chapter 9 where rating systems are discussed. This is how information in the book is reintegrated into later sections. The number **3** is referred to the third example shared in the section. In this case, the first two examples happen to be the LEED Rating System (13.9.1) and Passive House (13.9.2). As you become more comfortable with the book, the numbering system should be helpful to those who want to come back at a later date and retrieve an important link or a useful case study.

Figure 0.1b QR code for the Table of Sustainable Design Elements.

Although the "tree" metaphor has overall been very useful, there are a few downsides. First, there is no real meaning behind the right and left sides of the trunk. We tried to organize information differently on each side, but with not much success. Also, the tiles don't necessarily relate to each other vertically, hence the use of dashed lines between the rows. Some happy accidents included the location of climate change at the critical juncture between the "trunk" of nature and the "canopy" of humanity. If the climate fails, the whole top of the trees falls. To explore the "map" on your smartphone, simply use your smart-phone's QR code function and scan the image above.

You can zoom in and out and around the image to study the "Table of Contents" for this book. Or, the companion site at the web address below will provide a digital image of the table suitable for printing. On the companion website, other useful tools and resources are available that will augment your experience with this book.

Resource

Companion website: www.routledge.com/cw/fleming

Preface

Imagine being a passenger on the Titanic cruise ship. You are the only one on board who knows that the ship is headed for a catastrophic collision with an iceberg. You scream at the top of your lungs begging the crew to turn the ship. You run around the deck flailing your arms hoping someone will take notice. Alas, no one listens because the fog is so thick that there is no way to see what is ahead. No one "believes" you. Not because they distrust you, but because the captain told everyone onboard that the ship is unsinkable and that there was nothing to fear. The band played on, and champagne and caviar were served. Later that night, the ship did, indeed, crash into the iceberg, and, well, you know the rest.

The metaphor is quite simple. The ship represents nature, a strong foundation upon which to build a civilization. It's only in the last 60 years that nature was revealed to be far more vulnerable, and fragile than we thought. The "fog" in the "story" is our own mental state, our collective worldview that is so focused on obtaining extreme comfort, that we are blinded to the environmental threats ahead. We know deep down that the "iceberg" of human-induced climate change is right in front of us and that the longer we wait to take definitive action, the less time we have to change our ways.

This book will surprise you. Yes, there is plenty to fear about the future, but there are also many people and organizations working to "turn the ship." You will find yourself not only becoming hopeful, but also deciding to join in and help. Sustainable design offers not only a pathway to a healthy and resilient future, but also to finding purpose in life, a grand mission that will sustain your efforts far into the future.

A new kind of book

This book is horizontal. We chose to focus on the relationship between the parts of sustainable design rather than the parts themselves – which can be easily found on the internet. In other words, we will not be talking much about how to orient a building for passive solar gain, or how to select a rapidly renewable material. There are so many books out there already that accomplish these tasks quite well. This book is part essay, and part resource list. In many cases, we tried to provide useful links at the end of many of the sections in the book. Pictures are better viewed online rather than here in this black-and-white book. The information is also organized in a "nested" format, meaning that there may be moments when the information is repeated from an earlier chapter. This is not a mistake as there is an intention for each chapter to stand alone but also to tie into the larger set of chapters.

Target audience

This book is written for you, the student, the person taking their first class in sustainable design or beginning their journey to a new career. To that end, the book is written in simple, easy-to-follow sentences and is organized in a way that will allow you to find information quickly when needed later. For the advanced sustainability expert, or for the professor, you may find this book too simple. Try to keep in mind that the student is the target. Try to assume the perspective of your students, who want something easy to read and even easier to understand. As the professor, you can bring in your own perspectives and materials to enrich the learning experience of the students.

Disclaimers

Furthermore, our upbringing and education as Americans means that we have a rather provincial view. We ask that you translate the concepts to your own context and forgive our reliance on too many examples from the U.S. As authors, we believe that we are presenting the information in a neutral way without any bias, which, of course, is very difficult, however. The truth is our backgrounds as an interior designer and an architect will always influence how we think and write. We know, deep down, that landscape architecture, urban design, and planning are just as critical, if not more so, than building design, but, alas, you will notice the book remains building-centric. Nevertheless, the power of this book is the engagement of sustainable design at multiple scales: from the macro to the microscale, thereby opening the metaphoric doors and windows to new and different ways to think about design. Ultimately, this book will build bridges between the disciplines to discover new and powerful solutions to the challenges posed by climate change and other wicked problems of the 21st century. An urban designer must work with an interior designer. A city planner must collaborate with an architect. An industrial designer must work with medical professionals. Politicians must work with community planners. The problems and opportunities we see today are best attacked through deeply collaborative transdisciplinary design processes.

This book is written with the full knowledge of the current plight of many people all over the world and that reading about sustainability as a far off utopia might be painful. We seek a revolutionary worldview shift that will address basic fundamental needs in society like clean air and water and affordable food, now and in the future. For more information on this, see Chapter 10, which covers the sustainable development goals developed by the United Nations.

Finally, we have had the pleasure of teaching sustainable design to students from around the world, from different religious backgrounds, and from different cultures. This book relies on a secular approach to history with the full respect for religious worldviews that contradict science-based history. In the end, regardless of the background, it is hoped that this book will help to pave our way to finding a healthy, resilient, and sustainable future.

The companion website

The companion website will include, among other things, some video lectures and course materials such as quizzes and assignments, and there will be an opportunity to share feedback on the various points made in the book and to suggest corrections that can be implemented in the second edition. A full-size PDF poster of the Table of Sustainable Design Elements is available for download on the site. Also notice that the book is broken up into 15 chapters in order to correspond to a typical 15-week semester. Assignments and quizzes will align with this structure to create a good flow of learning and doing during the semester.

Introduction

Studying the earth over vast periods of time, we realize that the planet and its climate have always been changing. Four billion years ago, life appeared on earth. Only 300 million years ago, the great upheaval of the Pangea occurred, when a single landmass split apart to become the continents as we know them today. Sixty million years ago, a giant meteor struck the planet ending the long and glorious reign of the dinosaurs, ushering in the age of mammals. Early versions of humans first appeared *only* six million years ago and lived in trees. It was then that the climate changed causing the great forests to dissipate and the emergence of the great tall grass savannas of Africa. Humans began standing upright and walked on two legs, which allowed for better hunting and the use of their hands for manipulating tools. One million years ago, humans first started controlling fire, cooking their food, and their brains accelerated in both size and complexity. Eventually, humans evolved to become what we know today as the hunter-gatherer, humanity's first and most successful adaptation, occupying at least 90% of human history (Lee and Daly 1999). The nomadic hunter-gatherers, the original humans, citizens of the last ice age, intuitively practiced sustainable design because they had no choice. Their lives depended upon finding their niche in the local ecosystem in a cooperative partnership with the natural world around them.

Then came the Age of Agriculture. It was brought on by, of all things, climate change. As the planet came out of an ice age 12,000 years ago, the planet warmed, and the seemingly permanent massive glaciers retreated, creating the ideal climate and soil for growing food. Farmers settled down and began to "manage" the natural world through animal husbandry, irrigation, and hybridization of plants – the first genetic engineers. More importantly, with a sedentary lifestyle, the population explosion we see today was now fully underway – eventually causing great strain on the planet's resources.

Having mastered agriculture, the early industrialists developed wind and water mills, greatly increasing yields from flour and lumber mills. Sailing ships were used to map and classify the natural world, leading to amazing discoveries but also centuries of colonization, enslavement, and the beginnings of widespread resource extraction and pollution. Eventually, powerful and portable fossil fuels were discovered and placed into service to finally "conquer nature" – leading to riches beyond imagination and the establishment of the great modern cultures. At the same time, the first instances of contemporary ideals about rights begin to take shape. Slavery was ended, but the long journey towards the formation of a truly equitable society is still absent today as self-interested and short-term thinking continues to hold power.

It wasn't until Rachel Carson's *Silent Spring*, in the Information Age that the world started to see the damage wrought by industrialism. The 1960s saw the blossoming of the great empathic forces, causing the birth of the environmental movement as a response to the severe pollution wrought that occurred as the result of industrialism and materialism. By the late 20th century, human-induced climate change was becoming more and more evident. "Saving nature" became the mantra of progressively minded individuals.

And here we stand today, heavily influenced by climate change, at the dawn of a new worldview, in a time when the rise of renewable energy and social media are helping to build the platform of the next great leap in human consciousness. Sustainability, as we will see, is both the definer and driver of a new approach in thinking. We are seeing a new view of "integration" emerging, where humanity is striving towards forming a symbiotic relationship with the natural world.

At the very foundation of sustainability, we find ourselves in the middle of the eternal battle between self-interest and cognitive empathy – two invisible hands that govern our intuitive behaviors. The first invisible hand compels us to maximize profit, to do WELL. In this view, progress is connected to our drive for survival, security, and comfort. This driver is responsible for the quality of life that many first-world people experience. It catalyzes a culture of technical innovation, artistic wonder, and medical advancements. But the price to be paid for that "progress" is very high as undeveloped nations are the unwilling victims, having to deal with the legacy of colonialism: slavery, natural resource exploitation, and stolen land. In first-world countries, the underprivileged bear the brunt of progress as they toil long hours for low wages in poor conditions so that others can maximize profit. The second invisible hand, fueled by cognitive empathy, compels us to seek equity and justice in all situations, to do GOOD, and to be an agent of change for all stakeholders including the natural world itself. Empathy can do more, enabling us to take the perspective of those who are living far away and suffering from the effects of climate change, or even those future global citizens who have yet to be born. Empathy across time and space lies at the roots of achieving sustainability offering a second pathway for short-term profit seekers.

Traditionally, these two hands have been placed in opposition, a dichotomy that has held back authentic progress – a moment when both hands work together for a better future. Sustainability, as we will learn, has the potential to be a resolving force, a profoundly new paradigm for humanity. Sustainability began to take its definitive form in 1987 when the Brundtland Commission of the United Nations defined sustainable development in its groundbreaking work: Our Common Future. This document is a declaration of interdependence, unifying the motivation for profit (the first invisible hand) and the motivation for environmental protection and social equity (the second invisible hand). Eventually, John Elkington brought shape and form to the efforts through the development of the Triple Bottom Line, a powerful "accounting" mechanism for corporations and governments to measure their sustainability efforts. The three "P"s of sustainability – People, Profit, and Planet – are used as shorthand to define sustainability.

Accounting methods are useful, but they don't offer much inspiration to the artists and creative professionals – the people trying to use their imagination, not only to solve challenging social and technical problems, but also to provide an experience of delight, wonder, majesty, mystery, and social critique. The role of the designer and artist in the formation of sustainability is essential, if we are to create environments that people actually love, cherish, and sustain. Lance Hosey in his book *The Shape of Green* advocates for beauty in the sustainability bottom line. In this book, he states, "If it's not beautiful, it's not sustainable." Therefore, the Triple Bottom Line needs an additional fourth bottom line of "Place" to the triad of People, Profit, and Planet. Now, a more holistic quadruple bottom line for sustainable design is offered with a more exciting, thought-provoking, and engaging set of values for design.

Under the structure of Mark Dekay's Integral Sustainable Design, the Quadruple Bottom Line is placed into design practice by using integral theory as a basis for a structured, comprehensive, and more holistic approach to design. It uses a modified version of the four perspectives of Performance, Systems, Culture, and Experience as a way to help any person involved in a project to assume divergent points of view. It helps to provide a much-needed mental picture of sustainable design and a roadmap for transdisciplinary design, and also helps to build empathy among traditionally competing professions.

The Performance Perspective opens the way to include energy and resource performance as well as financial profit. The Systems Perspective allows for thinking about the interconnection between ecological, technical, and social systems in order to discover innovative solutions to difficult sustainability problems – such as climate change. On the subjective side, the Experience Perspective levels the playing field for artists and designers, who can now work side by side with engineers, builders, and accountants, to uncover uniquely beautiful and efficient solutions to all kinds of challenges. Finally, the Culture Perspective brings the ethical dimension directly into the heart of a project, generating important discussions around social equity and cultural expression.

Layered on top of the foundation, a range of new methods, concepts, and approaches have gained traction. It's a sign of the maturity of the sustainability movement. Bio-inspired design is now a major driver of innovation and more importantly a vehicle for co-creation with nature itself. Biophilic design, for example, starts with the hypothesis that according to EO Wilson, "humans possess an innate tendency to seek connections with nature and other forms of life." Biophilic designers seek to satisfy the basic human urge to affiliate with other forms of life. Biomimicry is an approach where the design team asks the question: What would nature do? Nature has been solving design problems for over three billion years. Many amazing sustainable products have been created by imitating models, systems, and elements found in nature.

Health and wellness has emerged as a second major driver of sustainable design. Design now becomes a healing process where evidence-based design methodologies are used to create spaces and places that improve human health and well-being – a radical change in focus for the design community. While human health is central to this approach, we are also tied to the overall health of the planet. Therefore, attacking human-induced climate change, reducing pollution, and working with nature is also a healing process – just at a larger sale.

Finally, many sustainable designers have shifted their focus from fighting climate change to preparing for the onslaught of environmental threats that are already happening and will continue to happen in a greater frequency due to human-induced climate change. Resilient designers ask how can we design buildings, sites, and communities to bounce back from the effects of hurricanes, flooding, and abnormally hot temperatures and other unknown "shocks" to our systems. Furthermore, this approach seeks to develop adaptive approaches to design so that even the long-term threats of sea level rise and drought are able to be dealt with in a nonemergency function.

All of these new approaches along with traditional sustainable design methods are challenging and require new design processes to help us think and design differently. The integrated design process – also known as concurrent design, or integrative design, or participatory design – is the key to unlocking the full potential of sustainable design. The stereotypical designer working alone and unleashing their design masterpieces is replaced by a team of designers from all disciplines, and all stakeholders who might be impacted by the design. In addition, the natural world itself is recognized as a stakeholder so that design decisions are understood within an environmental context. The all-important shared sustainability values are facilitated by the sustainable designers who organize and deliver *design charrettes* to collectively tackle the most pressing problems and the greatest opportunities on a design project.

The types of projects attacked by sustainable designers range from the global scale to the microscale where diseases are prevented through a health and wellness approach to design. At the global scale, climate change, sea level rise, desertification, and deforestation are all targets of sustainable development. Sustainable designers work hard at this scale to bring about real change. The United Nations released its 17 sustainable development goals, with "fighting poverty" as one of the key drivers of sustainable design, underscoring the importance of social equity in the overall movement towards a sustainable future.

At the regional scale, we reimagine entire cities as holistic self-supporting urban ecologies by redesigning transportation, energy infrastructure, building design, and urban parks. We try to imagine an entire city as self-sufficient where all its resources and waste are procured and recycled locally.

At the district scale, sustainable design is used to help communities to share water and solar resources, while also reinforcing the local economy by specifying local materials and training local laborers. Entire communities can become a microcosm of the sustainable city, where all of its needs for food, water, energy, open space, and more are met within walking distance. Third-party standards and rating systems help design teams and stakeholders achieve these goals. EcoDistricts, 2030 Districts, and LEED for Neighborhood Development are a few examples. Sustainable designers are good at creating and using these systems.

At the site scale, holistic frameworks of sustainability values are placed into practice to create sites that are ecologically resilient as supported by technical infrastructure and restorative ecological practices. But the site is also the place where we directly interact with the natural world offering an experience vital to our physical needs for health and our social needs for a sense of place. Rating systems such as the Sustainable Sites Initiative are used to help guide the design effort and check to make sure that we reach high levels of performance.

At the building and interiors scale, sustainable designers focus on creating "living buildings" that require no outside energy, water, or materials to support its function – the first examples of a co-creative partnership with nature itself to provide for all the needs of the building users. Other sustainable designers use the LEED Rating System or Passive House to guide their efforts towards high-performing projects.

At the scale of the human, eco-design strategies help to create better products. Sustainable designers work on new kinds of energy systems or new water purification systems, or develop and use rating systems like Cradle to Cradle to make sure that the users of our buildings are not exposed to toxins and that our materials do not end up in landfills.

Lastly, we zoom in further going inside our own bodies, and we start to think about how design can contribute to making us healthy and strong. We can use design to achieve health metrics by increasing access to fresh air, daylight, healthy food, and biophilic elements. We begin to make life-enhancing design decisions. The Well Building Standard is a tool to help make sure that all our goals are met at very high level.

So that's it, the entire spectrum of sustainable design and the entire introduction to this book. If you have read this far, there is no turning back now – you have been invited into the rarified realm of sustainable design. On one level, it's not any different from regular design. There are still drawings and models to be made, ideas to be explored, and design solutions to be developed, but this time there is an intentionality to the work – a conscious effort to equally represent all of the values of sustainability: environmental, social, economic, and experiential. Empathy lies at the very core of this approach – making decisions that affect people far away and far off into the future. Ultimately, this requires a great leap in consciousness, and hopefully this book will provide a much-needed springboard into the sustainable future.

Reference

Lee, R.B. and Daly, R. Eds., 1999. *The Cambridge Encyclopedia of Hunters & Gatherers*. 1st edn. Cambridge: Cambridge Press.

1 Space, time, and sustainable design

Sustainable designers think very deeply about the design process. This is no surprise since the goal of the sustainable designer is to seek an alternative reality – a new operating platform for the human race. This new platform must be radically different from today's models of design practice if we are to find our way out of the current climate predicament. After all, it would be unrealistic to expect that we can solve today's problems with the same knowledge base and worldviews from the past. Discovering a new model for design requires a deep dive into the foundations of reality itself. There we find the space-time continuum, the basis of human perception in physical space. We begin to discover new ways of seeing the world, and ultimately to developing new design skill sets that will empower sustainable designers to better attack the world's most pressing problems. This chapter will begin by studying space and then time to uncover useful principles for sustainable design and to develop a deeply important "mental model" of reality – a foundation for transformative sustainable design.

1.0 Space

Thinking across space is one of the fundamental skill sets of a sustainable designer. The simplest way to think about space and sustainable design is through the simple, but meaningful phrase: Think globally, act locally. The desire to reduce global pollution and to specify local materials indicates an ability to understand how specific design decisions impact the planet at different scales. The following passages are intended to break down the salient aspects of each step in the scale of space.

1.0.1 Cosmic scale

Cosmic space is infinite, the product of an ever-expanding universe, unfolding, evolving, always in a state of becoming. It is a miracle to be sure, worthy of a starting point in the pursuit of learning about sustainable design. Imagine the miracle of intelligent life. We are the only sentient beings within millions of miles of space. The chances of our existence are 1 in 100 billion (Siegel 2017). We are a speck of dust on a speck of dust on a speck of dust somewhere in the vastness of the galaxy.

Think for a moment about an alien race arriving on planet earth. After traveling millions of light years, they would be overjoyed to discover a planet with sentient life. They would see the verdant beauty of the planet and marvel at the azure waters. They would see rich and beautiful cultures, amazing monuments and buildings – a truly progressive civilization. The aliens would be impressed at first and quite proud of the human race. Upon closer inspection, they would see the dark underbelly of progress: the destruction of the natural landscape, the extinction of millions of species, and the climate altered in significant ways. And they would see countless examples of humanity's sad legacy of enslavement, oppression, and discrimination. This is the scale of space that sustainable designers occupy from time to time, seeing the planet in its entirety as one organism, one place where everything is connected. The magnitude of even the smallest design decisions, like using a plastic straw, when multiplied by the millions of other similar decisions, have significant impacts at this scale. Today, the aliens would see our oceans destroyed by the careless use of disposable plastic – the ultimate convenience for us, but a death sentence to the oceans, and ultimately to our own species.

1.0.2 Macro space

In 1968, astronauts looked back on the planet earth for the first time, and saw a big beautiful blue marble. They saw the planet in its entirety, not as a lifeless collection of minerals, but as a complex, interdependent, lively ecosystem. This is the macroscale of space, and it is critical for sustainable designers to be able to conceive of the vastness of the planet and predict the impacts of design solutions to far-away places. Climate change is apolitical: It doesn't observe borders. Rivers link different countries together and form the borders between others. Imagine the Colorado River, originating in the Rockies and slowly finding its way towards Mexico and the Pacific Ocean. Along the way, millions of gallons of water are extracted to irrigate crops, dammed to produce energy, and used for recreation. The Colorado River rarely reaches the Pacific Ocean anymore (Postel 2014), dying a slow death after it crosses the U.S. border into Mexico in a small, sad puddle of muddy water.

At the north pole, unusually warm temperatures are melting arctic ice at a record pace, causing village structures to literally sink into the ground and polar bears to move south in search of food (Derocher et al. 2018). In the rainforests of Brazil, the forests are burned daily to make space for cattle ranching, replacing the oxygenation process of photosynthesis with carbon dioxide emissions and eventually methane emissions from millions of cows soon to be slaughtered for steak dinners. The great rivers of the Yangtze in China and the Ganges in India are fed by the spring melt of winter snow packs in the Himalayas. But the glaciers are retreating and the snow packs are diminishing, threatening the water supply for millions of people. The macroscale enables the designer to imagine the impacts of their decisions at the global scale, a frightening but empowering point of view. While it's true that the negative impacts of our design decisions are multiplied by the thousands each day, plunging the global ecology into a downward spiral, there are literally millions of people around the world working to turn the table. By joining together to develop new frameworks, new design approaches, and a whole range of solutions, sustainable designers have clear pathways to engage and attack global issues at the scale of their projects, thereby changing the very face of the planet. Passive House (13.9.2) is one of the many rating systems that are transforming design and construction practices to align with global solutions for human-induced climate change. Resilient design has also emerged, asking design teams to consider how projects can adapt to changing conditions and bounce back from shocks like hurricanes, forest fires, and floods.

1.0.3 Mezzo space

The Mezzo scale is *our* space, the scale of the human. We perceive space all around us, and we are shaped by it in a myriad of ways. Our experiences and ultimately our happiness and survival depend on our manipulation of objects in the human scale for shelter, food, clothing, and all the things that make life worth living. The shape, form, and materiality of our buildings, interiors, textiles, and products are the result of countless hours of effort by the designer seeking to maximize the experience for end users. Artistic expression in many forms is embedded in the process, and without that, we are left with a life devoid of joy, mystery, and happiness. Design matters, but sustainable design matters more. The macro and microscales compel us to think beyond the human scale, to predict the impacts on human health now and in the future. Regional ecologies, the backbone of our living systems, rely upon us to think at the macroscale to make decisions that are restorative rather than destructive. And yet, many of us remain fixated on the human scale spending out time shaping and making beautiful spaces and objects to deepen the human experience. What if we could figure out a way to keep the traditional, aesthetic goals of design at the mezzo scale, but overlay the awareness of macro- and microscales to find the ways to address the big problems of today and tomorrow? That is exactly what sustainable design is all about.

1.0.4 Micro space

Micro space is all around us, and perhaps more importantly inside us. Just as the planet is made up of nested ecosystems, inside our bodies there are literally millions of tiny ecologies all working together to make us a whole and healthy person. Design impacts this scale in so many important ways, from the psychology of how we experience space, to the quality of the air we breathe, and to the types of nutrients that enter our digestive system. At the microscale, life itself depends on the quality of interactions that occur inside our bodies and in the environment around us. Doctors and scientists spend their lives understanding this scale, using microscopes to reveal typically invisible ecologies that form the basis for human health. The union of design and science is the key to unlocking this scale as a new frontier for the creative community, and new partnerships with the medical professions are now available. Zooming in to an even finer scale, we find ourselves in quantum space, a molecular reality of electrons, protons, and quarks. These form the building blocks of life itself. Surprisingly, designers are at work at this scale, especially in the way light is manipulated to create a "vibration" in the spaces and places we inhabit. The more we zoom in and the deeper we consider this scale, the more we discover nonlinear and, as yet, inexplicable aspects of reality.

1.0.5 Virtual space

This type of space is going to become critical in our movement towards a sustainable future. With the number of humans on the planet increasing, living space dwindling, and a failing environment, more and more of our existence will take place in virtual space. Whether this is a good or bad thing from a moral perspective is a larger discussion, but we can already see a future where entire virtual buildings and communities will be used for human purposes such as meetings, education, the creation of art, and more.

It's hard to imagine just how far the digital world has come and how quickly innovation is now occurring. The great cacophony of data streaming in cables and in the air holds the key to an "information ecology" that will transform life as we know it and that holds the very real potential to fighting climate change by dematerializing physical matter typically associated with all the goods and services we need to survive.

The amount of energy and technology used to support virtual space is still quite high, and the devices themselves we use to interact digitally have their own ethical issues and environmental impacts, especially the rare earth metals required for fabrication. But virtual space, and by default virtual time, can reduce our impacts on the planet in very real ways – one of the many pathways to a sustainable future. Buckminster Fuller, one of the first contemporary sustainable designers, promoted a basic axiom to "do more with less." Mr. Fuller understood the opportunities to deliver goods and services with a greatly reduced impact, something he called ephemeralization, a large word for a special kind of alchemy that defines one of the most powerful sustainable design approaches. At a more pragmatic level, virtual space allows us to "simulate" proposed sustainable buildings and measure their performance over virtual time. This allows for the prediction of a proposed design project's environmental and energy performance. Finally, with the advent of 3D printers, the connection back to physical space and objects is reached, creating a blurred world between physical, augmented, and virtual reality.

Thinking across space is a critical foundational skill set necessary for effective sustainable design. Scales are nested within each other, all interacting at all times to define the quality of our reality. Sustainable design relies upon that understanding in its formation and application. Right now, many see sustainable design as a compliance path, as in, "I have to do these things to get my new project certified" as green or sustainable. That is a necessary but limiting approach and typically leads to resentment and backlash against sustainability. It's better to learn how to see the world at all of its scales and understand how your decisions fit into the great picture. That is what makes sustainable design so exciting.

Additional resources

2030 Palette, http://2030palette.org/
Architecture 2030, http://architecture2030.org/
The Power of Ten: Charles and Ray Eames, www.youtube.com/watch?v=0fKBhvDjuy0

1.1 Time

There is a special relationship between time and sustainable design. The environmental impacts from the past are felt in the present now more than ever. The future looms over us, threatening us with images of dystopia, climate collapse, and a lower quality of life. The present offers us a unique opportunity to dwell in the time continuum and consider how we got to this point in history, and it also allows us to begin to chart a course towards the sustainable future. Sustainable designers are very good at looking into the distant past to uncover tried and true passive strategies for cooling, heating, and more. Thinking over the long term is also a specialized skill set with the impacts of a single decision or performance simulation of an entire building.

Worldview shifts over time

We can view the past as a series of hyper-accelerated jumps in evolution, as shifts in worldview, or in how the collective or dominant culture views itself in the cosmos. This has great importance to how we understand sustainability, because the way we frame our views to nature, and the way we use technology, and the way we communicate determine the essential aspects of a worldview.

Perception of reality changes over time

Ken Wilber, the founder of Integral Theory, one of the basic frameworks that will be used in this book (1977), believes that consciousness is constantly evolving, and that as a person becomes more inclusive in one's understanding, consciousness will rise to a higher level. The same is true of entire societies. As societies evolve, their consciousness changes and with that comes new worldviews. Wilber (1977) argues that society is heading towards a period he calls "the integral consciousness period." For this book, we are using the term Age of Integration to reflect that. Dramatic changes in climate, new energy sources, and new communications have been shown to drive upward shifts in human consciousness and societal organization (Rifkin 2009).

Nested time

Like space, time is also "nested." The transition from one worldview to the next does not mean an end and beginning. Rather, the previous worldview is carried forward and nested inside the next worldview, sometimes playing a prominent role in shaping society and sometimes lying dormant waiting to be rediscovered by a future generation. The hunter-gatherer worldview, with its connection to nature and integrated model of living, is just now finding its way back into the corners of mainstream thinking about sustainability, especially through the application of bio-inspired design. The Age of Agriculture, dominated by monotheism and "traditional" values, continues to play a critical role in national politics in many countries, even though the world consciousness has moved on through industrialism, information, and soon into the next worldview of integration. Understanding nested time is a critical foundational skill set for sustainable designers because the necessity of collaboration across disciplines and stakeholder engagement to meet ambitious environmental goals on a project is often complicated by team members operating out of different worldviews. Some members are still transfixed with the industrial motivation of maximizing profit, thereby finding every way to save money on a project without considering the ecosystem or the well-being of building users. Others are focused on the social equity of a project, making sure that everyone benefits, while still others are focused on the integration with natural world, finding ways to build a regenerative project. The complex interaction of all of these competing fundamental drivers of behavior is what makes a project so interesting and so difficult to manage. Sustainable design forces people who are still operating out of old worldviews into a very uncomfortable position of having to think differently – and that, forms one of the great challenges for sustainable design practice in an emerging worldview.

Time scale: seasons, days, and hours

Sustainable designers are interested in understanding the passage of time via different seasons. Landscape architects already think across seasons because the plants they specify have cycles that change with the seasons. But architecture and interior design are often thought of as "frozen music" (Goethe and Eckerman 2010) as in a fixed solution for a fixed moment in time. Because we want to reduce our reliance on mechanical solutions for heating, cooling, and lighting buildings, we have to understand how the climate changes over the course of a year and find ways to harvest resources such as sunlight, wind, and water. We can then construct our buildings in ways that take advantage of those changes in temperature, sun angle, wind direction, precipitation, and more. Finally, by analyzing a project as it passes through all its daily and seasonal cycles, we allow unique opportunities to be revealed which allow us to harvest more energy, integrate with natural systems at a deeper level, create more cultural vitality, and inspire a sense of beauty and place. Thinking across time leads to new discoveries in the design process – especially opportunities to work with nature, save energy, and create connections.

1.1.1 Prehuman history

The Pangea

The current geographic makeup of the continents has not always been this way. A quick look through history reveals that *only* 175 million years ago, the continents as we know them were completely joined into the Pangea (Our Planet 2018). The word "quick" is used to denote the relatively short amount of time in relation to the 3.5 billion years ago when the planet was first oxygenated.

The joining and splitting apart of continents is part of a very long cycle that repeats roughly every 300 million years. The Pangea broke apart to form the continents as we know them today in three stages. Why does this matter for sustainable designers? After all, it was a long time ago by human standards, but evidence suggests that the impact of the event led to widespread extinction of species and is rated as one of the top five extinction periods in the planet's history (BBC 2018). 90% of marine life and 70% of terrestrial species were lost due to changing climate including higher CO_2 levels, volcanic eruptions, and ocean acidification (Sahney and Benton 2007). The earth then is not as permanent or as invincible as one might think, and it can, and does, go through catastrophic events that alter the geologic, biologic, and climatic history in significant ways.

The K2 giant meteor strike

Similarly, it was *only* 66 million years ago the dinosaurs experienced extinction. Imagine the time back then. It was hot, very hot by today's standards, and CO_2 levels were at least five times higher than they are now (Moskvitch 2014). The current theory states a giant meteor K2 hit the Yucatan Peninsula in what is now Mexico. It caused an intense cooling period because the soot from the impact blocked the sun interrupting our normal greenhouse gas cycle. The planet cooled by as much as 26 degrees Fahrenheit in five years (Brugger et al. 2016). Plants died first, and then the animals soon followed. It is estimated that 75% or more of all species on earth vanished. Here again, like the breakup of the Pangea, we see that the planet is vulnerable to random but powerful events and that it can change, in a geologic wink of the eye, to form a new climate with new species in a very short time period. Of course, the end of the dinosaurs ushered in a new geologic age that would be more favorable to the survival of mammals, and therefore humans as well. In working through this exploration of time and changes to the earth, there is a pattern of death and rebirth, a cycle of life that is repeated at many scales and far more often than one might imagine.

Humanity

And then we came along…

Only six million years ago, early humans (hominids) began to walk upright. Early climate change during that period forced the great forests of the period to die off, replaced by the savanna of tall grasses as the most common form of landscape (Smithsonian National Museum of Natural History 2018). The primates, who had adapted to living in the trees, slowly over the course of generations began to move to the plains. Some scientists argue that this meant that they could not see around them in the tall grasses, making it difficult to hunt and also more vulnerable to predators. So, they eventually, over many generations, began to walk upright. Other scientists argue that humans began walking upright to free up their hands to do work and use tools like spears for hunting. Either way, walking upright became the next big leap in evolution for humans, and it occurred over a very long period of time in human years and most likely as a response to a changing climate. The relationship between humankind's evolution as a species and radical changes in climate will be a recurring theme throughout this chapter.

In Figure 1.1.1a (below), it is quite clear that the temperatures have always been shifting between extremely cold ice ages and warmer periods called interglacials. Interglacials are unique periods in geologic history when the planet warms and the ice sheets melt and the glaciers retreat, and sea level rise occurs. The current Holocene interglacial is circled. Clearly, there has been a relatively long warm period, the perfect conditions for the flourishing of human life.

Time scale: thousands of years

Only 200,000 years ago, human beings as we might recognize them began to roam the planet and become the dominant species. Over that period, the climate changed quite often moving between ice ages and interglacials. Humans with their advanced brains, opposable thumbs, and upright mobility adapted to a variety of warm and cold climates.

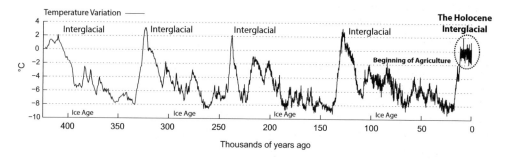

Figure 1.1.1a Variations in CO_2, temperature, and dust from the Vostok ice core over the last 450,000 years – redrawn by the authors.

Source: By Vostok-ice-core-petit.png: NOAA derivative work: Autopilot (talk) – Vostok-ice-core-petit.png, CC BY-SA 3.0, https://commons.wikimedia.org/w/index.php?curid=10684392

Time scale: hundreds of years

Next, we see the next big impact, the next metaphoric meteor – humanity and with us comes new climatic period known as the Anthropocene. As we saw before, radical unexpected changes in climate led to radical changes in the geography of the planet and by default radical changes to life on the planet – especially massive extinctions. The impact of human expansions and human-induced climate change has the potential to be just as devastating as the K2 meteor or the breakup of the Pangea. Sea level rise, ocean acidification, temperature rise, and the shifting of ocean currents were not only predicted years ago by scientists but are actually occurring in real time right before our eyes. We are in the middle of one of the most impactful periods of climatic change. CO_2 levels are at their highest point since 800,000 years ago – 410 parts per million (Kahn 2017). We are in one of the six largest extinction events in all of planetary history. While humanity thrives, an apocalypse for plants and animals is in full effect. Our species is still expanding at a very fast pace with all the earth's resources being used at an alarming rate. In short, just as a giant meteor wiped out the dinosaurs, we are the "giant meteor" that is killing everything around us except for our own species. So, where did we go wrong? After all, we are intelligent, compassionate species. The chapter will end by trying to answer this question. We will move through history briefly examining how our views of nature changed depending on our worldview – for better or worse. Ultimately, our views, or values, drive our behaviors (Figure 1.1.1b).

Resources

The real reasons why we walk on two legs, and not four, BBC www.bbc.com/earth/story/20161209-the-real-reasons-why-we-walk-on-two-legs-and-not-four

World Population Clock, The United States Census Bureau, www.census.gov/popclock/world

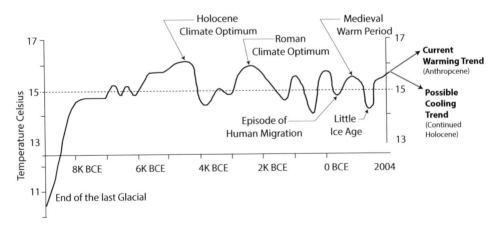

Figure 1.1.1b The Holocene climatic period. Average near surface temperatures of the northern hemisphere during the past 11,000 years complied.

Source: R. Fleming (2013).

1.1.2 *The Age of Hunter-Gatherer*

Climate change

The impact of the end of the last ice age for the human race is of paramount significance. Rising temperatures led to the glaciers receding and the emergence of more hospital climate and geography. The thawing of planet earth, coupled with the innovations listed below, meant that humanity was now poised to thrive as a species (Table 1.1.2a).

Energy

The benefits of controlled fire as an energy source are obvious. But try to begin to imagine just how much this "technology" changed conditions for humankind. Fire allowed for the cooking of food, thereby allowing for better digestion of the food, better nutrition, and the rapid development of the human brain. Fire allowed for protection from predators and insects, and allowed humans to expand their usable hours into the evenings. With more time, more security, and a more developed brain, humans could now begin to free themselves from the day-to-day struggle to survive to consider activities beyond mere day-to-day survival. Eventually, humans would develop language, art, and technology – the foundation for civilization as we know it.

Communication technology

The development of spoken language and drawings offered the first steps towards a more abstract and by default a more objective view of the natural world. The assignment of sounds to describe elements found in nature signaled the beginning of a more formal ordering of the natural world and perhaps the beginning of the end of a completely connected relationship between humankind and the natural world – the dawn of a human consciousness. The ability to make the leap in abstract thinking to draw an image of something found in the nearby environment signifies a new level of human mental development. The natural world was now being portrayed as something "other" than humankind, thereby further laying the groundwork for a new, more separated relationship between humankind and the natural world.

Table 1.1.2a Overview of the Age of the Hunter-Gatherer

Worldview	The Age of the Hunter-Gatherer
Time frame	80,000 BCE–12,000 BCE
Space	The world is so big; it can't be comprehended by human consciousness
Time pace	Slow
Climatic context	Near the end of the last ice age
Energy	Food and fire
Communications	Spoken language
Relationship to nature	Integrated
Design	Intuitive, organic, closed loops

Source: Created and drawn by the authors.

Worldview shift over time and views to nature

Animism is the belief that the natural elements, air, water, earth, fire, and plants, have a soul. This belief system suggests a deeply connected, respectful, and perhaps reverent attitude towards the natural world. Therefore, the behaviors of early humans are considered to have been in unison with the natural world (Peoples et al. 2016). However, evidence suggests that nature was altered by controlled burning of forests and other alterations to improve hunting, for example (Peoples et al. 2016). The temptation to romanticize the hunter-gatherers as the ultimate example of sustainable living, and a model for how we might live today should be avoided. However, it is more important to capture the spirit of the time rather than to literally go back to living in that manner. But the long story of humanity and its relationship to nature starts here.

1.1.3 Age of Agriculture

Early climate change

The onset of the Holocene (Figure 1.1.1b) provided a warmer and more hospitable climate for human development. The end of the long and harsh ice age, thanks to early, *natural*, global warming, aligns well with early agricultural practices and the rise of a higher-level intellectual capability of humankind. This unique climatic period marks the beginning of what is commonly referred to as "civilization." It can't be stated strongly enough that the rise of the human race and the change in climate are directly related. It is hard to imagine the great wonders of the Age: the Pyramids, the Hanging Gardens of Babylon, or the Great Colossus could have ever occurred if the earth had not warmed (Table 1.1.3a [below]).

Energy: surplus food

The warm climate also meant that growing food would become easier and the yields more bountiful. Surplus food is, indeed, energy and, along with fire, became the primary energy source in the Age of Agriculture. The innovations of irrigation, and the granary were central to the continued rise of the agricultural worldview. The evolution from basic horticulture, food gathering, and hunting led to an opportunity to live in a fixed location as opposed to the previous nomadic lifestyle of the hunter-gatherer. In addition, the innovation of animal husbandry created an even greater abundance

Table 1.1.3a Overview of the Age of Agriculture

Worldview	*The Age of Agriculture*
Time frame	12,000 BCE–1,750
Space	The world is flat and finite
Time pace	Slow
Climatic context	The Holocene – a warm planet
Energy	Surplus food
Communications	Writing
Relationship to nature	Managed
Design	Vernacular and monumental

Source: Created and drawn by the authors.

and consistency of food sources and, ultimately, a food surplus. The formula for accelerated population growth was now in place, and humanity began to flourish in earnest and expanded across the planet. But it was more than just "excess energy" contained within the food that changed society. Increased food security meant less day-to-day worry about meals, which would allow humans to pursue a more precious commodity – surplus time. Now, the freedom to pursue non-survival-based activities contributed to the establishment of class structure, more invention, new metaphysical explanations of nature, larger expressions of culture, and a new conception of the cosmos – in short, a new level of human consciousness – a new worldview.

Communication: writing

Trade routes, oral traditions, and drawings were the primary method of communication at the beginning of the Age of Agriculture, but with emergence of writing, a new level of subject/object relationship between humans and nature was formed. The number of lateral thought processes and the capacity of the human brain to think abstractly were greatly heightened. Now, an element of the natural world, like a certain kind of tree for example, is understood through a series of visual symbols, which refer to specific sounds, which, when combined, trigger a synapse in the brain that signals a recognition of that certain kind of tree. This great leap in cognitive ability also meant the next step in the objectification of the natural world. The addition of writing and the development of math were necessary to manage an increasingly complex and larger society with more diversified trade. The rise of money as a proxy for the bartering system meant that plants and animals were viewed by some as a commodity, rather than a sacred part of the ecosystem – a transition away from Animism as the guiding principle of societal morals.

Views towards nature

While the ebbs and flows of nature via the passage of the sun and moon, yearly flooding patterns, and the coming and going of the seasons served as consistent connections to nature, the Age of Agriculture marked the beginning of a new relationship between humankind and the natural world. Nature, still viewed as an all powerful force controlled by the gods, would slowly become demystified though fledgling scientific efforts. The communication methods of writing and the abstract mathematical processes meant that nature was as much a mystery to be explained or a commodity to be bought and sold, as it was a magical force imbued with spirit. The transition from an animistic, integrated conception of the natural world to an abstract one, where humans and nature exist in a duality, formed the basis of our current-day politicization of nature in the climate change debate. In any case, the Age of Agriculture firmly entrenched humans as "stewards" of nature from a positive sense, but also as future destroyers of nature. Furthermore, in an effort to expand knowledge and ownership of territory, the "heroic" colonial "explorers" inadvertently spread disease and death to the entire nations of indigenous people, as well as subjecting many others to slavery and or genocide. This underscores that a "jump" in worldview does not necessarily mean a jump in ethical behavior. In fact, the separation from nature, along with newfound technological prowess, laid the groundwork for widespread destruction of the earth in the search for resources to be used in the production of material goods.

1.1.4 Age of Industry

The Age of Industry or the Industrial Revolution as it is commonly referred to happened in two phases. The early phase mostly referred to as the Age of Enlightenment, and the later phase typically referred to as the Modern Era. Remember that each new worldview does not replace the previous one. While the early Age of Industry was blossoming, the vast majority of humankind still lived in either an agrarian society or a hunter-gatherer existence. Both forms of living still exist today (Table 1.1.4a [below]).

Energy sources

In the early part of the Age of Industry, the ability to think abstractly led to the development of bigger and better machines to harness the energy potential of wind and water through the use of windmills, sailing ships, and water wheels. The total number of wind-powered mills in Europe rose to over 200,000, and the number of waterwheels is estimated to have reached 500,000. In some cases, this increased the production at lumber or grain mills by 20-fold. Sailing technology rose dramatically, making it possible to travel longer distances at greater speeds, thereby further opening up the world to trade, colonization, imperialism, and eventually to death and exploitation of indigenous cultures around the world. This period, which predated the use of fossil fuels, led to unprecedented increase in the rate of production which led to wealth and comfort for many. An explosion of intellectual growth supported by the scientific methods and mathematical progress soon followed. The scientific method and algebra, developed in Persia, helped to reveal all of nature's secrets either through scientific curiosity or for the more pragmatic desire for profit.

The second phase of the Industrial Revolution featured the discovery of fossil fuels, and the further discovery of the steam engine meant that the agricultural consciousness based on food surplus was now giving way to a new worldview, not built on the backs of millions of slaves, but on the gears and wheels of fossil fuel-powered machines. Humanity's fundamental conception of their role and place in the world was changing dramatically – leading to a new consciousness and a new worldview.

Perhaps the U.S. Civil War best reflects the transition between one worldview and another. The South desired to continue using the "old ways" of a slavery-based production and agriculture system, whereas the North sought a "new way" based on fossil fuels and machines. This was a violent clash of worldviews, with the old agrarian worldview losing to the emerging industrialized worldview. It was only ten years after the civil war that the U.S. held the 1876 international exposition in Philadelphia which featured technological machinery from all over the world.

Table 1.1.4a Overview of the Age of the Industry

Worldview	The Age of Industry
Time frame	1750–1960
Space	The world is round
Time pace	Fast
Climatic context	The Holocene – a warm planet
Energy	Fossil fuels
Communications	Printing press, radio, telephone
Relationship to nature	Destructive
Design	Linear, open loop, toxic

Source: Created and drawn by the authors.

Communications revolution and human consciousness

While new forms of energy laid the foundation for an emerging consciousness, it was the communications technologies of the time that spread information in increasingly faster ways. The invention of the Gutenberg Press was built upon the work of over a thousand years of printing innovation in China and Korea. It freed people from the slow and laborious task of hand writing each book and opened the door to the first type of mass communication in history. The press was widely adopted and replicated. Works of fiction, religion, science, and politics were widely printed. As the cost of books fell, literacy levels rose and the spread of news, ideas, and philosophies accelerated at unprecedented rates leading to the establishment of a new worldview. Later in the early 20th century, the advent of the telegraph, telephone, and radio exponentially expanded the reach of communication technology and increasing the speed at which new ideas circulate across the planet.

Views towards nature

Up until the late in the Age of Industry, nature and religion still held considerable sway in impacting people's thoughts about nature and the meaning of existence. Even Newton, the founder of modern physics and proponent of a mechanistic view of nature, identified God as the force that put the whole machine in motion (Newton and Thayer 2010). But it wouldn't be long before nature and religion would come under direct attack. René Descartes, fueled by the three pillars of the Enlightenment: reason, individualism, and skepticism, took the mantle of reductionist thinking to new levels – especially with respect to the natural world where animals were reduced to mere machines (Cottingham 1978) – a far cry from the animistic views of animals as having a soul.

   The complete decoupling of the spirit and value of nature meant that any moral or ethical shackles were now removed and the race towards the eventual domination of nature was now in full swing. Economic theories began to shift, further solidifying the belief that the supply of natural resources was endless and that growth was now the primary goal. The use of coal and later oil and gas would once again exponentially increase the rate at which humanity could plunder the resources of the natural world. The simple equation of the Industrial Age – plunder resources from the earth, place them into production in the form of energy and products, consume those products, and ignore the impacts of waste – has come to represent an acceptable level of damage necessary to deliver progress.

1.1.5 Age of Information

By the end of the 1950s, the destructive impacts of the Industrial Revolution were not only increasingly visible but clearly identified as the cause of a deteriorating environment. Rachel Carson's groundbreaking book, *Silent Spring*, published in 1960 included one of the first studies (Griswold 2012), to clearly link the use of pesticides such as DDT in agricultural practices to a larger ecosystem problem. The idea that birds would eat bugs who held the DDT poison inside them, and would then also die, brought home to large segments of the population the interconnectedness of ecological systems. In 1968, the first photographs sent back to earth from the Apollo 8 mission left the world in a state of wonder. Never before had the earth been viewed from afar, and it immediately impacted humanity's sense of scale – a deeper understanding that humanity really was all connected by the common space on the sphere of earth. The great level of variation of color and texture indicated that the planet was far more differentiated and beautiful than many would have believed and

Table 1.1.5a Overview of the Age of the Information

Worldview	The Age of Industry
Time frame	1960–2005
Space	The world is shrinking
Time pace	Faster
Climatic context	A warming planet
Energy	Nuclear and hydro
Communications	Television + internet
Relationship to nature	Savior
Design	Green design

Source: Created and drawn by the authors.

helped to solidify the growing conception of the planet as something complex, diverse, and perhaps even magical. At the same time, scientists were beginning to confirm the obvious: the legacy of the industrial model of progress formed during the Industrial Revolution was not only polluting the environment but changing global temperatures. First, global cooling was predicted due to the heavy smog cover blanketing the planet blocking the sun. Quickly through, scientists began to uncover the more threatening phenomena of global warming. Even though global warming was predicted long before, it was slowly dawning on a small but influential segments of society that humanity held the power to fundamentally alter the planet's ecosystem – and not for the better (Table 1.1.5a [above]).

Energy: nuclear and hydro

As early as 1970, the largest oil companies understood the threat of global warming as an effect of burning fossil fuels but elected to cover it up and maintain a business as usual approach (Gleiser 2016). While gas, oil, and coal continued to dominate as the main energy sources of choice for the developed world, the rise of nuclear and large-scale hydro power used to make electricity would become increasingly important as society was beginning to adopt an overall information-based worldview that relied upon electronic communication. A new global consciousness was emerging that saw the world as more connected. Space was shorter and nature was once again relevant.

Communications revolution

The formation of the internet underscored this emerging worldview. Information could not only flow more quickly and more freely, but it was also decentralized, giving power to individuals to shape reality – a significant departure from the centrally controlled media of radio and television which had already been compromised by drive to maximize profit. Like the growing perception of nature as an interconnected web or network of information, technical infrastructure now has its own information ecosystem.

The newly formed global consciousness in the 1960s reflected the long pent-up frustration by large segments of society who were still denied basic rights and economic opportunity. The legacy of racism, discrimination, and oppression, present since the dawn of humanity, had finally reached its boiling point in many parts of the world. The fight for social equity was in full force, with many sacrificing their lives in pursuit of a better quality of life. Furthermore, anti-war protests were quite common. In short, the Age of Information was a revolutionary period in world history and it was televised in real time – a first. New media played a key role in encouraging a great empathic flowering among the privileged sections of society. The overwhelming belief that a new world order based on love, compassion, fairness, empathy, and respect for nature reached its zenith towards the end of the 1960s. Here, we see the seeds of a sustainable future. The dual priorities of "saving nature" and building an *equitable society* are still yet to be achieved, but they were firmly established in the Information Age and would continue to grow – leading to the next worldview of the Age of Integration.

1.1.6 The Age of Integration

The transition

It's impossible to pinpoint the exact year when one worldview shifts to another, but 2005 will be a good starting point. A series of events occurred that marked the beginning of a transition from the Information Age to the Age of Integration. Human-induced climate change came into full focus, as record heat waves swept across the globe, cataclysmic hurricanes ripped thought Gulf of Mexico, gas prices hit record levels, and the Indian Ocean tsunami wrought a destructive path. High gas prices slowed consumption but opened the door to the ecologically devastating processes of tar sands removal in Canada and the risky reality of unevenly regulated natural gas fracking industry. Al Gore released the book and movie *An Inconvenient Truth*, which linked human activities, to CO_2 emissions to global warming to climate change. The public began to "see" the harsh reality of climate change and began to take serious action to address the problem. Sustainable development and the Triple Bottom Line became the default goals and frameworks that both governments and corporations used to organize and measure environmental activities. The 13 years that followed featured historic forest fires, an accelerated melting of the ice caps, more sea level rise, record droughts in the western U.S. and India, and devastating hurricanes in the Philippines, India, New York City, Houston, Florida, North Carolina, and Puerto Rico. By 2018, it became clear that the earth's ability to support human society at the current pace of development and growth was not possible (Table 1.1.6a [below]).

Climate change and new energy regimes, normally a predictor of massive change in civilization consciousness, have yet to deliver the full transition to the next Age of Integration. Instead, we live in limbo, numb to dramatic weather disturbances and to the fact that we are now living in the first ever human-induced climatic epoch – the Anthropocene. We are blind to the massive transition to renewable energy and electric cars – choosing to cling to our fossil

Table 1.1.6a Overview of the Age of the Integration

Worldview	The Age of Industry
Time frame	2005–
Space	Dematerialized
Time pace	Instantaneous
Climatic context	The Anthropocene
Energy	Renewable energy
Communications	Social media, distributed networks
Relationship to nature	Co-evolution with nature
Design	Sustainable design

Source: Created and drawn by the authors.

fuel-powered vehicles that sputter and lurch, spitting out CO_2 and quickening the pace towards climate destruction. Even the advent of new communication tools like social media and the smartphone has failed to deliver the goods on a promising new way of seeing the world.

Human motivations

At the core of the hesitation toward a sustainable lifestyle is the age-old conflict between self-interest and empathy. Covered in Chapter 3, self-interest as a biological motivation plays out in a myriad of ways from climate denial to reactionary elimination of environmental regulations. It can also be seen in the rise of racism and nationalism as a competing pathway as we enter the Age of Integration. In the USA, the promise of social equity for people of color and women in a society dominated by white males that peaked in the 1960s is still waiting to be realized. The human tendency towards bias and tribalism is as strong as ever, rising quickly by turning a blind eye to recent history.

The sustainable future?

Ultimately, humanity will move towards the utopian vision of sustainability either by conscious choice through a complete restructuring of the world's energy system, a serious mobilization to fight climate change, and the establishment of completely new and ambitious regenerative design methodologies. Or, humanity will move towards a sustainable future out of desperation, a sad, non-resilient form of sustainability, involving deep levels of change and sacrifice to maintain survival with a much lower quality of life than many experience today. The rest of the book explores the first option, an intentional transition into the Age of Integration, by looking at the values, frameworks, strategies, and then the application at various scales of an authentic sustainable design, seeking to improve our quality of life and the ecosystems that have suffered damage.

1.1.7 Virtual time

It's hard to believe, but we are now seeing the dawn of virtual time – which is time that occurs in virtual space. While the physical world deteriorates and society hesitates to take definitive action, the virtual world is growing at an exponential pace. Just as virtual space can be inhabited through computer technology, we also now have the ability to run simulations over virtual time to see the results of our choices in design. These simulations can be repeated many times and in parallel, thereby finding and exploring multiple pathways towards a desired result. Our climatic future can be projected over and over again to see the possible scenarios for human flourishing or demise. This is important to sustainable designers because we can now design buildings, infrastructure policies, and initiatives to attack global issues and simulate their performance over the course of 30–50 to 100–1,000 years. We can make one virtual change to the design and run the simulation again. This allows us to begin to predict how actual physical buildings will perform in the real world. Mastering virtual time is a critical skill set for a sustainable designer.

1.1.8 Conclusion

Space and time are frameworks for design, which enable us to think about projects in a holistic way. We can also begin to understand how our decisions will impact distant parts

Table 1.1.8a Summary of worldview shifts

	Age of Hunter-Gatherer	Age of Agriculture	Age of Industry	Age of Information	Age of Integration
Attitudes of development	Tribal	Traditional	Modern	Postmodern	Integral
Energy type	Fire	Surplus food	Fossil fuels	Nuclear/hydro	Renewable
Climate	Ice age	Holocene	Holocene	Holocene	Anthropocene
Communications	Speech	Writing	Printing press	Internet	Social media + smart phone
Relationship to nature	Integrated with nature	Nature managed	Nature destroyed	Nature saved	Co-evolution with nature
Architectural forms	Organic architecture	Vernacular + ceremonial architecture	Classical + modern architecture	Postmodern architecture	Sustainable architecture

Source: Created and drawn by the authors.

of the planet. The most obvious examples are the polar ice caps. They were already melting due to the interglacial period, and the additional melting due to human-induced global warming varies depending on which scientific reports are used. The relationship between our greenhouse gas emissions and significant changes to ice caps and the rest of the world for that matter signals the beginning of an environmental catastrophe not seen since the extinction of the dinosaurs. For sustainable designers, thinking across time is an essential skill set. Figure 1.1.8a summarizes the shifts in attitudes of development, energy use, communications technology, relationships to nature, and forms of architecture (Table 1.1.8a).

Form does follow worldview, and we are entering a new worldview of integration. Seeing the long-term patterns brings into focus the essential challenges of our time. The Anthropocene climatic period is not unlike the devastating impacts of the K2 meteor, or the breakup of the Pangea, but in this case, we also have the necessary tools to deal with the threats. First and most important is an emerging attitude of development that seeks partnership with nature – bio-inspired design, which is covered in Chapter 6. The rise of renewable energy is an obvious step forward followed by a complete transformation of the communications systems, allowing for ubiquitous and instantaneous sharing of information – the perfect tool to help change people's mindsets towards an attitude of development that favors nature. Finally, interior design, architecture, landscape architecture, urban design, city planning, and engineering along with industrial designers and fashion designers are all professions that have the knowledge to transform their practices and the projects they design. To that end, the next chapter will focus on environmental literacy in order to help the reader better visualize and understand the environmental threats we face today.

References

BBC, 2018. *Big Five Mass Extinction Events*. BBC. www.bbc.co.uk/nature/extinction_events edn.

Brugger, J., Feulner, G. and Petri, S., 2016. *Baby, Its Cold Outside: Climate Model Simulations of the Effects of Asteroid Impact at the End of the Cretaceous (2017)* https://agupubs.onlinelibrary.wiley.com/doi/pdf/10.1002/2016GL072241 edn.

Cottingham, J., 1978. *'A Brute to the Brutes?': Descartes' Treatment of Animals: Discussion* https://philpapers.org/rec/COTABT.

Derocher, A.E., Lunn, N.J. and Stirling, I., 2018. *Polar Bears in a Warming Climate*. Oxford. https://academic.oup.com/icb/article/44/2/163/674253 edn.

Gleiser, M., 2016. *ExxonMobil vs. The World*. NPR. www.npr.org/sections/13.7/2016/11/30/503825417/exxonmobil-vs-the-world edn.

Goethe, J.W.V. and Eckermann, J.P., 2010. *Conversations with Goethe in the Last Years of His Life*. Boston: Nabu.

Griswold, E., 2012. How Silent Spring Ignited the Environmental Movement. *The New York Times*.

Kahn, B., 2017. *We Just Breached the 410 Parts Per Million Threshold*. Climate Central. www.climate central.org/news/we-just-breached-the-410-parts-per-million-threshold-21372 edn.

Moskvitch, K., 2014. *Dinosaur Era Had 5 Times Today's CO_2*. Purch. www.refworks.com/refworks2/?r=references|MainLayout::init edn.

Newton, I.S. and Thayer, H.S.E., 2010. *Newton's Philosophy of Nature: Selections from His Writings, p. 42, ed. H.S., NY, 1953)*. New York: Dover Publications.

Our Planet, April 21, 2018, Ancient Earth (240 Million Years Ago) – Pangaea [Homepage of Word Press], [Online]. Available: https://ourplnt.com/ancient-earth/ancient-earth-240-million-years-ago/ [March 5, 2018].

Peoples, H.C., Duda, P. and Marlowe, F.W., 2016. *Hunter-Gatherers and the Origins of Religion*. Springer US. https://link.springer.com/article/10.1007%2Fs12110-016-9260-0 edn.

Postel, S., 2014. *A Sacred Reunion: The Colorado River Returns to the Sea*. National Geographic. https://blog.nationalgeographic.org/2014/05/19/a-sacred-reunion-the-colorado-river-returns-to-the-sea/ edn.

Rifkin, J., 2009. *Empathic Civilization: The Race to a Global Consciousness in a World in Crisis*. 1st edn. TarcherPerigee.

Sahney, S. and Benton, M., 2007. *Recovery from the Most Profound Mass Extinction of All Time*. The Royal Society. http://rspb.royalsocietypublishing.org/content/275/1636/759 edn.

Siegel, E., 2017. *The Odds of Your Unlikely Existence Were Not Infinitely Small*. Forbes. www.forbes.com/sites/startswithabang/2017/05/16/the-odds-of-your-unlikely-existence-were-not-infinitely-small/#255ae3e840b0 edn.

Smithsonian National Museum of Natural History, 2018. *Climate Effects on Human Evolution*. Smithsonian Institution. http://humanorigins.si.edu/research/climate-and-human-evolution/climate-effects-human-evolution edn.

United States Census Bureau, July 24, 2018-last update, World Population Clock [Homepage of US Department of Commerce], [Online]. Available: www.census.gov/popclock/world [April 4, 2018].

Wilber, K., 1977. *The Spectrum of Consciousness*. 2nd edn. Wheaton, IL: Quest Books.

2 Environmental literacy for the sustainable designer

This chapter represents the absolute minimum level of knowledge needed to begin to see the state of our environment and to form a foundation for sustainable design. In the chapter, additional resources are offered to encourage the reader to delve more deeply into the topics that are most interesting or alarming. The responses to the environmental problems will be covered in the majority of the book starting in Chapter 4. If you are already environmentally literate, this will provide a basic review.

2.0 Introduction

The earth is already four billion years old and has experienced numerous and significant geologic and climatic changes. We saw how during the great Pangea, the land mass was split apart to form the continents as they are today, causing the extinction of millions of species. We also looked at how a giant meteor wiped out the dinosaurs and a host of other species in just a few short decades. We saw the humans, as we might recognize them today, just arriving at the end of the last ice age, 14,000 years ago. Since then, humanity flourished benefiting from the latest interglacial period – the Holocene. But with the rise of humanity also comes the next great apocalypse – a radical transformation of our climate, our geography, and a widespread extinction of species through habitat destruction, human-induced climate change, and extreme pollution.

We also saw how humanity has changed its views of nature over time, from the hunter-gatherers who lived in a connected way with nature, to the agriculturalists who managed nature, to the industrialist who conquered nature for their own profit. During this time, we became separated from the natural world, which became a commodity to be exploited. It wasn't until the Age of Information in the 1960s that we thought we needed to "save" nature and build a more positive relationship with the planet earth.

Currently, we see the conditions for the next great leap in human development: decentralized renewable energy systems, distributed social media networks, and the first human-made climatic period ever – the Anthropocene. We now stand at the proverbial crossroads. One path takes us towards a dystopian future of water scarcity, intense weather, and human suffering. The other road offers the possibility of a sustainable future, where we can raise the quality of life for all humans through a co-evolutionary relationship with nature.

By now, you can see that the language and tone of the book so far stems from an admittedly humanistic point of view. After all, nature was here before us, and it will be here long after we are gone. It has its own arc of evolution, now intertwined with our own pathway of development. As will be argued later, sustainability is a decidedly human construction, an artificial vision of what the future could or should be, and nature plays the central character of the story.

We must work harder to understand nature better, or at least become more environmentally literate. While there is not the time and space to cover everything here, this chapter will lay out the basics and study the interrelationships to aid in building a connected mental map of the most significant ecological issues threatening our future. The first step in starting on the path towards environmental literacy is to establish a "working agreement" on the *nature* of nature.

Nature versus natural

According to the Oxford dictionary, the definition of "nature" is as follows:

> The phenomena of the physical world collectively, including plants, animals, the landscape, and other features and products of the earth, as opposed to humans or human creations. Synonyms; the environment, the earth, the universe, the cosmos, natural forces.
>
> (Oxford 2018a)

Note the contrast above of "... the physical world collectively ..." and "... as opposed to humans or human creations." Humans are mammals, we are a phenomenon, or an occurrence, of and in the natural world, but yet the definition says that we aren't. If we look at the definition of "Artificial," we see:

> Made or produced by human beings rather than occurring naturally, especially as a copy of something natural ... not existing naturally; contrived or false. Synonyms; synthetic, manufactured, machine-made, fabricated.
>
> (Oxford 2018b)

Also of interest is that these definitions were created in, and reflect the Industrial Worldview. We saw in Chapter 1 that worldview does affect actions. This is quite a challenge. As a response, we could ask the following question:

> If human beings are animals and part of the animal kingdom,
> then
> is everything we make and do by default, natural?

For example, is there a difference between an untouched forest and a tended garden? Is Central Park in New York City natural? What about buildings or products that contain only natural materials? Plastic is made from oil, which is a natural resource. What about nuclear fission or the atomic bomb? These are deep and complex questions that could take a lifetime to answer with no satisfying result. It is not possible, nor necessary, for us to come to a philosophical agreement of what is nature in this section of the book.

If the Age of Integration is about returning to the understanding that we are part of the natural world, then we need a greater understanding of the elements of nature and their interrelationships. By studying the balanced and biodiverse systems that exist without human intervention, we can learn many lessons to drive sustainable design. We will see natural systems not destroy their source of survival, precisely because the predator, prey, and biodiverse relationships create balance. They do not take more than they need for survival, and their waste is food for something else.

In sustainable design, we look to harness this creative power and work in a restorative way to comprehensively evaluate the long-term consequences towards better decisions.

Ecosystems across scale

Nature is often defined as a great ecosystem or as a series of ecosystems connected at different scales. An ecosystem is an interrelated community made up of living organisms and nonliving components such as air, water, and mineral soil (Molles 2015, p. 485). Ecosystems come in many scales with the global ecosystem being the largest normally considered in sustainable design, but even that is affected by the cosmic scale with the sun, moon, and the occasional meteor strike having a great impact. Ecosystems also exist inside our bodies with millions of complex interactions that govern our health and well-being.

Ecosystem goods and services

Ecosystem goods and services are based on the essential elements of the earth: air, water, earth, and fire, as well as all living things – flora and fauna. *Ecosystem goods* are the physical things we take from the earth to create immediate value, like iron for steel, trees for building materials, gold, and plants or cattle for food. Historically, these have been referred to as natural resources. *Ecosystems services* are the processes and systems that nature provides to support life including the regulation of our climate, water purification, plant pollination, enrichment of soil, and even medicine. Ecosystem services also include intangible benefits such as natural beauty. Lastly, ecosystems provide a cultural service in their ability to support communities that have developed spiritually intrinsic relationships with the natural world – especially sacred sites. Many hunter-gatherer communities still exist today and still perform rituals on sacred sites considered to have augmented spiritual value.

All of these ecosystem services interact in complex, interdependent systems which create the conditions for life. If one of the elements like water, for example, is compromised, the entire ecosystem is threatened. For example, as global temperatures continue to rise, the polar ice caps melt. Less ice means that polar bears need to travel farther for food expending more energy in the process, leading to less body fat to sustain successful pregnancies, ending up in an overall decrease in polar bear population (Derocher et al. 2018, p. 163). This doesn't sound like a huge problem, until you look at the balancing effect polar bears have on the seal population. By eating seals, polar bears help to keep the seal population down, meaning they eat less fish, leaving more fish for human consumption. From the really big picture, species go extinct all the time. The disappearance of the polar bear may allow for a new species to emerge, or they may adapt by moving further south, which would then have an effect on the balance of another ecosystem.

The ability to support human life depends upon ecosystem goods and services. The ability of the earth's ecosystems to support life is referred to as carrying capacity. There are those that argue that we have exceeded the carrying capacity in several significant ways including loss of biodiversity, climate change, high levels of nitrogen and phosphorous, and land conversion. In the rest of this chapter, we will briefly review the primary ecosystems goods and services, and then explore their relationship through a brief study of energy and climate systems.

Additional resources

Feedback Loops: How Nature Gets Its Rhythms – Anje-Margriet Neutel, TED Talk, www.
   youtube.com/watch?v=inVZoI1AkC8
The International Union for Conservation of Nature Website, www.iucn.org/

2.1 Air

There is nothing better than waking up in the morning and taking a deep breath of "fresh air." We take clean air for granted and assume it will always be there for us. Most of us breathe without really thinking about it or even realizing we are doing it because, from the second we are born, it has been part of our life.

The air, or the atmosphere, is composed primarily of three main gases: nitrogen making up 78%, oxygen at 21%, and argon at 0.9%. The other 0.1% of gasses include CO_2, methane, ozone, and helium in varying amounts (NCCO 2018). There is also water vapor in the atmosphere as well. This can range from 4% in tropical areas to 1% in more arid or desert areas.

The reason it is important to understand the composition of the atmosphere and how it changes is it "… determines the atmosphere's ability to transmit sunlight and trap infrared light, leading to long term changes in the climate" (NCCO 2018). The greenhouse gases (GHGs), which make up a very small percentage of the atmosphere and can vary with the seasons, still greatly affect the temperature of the earth and its global energy balance (NCCO 2018).

2.1.1 GHG emissions

We inhale oxygen and exhale carbon dioxide. Plants have a reverse process using photosynthesis to take in carbon dioxide and emit oxygen. This system of interactions between air, plants, and animals is necessary for survival and provides a balancing feedback loop. When one of the elements in the feedback loop changes, the system is placed into stress. Today, the buildup of GHGs such as carbon dioxide forms a layer in the atmosphere that blocks heat from escaping into space, thereby reflecting heat back into the planet's atmosphere – see Figure 2.1.1a (next page). This is the same thing that occurs in a greenhouse. Heat enters the greenhouse but can't escape leading to a buildup of heat. For our environment, the result is a warmer-than-usual planet – global warming. In Chapter 1, we saw that the planet has already been warming due to the Holocene interglacial period and now, by altering the GHG layer, we create even more global warming. Furthermore, as the planet warms, polar ice cap melt causing the planet to absorb more heat, which in turn leads to more warming. This is a positive feedback loop that accelerates and amplifies the effects causing the overall system to become more unstable.

The prominent GHGs are as follows: Carbon dioxide (CO_2), Methane (CH_4), Nitrous oxide (N_2O), and Fluorinated gases.

2.1.1.1 Carbon dioxide (CO_2)

Carbon dioxide (CO_2) is produced by burning fossil fuels, trees, or wood. CO_2 is removed from the atmosphere, or sequestered, when it is absorbed and stored by plants and the ocean. According the EPA, CO_2 accounted for 81% of the GHG emissions in 2016 (EPA 2018a). CO_2 is measured in parts per million (ppm). In 2013, CO_2 levels rose to 400 ppm

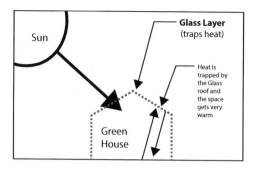 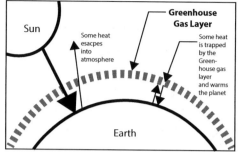

Figure 2.1.1a GHGs and global warming.
Source: Created and drawn by the authors.

significantly higher than the levels in 1950, 300 ppm, or during the last interglacial periods where they were approximately 280 ppm (NASA 2018). Just four years later, CO_2 levels reached 410 ppm, a new record (Kahn 2017). As CO_2 levels rise, more heat is trapped in the atmosphere.

2.1.1.2 *Methane (CH$_4$)*

Methane (CH_4) is emitted during the production and transport of coal, natural gas, and oil (EPA 2018a). It is also produced when organic material decomposes, permafrost thaws, or during agricultural processes. Methane emissions comprise 10% of GHG (EPA 2018a). Researchers are also speculating that as temperatures continue to rise, methane emissions will increase from the sediment and freshwater wetlands (Princeton 2014). In addition, the melting permafrost in Northern Russia also melts and emits even more methane, thereby accelerating the negative effects of global warming.

2.1.1.3 *Nitrous oxide (N$_2$O)*

Nitrous oxide (N_2O) is emitted when fossil fuels and solid waste are burned, as well as from certain agricultural and industrial processes, and accounts for 5% of the GHG emissions (EPA 2018a).

2.1.1.4 *Fluorinated gases*

Fluorinated gases are emitted from industrial processes in smaller quantities, but because of their high potency to increase global temperature, they are referred to as High Global Warming Potential (GWP) gases (EPA 2018a).

A gas's *warming potential* is determined by three factors (EPA 2018a): How much of the gas is in the atmosphere? How long do they stay in the atmosphere? How strongly do they impact the atmosphere?

Each GHG has been tested and given a GWP number that reflects how long the gas remains in the atmosphere and how strongly it absorbs energy and promotes warming (EPA 2018a). The higher the GWP, the more they contribute to global warming. As we mentioned above, even a small amount of these gases can cause significant changes to the earth.

As a response, many clients who commission design projects, or government organizations that oversee the development of cities are all concerned about global warming and are anxious to find ways to attack the problem. Building codes, zoning ordinances often have energy-reducing requirements. Most green building rating systems have a significant focus on reducing GHGs emissions.

Additional resources

The Holocene Epoch: University of California Museum of Paleontology, www.ucmp.berkeley.
   edu/quaternary/holocene.php
Is There A Ticking Time Bomb Under The Arctic?: NPR, www.npr.org/sections/goats
   andsoda/2018/01/24/575220206/is-there-a-ticking-time-bomb-under-the-arctic
Present: Climate Forcing Factors: NSF, www.plantsciences.ucdavis.edu/plantsciences_faculty/
   bloom/camel/present.html
Overview of Greenhouse Gases: EPA, www.epa.gov/ghgemissions/overview-greenhouse-gases

2.1.2 Air pollution

2.1.2.1 Particulate matter

PM2.5 refers to any particulates in the air that have a diameter of less than 10 μm (EPA 2018b). Particles of this size are particularly dangerous because they are able to get deep into your lungs and even into the bloodstream. Exposure to PM2.5 has been shown to cause nonfatal heart attacks, irregular heartbeat, asthma, increased respiratory issues, or even premature death in people with existing heart or lung disease (EPA 2018b). PM2.5 is at the root of much of the reduced visibility and smog seen over cities and even some national parks. It also causes environmental damage by "… making streams and rivers more acidic, changing the nutrient balance in coastal waters and river basins, depleting soil nutrients, and damaging sensitive forests or farm crops" (EPA 2018b). In 2014, marathon runners needed to wear respirators and masks because the toxic smog was so thick (Asia Pacific 2014). PM2.5 in the air we breathe has increased significantly since pre-industrial times. However, many countries have been able to significantly reduce PM2.5 back to safe, breathable levels through regulation and technological innovation.

Other forms of air pollution

There are also a host of other types of air pollution including carbon monoxide, lead, nitrogen dioxide, ozone, and sulfur dioxide. Significant indoor air pollution threats include volatile organic compounds (VOCs), formaldehyde, and others. All of these sources of air pollution along with PM2.5 and GHGs combine to create a significant threat to human health both in short and long-terms.

As a response, the focus on air quality in design has risen dramatically in recent years, with indoor air quality becoming increasingly important since people spend so much time indoors. Chapter 8 will address the relationship between health and design.

Additional resources

Air Pollution, Causes, Effects, and Solutions: National Geographic, www.nationalgeographic.
   com/environment/global-warming/pollution/

2.2 Water

2.2.0 Introduction

Water is an astounding substance. Like air, it is the basis of all life as we know it. We can live weeks without food, but only days without water. The human body is approximately 60% water, and the 71% of the earth is covered with water. Water molecules are the same as they were billions of years ago. In other words, we are, in effect, drinking the same water as Aristotle, Imhotep, and Cleopatra. Water quality changes over time, and what we think of as "pure" is really just the type of water with the proper acid and pH levels to support life and biodiversity as we know it.

Water has been integrated into cultural practices all over the world for millions of years instilling beauty, peace, and pleasure. Fountains, pools, and water features along with rivers or oceans have long been gathering places and sought-after elements in the built environment.

Water is precious. Water is a vital element to all living systems – without it, life, as we know it, ceases to exist. From that point of view, we need to understand that the trillions of decisions, we make each day, have a direct impact on the quality and quantity of our water supply.

2.2.1 Water pollution

Water pollution refers to changes in the quality or composition of water and water bodies. Pollution can occur anywhere in the water cycle. It can come from dumping toxic waste into water bodies, as it filters through the atmosphere to create acid rain, or through the earth into water veins that supply wells, and springs.

Major sources of pollutions are industrial waste, agricultural processes, oil industries, radioactive waste, and consumer waste (Ukela 2017). Industrial and radioactive waste that ends up in rivers and streams still happens regardless of established regulations. Oil industries are responsible for petroleum products being spilled directly into rivers and the oceans as well as spills on land that find their way into water supplies. Consumer waste is the result of direct dumping into waterways and trash that is discarded elsewhere that ends up in the water because of wind or flooding. Plastic and trash in the oceans is becoming a huge problem in the environment especially the health risks posed to all species.

One of the major pollutants of water is chemical and organic fertilizers used in agriculture. These come in the form of products, both natural and synthetic, that are applied to the ground to enhance crops, or the waste products from all types of livestock that are raised for food. Nitrogen and phosphorus are natural products that when they make their way into the water lead to increased plant life, especially algae. When there is too much algae, sun is blocked and slows plant growth under that water which other species like fish or ducks or crabs depend on. When the algae die and decompose, they use up a lot of oxygen, which leads the fish and other underwater species to suffocate (US Fish and Wildlife 2018). The agricultural industry is responsible for releasing literally millions of tons of organic pollutants into the water each year (EPA 2018c).

2.2.1.1 Ocean acidification and the coral reefs

Ocean acidification is the ongoing decrease in the pH of the earth's oceans, caused by the uptake of carbon dioxide from the atmosphere. pH is a measure of acidity. As CO_2 levels have always shifted in past millennia, so too have pH levels shifted – changing the conditions of life in the oceans and by default changes in the evolution of species on the planet. Today, pH levels have shifted by as much as 30% due to increased CO_2 in the atmosphere. Air and water pollutions are interlinked problems.

Increasing acidity is thought to have a range of potentially harmful consequences for marine organisms, such as depressing metabolic rates and immune responses in some organisms, and causing coral bleaching. Coral reefs provide habitat for trillions of creatures as part of a complex marine ecosystem that ultimately supports human life. The Great Barrier Reef, for example, has lost 50% of its coral over the past 20 years (Cho 2018). The impacts to the local ecology and economy are grave as fisheries shut down and tourism declines (Cho 2018).

2.2.1.2 Plastics in the water

Plastic is a durable material that does not degrade naturally, which makes it a useful product for humans, but once its useful life is over, it has the potential to become a nearly permanent source of environmental contamination. It is estimated that by 2050, humans will have produced nearly 34 billion tons of plastic, which is 100 times more than the combined weight of all the people currently on the planet (Geyer et al. 2017). Although plastics can be recycled or converted to energy through incineration, it's estimated that nearly half the plastics we produce currently wind up in landfills or contaminating the environment. Sadly, some of the plastic that is not recycled ends up in the ocean. Massive gyres of plastic are trapped in the ocean by currents, forming floating "islands" that can reach more than a million square miles in size, not including the denser waste that sinks to the bottom.

These plastics have a negative impact on all life in the aquatic system – blocking the light which algae and microorganisms need to grow, choking or starving marine animals and birds that eat the small plastics, and drowning animals which get tangled in plastic debris. This has led to the unnecessary death of millions of animals and the destabilization of a major human food source. Eventually, fish contaminated with plastic waste can be caught and eaten by humans, too. A recent study found that tap water around the world contains plastic microfibers, including in 94% of samples taken in the U.S. The health impacts of ingesting plastic microfibers are still unknown (Carrington 2017).

Other major water pollution sources include acid rain from coal-burning plants, point source pollution from factories, wastewater from fracking, and storm water runoff from developed sites and agricultural lands. Sustainable designers are focused on attacking problems associated with water, by finding ways to capture and purify water on site, reducing GHGs which will protect oceans from acidification, and by specifying water-saving technologies in a design project.

Additional resources

Water Topics: EPA, www.epa.gov/environmental-topics/water-topics

2.3 Earth/materials

The earth through its slow shifts and changes, holds a record of everything that has happened throughout history. We can study geology to make guesses about what life on earth has been like in the past and what might happen in the future. The quality of the earth itself is directly tied to complex interactions between the other elements of water and air and with flora and fauna. A portion of the earth itself includes trillions of fossils and therefore is the source of much of the world's energy. In addition, all of the "precious" minerals come out of the earth, ecosystem goods, are processed into materials and eventually products that help to meet the needs of humans and increase our quality of life.

2.3.1 Soil

Soil, or "dirt," often has a negative connotation in our language. Even the word "humble," meaning low in dignity or importance, gets its roots from "humus", the decayed plant matter that makes up the majority of fertile soil. However, soil is an extremely valuable resource – all life depends on it in some way. It serves a vital role in providing habitat and nutrients for plants and animals, recycling dead organic matter, filtering and distributing water, and supporting human buildings and infrastructure. In nature, it can take up to 3,000 years just to form an inch of biologically active, fertile topsoil (White 2015). Apart from the critical role soil plays in supporting flora and fauna on the planet, recent studies have shown that certain bacteria present in soil have an antidepressant effect on humans, promoting a healthy immune system and stimulating the release of serotonin in the brain (Lowry et al. 2007). Unfortunately, soil quality is in rapid decline through human land use practices. Soil erosion, disruption, and exhaustion stemming from a myriad of human forces including development, deforestation, and agriculture. We also forget that millions of tiny life forms exist in each square yard of soil. These are mini-habitats that form the basis of the larger ecosystems that we depend on.

The design and construction industries are a direct threat to soil, especially when we compact soil, or remove it, the soil is disturbed and often destroys the delicately balanced ecosystems. We are quick to identify sand, clay, and silty soil, but we almost never consider the interrelationship between soil type, local geography, local weather, water patterns, and biodiversity. Those macroscale considerations and longer time periods are invisible to most, leading to short-term decisions that often compromise the long-term health of the land.

2.3.2 Raw materials

Raw materials are "crude" materials that can be converted through manufacture, processing, or combination into new and useful products. The earth provides us with thousands of raw materials – some of which are renewable over a human life span, but many of which are not. We use some raw materials to produce energy in order to process *other* raw materials – like the use of petroleum in the making of steel. We tend to think of our supply of materials is unlimited, and that there will always be more to be found or we will find new substitutes. The endless extraction, processing, transporting, use, and disposal of materials reflect a very high amount of energy and therefore amount of pollution. For example, the Belo-Monte dam that is under construction in the rainforests of Brazil is designed to provide energy for iron-smelting plants that convert the raw material of iron ore into useful products for different manufacturing industries. It is destroying the local ecology and threatening the way of life for thousands of indigenous people (Hurwitz 2011).

Your smartphone contains over 50 different kinds of metals. The impacts of the extraction and processing of these materials include a swath of environmental destruction, a drop in human health, and the loss of lives through civil wars (Nogrady 2016). So, as a sustainable designer, the selection of materials for use in built projects and products is not to be taken lightly – especially considering the levels of waste currently in effect all over the world.

2.3.3 Waste

Waste is a funny word, since there is technically no waste in a natural ecosystem. The waste from one organism is food for another organism. William McDonough coined the phrase "waste = food," and yet we tend to think of waste as something to be removed and buried in the earth where it rots – sending millions and millions of tons of CO_2 and methane into the atmosphere. We now create about 2.12 billion tons of solid waste globally per year that is kept in landfills or dumped directly into the environment (World Waste Facts 2018). There are many types of waste:

- **Plastics and product waste.**
- **E-waste:** 44.7 million metric tons of e-waste produced annually around the globe = 4,500 Eiffel Towers. That's 6.1 kg/person/year (Baldé et al. 2017).
- **Textile waste:** 25 billion lbs/year in the U.S. alone, 85% goes to landfills (Harmony 2018). That's just measuring postconsumer waste. About 15% of fabric is discarded in the process of converting it to clothing. It takes more than 5,000 gallons of water to manufacture just a T-shirt and a pair of jeans. Second to oil, the clothing and textile industry is the largest polluter in the world (Sweeny 2015).
- **Food waste:** Roughly one-third of all food that is grown is ultimately spoiled or discarded, which is an unbelievable statistic when you remember that over a billion people on the planet are suffering from malnourishment (FAO 2018).
- **Toxic waste.**
- **Construction waste:** 534 million tons of construction and demolition waste generated in the U.S. in 2014.
- **Human waste.**

Sustainable designers worry about waste all the time, not just because it pollutes our air, earth, and water, but also because it is unnecessary. It is entirely possible to close the materials loops that lead to waste by mimicking nature's processes. Zero waste can generate enormous benefits to an ecosystem and to society in general. Sustainable design, especially in the form of biomimicry and through life cycle assessment, is targeting a zero-waste future.

2.4 Fire (energy)

There's nothing better than sitting by a warm fire on a cold day. It's not just the practical benefit of staying warm. It's one of the most primordial human experiences, and it's associated with cooked food, safety, and light. As a classical element, using the word *fire* fits so well when grouped with air, water, and earth. Functionally, we use the term "energy", which fits better as a 21st-century word. Energy, like the other elements, is embedded in all ecosystems. All of life is activated by energy. Energy is eternal. It changes its form but never actually disappears.

Energy across space and time

Over the span of human history, we have seen the types and power of energy change, and with each change, a new worldview or level of consciousness emerged. In Chapter 1, we saw how people in the Age of Agriculture began to settle in one place because food sources became more predictable and abundant. Or, we saw the explosion of communication, travel, and commerce that occurred after the emergence of fossil fuels. Today, we see the beginning of another shift in human consciousness as renewable energy offers decentralized energy for anyone with the capability of harvesting it on-site. And, renewable energy emerged just when it was needed most, a low-carbon option to fossil fuels, thereby offering a pathway out of the threat of global climate change.

Today, we have many innovative ways of using earth's natural resources to create, harness, distribute, and use energy. The terms *stocks* and *flows* are used to define the type of energy in use. Stocks are those resources that are used once. Some stocks, such as firewood, can be renewed by regrowing the forest. Therefore, firewood is a renewable stock. Oil, on the other hand, is a nonrenewable stock because it would take millions of years to restock that resource. Flows are readily available energy sources such as solar, wind, or water which can be harnessed over and over again when available. A consistently flowing river offers a source of energy that is harnessed via a dam. Solar power is another example. Energy from flows is most often referred to as renewable energy.

The goal of the sustainable designer is to be able to understand the advantages and disadvantages of each type of energy and know which types of energy should be used in a given situation. Of course, the goal is to eliminate the need for fossil fuels and nuclear power through the expansion and evolution of renewable energy sources, but the reality is that in some situations, the use of fossil fuels is unavoidable at this point in time. Knowing which fuel to use and making a design project as efficient as possible is the key to successful sustainable design. Each energy source comes with advantages and disadvantages, but a choice must be made nonetheless. The issues that shape the choices are listed below:

- Worldview: Which worldview is the energy source associated with? Is it a source of energy that is in decline?
- The amount of power generated is more important than one might think. While renewable energy sources are generally "clean," they generate low amounts of power which may not be suitable for a range of project types such as hospitals and factories.
- 24/7 availability speaks to the ability of an energy source to run consistently day and night to provide electricity on a consistent basis. Obviously, wind and solar are not consistent sources of energy.
- Portability: One of the great advantages of fossil fuels is the ability to transport large quantities of energy via tanker, train, and truck. Until battery technology is further developed, renewable energy is not portable.
- Resources remaining: Fossil fuels are finite. It gets more difficult to obtain finite energy sources. Tar sands and fracking are complex systems for mining finite fossil fuels. They generate varying levels of damage depending on the ability of local government to regulate the industries, and by the companies themselves in their willingness to follow proper procedures in mining the resources. Renewable energy is perpetual, always available in great quantities.
- Pollution levels: The amount of pollution generated per unit of energy varies greatly between different sources of energy with fossil fuels generating tremendous amounts of pollution, while renewable energy sources are very clean.
- Short-term danger: Different energy sources carry varying levels of danger to human life and ecosystems. Coal mine explosions, oil rig explosions, oil tanker crashes, nuclear meltdowns, and dam failures are all real risks in the use of energy.
- Long-term danger: Nonrenewable fossil fuels pose significant long-term risks from CO_2 emissions, methane burn-off, acid rain, and climate change itself.

2.4.1 Nonrenewable energy

Coal, petroleum (oil), and natural gas are fossil fuels created from the remains of plants and animals that existed on the planet millions of years ago. After processing, they can all be burned directly to make heat and energy, or burned in a power plant and converted into electricity. Early in the sustainable design process, in the goal-setting process, different sources of energy are discussed regarding their use on a project.

2.4.1.1 Coal

It's easy to look at all the negative impacts of coal, but coal is a remarkably powerful fuel source that ignited the industrial revolution. Today, coal power plants generate terra watts of energy for human civilization. The levels of comfort and convenience enjoyed by millions come from the benefits of coal. Unfortunately, the coal-mining process is fraught with huge risks to the society including the leaching of radioactive sludge into the groundwater and nearby rivers. Coal miners lose their life when explosions occur, and many die young as the result of black lung disease. Table 2.4.1.1a [below] outlines the different aspects of coal power. Notice that coal comes with both short- and long-term dangers.

Table 2.4.1.1a An overview of coal energy for sustainable designers

Evaluation	Coal energy
Worldview	Industrial
Power (power generated per unit of energy)	High
Availability (24/7 availability to deliver electricity)	Yes
Portability (ability to transport energy source)	Yes
Decentralization (ability to generate energy on site)	No
Resources (remaining)	Finite
Pollution (pollution generated per unit of energy)	High
Short-term danger	High
Long-term danger	High

Source: Created and drawn by the authors.

2.4.1.2 Oil

Oil, like coal, is a tremendously powerful source of energy. One barrel of oil (42 gallons) is able to generate energy to accomplish all of the tasks listed below (JWN Staff 2016):

- Enough gasoline to drive a medium-sized car over 450 km (280 miles).
- Enough distillate fuel to drive a large truck for almost 65 km (40 miles). If jet fuel fraction is included, the same truck can run nearly 80 km (50 miles).
- Nearly 70 kWh of electricity at a power plant generated by residual fuel.
- About 1.8 kg (4 lbs) of charcoal briquettes.
- Enough propane to fill 12 small (14.1 ounce) cylinders for home, camping, or workshop use.
- Asphalt to make about 3.8 L (1 gallon) of tar for patching roofs or streets.
- Lubricants to make about a 0.95 L (1 quart) of motor oil.
- Wax for 170 birthday candles or 27 wax crayons.

Despite all of the negative effects of oil, which will be covered shortly, it is necessary to acknowledge the societal benefits that oil has provided for many years. The statistics above are quite impressive when you consider that millions of barrels of oil are extracted from the earth each day. That is a lot of power. Oil is portable, meaning that it is relatively easy to transport to different parts of the world. Of course, that comes with risks. Pipelines leak. Train cars carrying oil crash leading to loss of life. Oil rigs explode leading to ecological destruction and loss of life. Oil tankers crash, spilling millions of gallons of oil into pristine ecosystems. The tar sands in Canada produce oil through a highly toxic process leading to increased cancer deaths of local residents and the ruination of local ecosystems (NRDC 2018) (Table 2.4.1.2a [below]).

Lastly, the burning of oil releases many chemicals and toxins into the air, most notably CO_2 which contributes to global warming. Based on these and other negative aspects of the use of oil, sustainable designers strive to either find alternatives to using oil on projects or try to eliminate fossil fuels all together by using solar or wind power. Fossil fuels are still extremely profitable. However, extraction of fossil fuels is becoming more and more expensive and inconvenient as we approach "peak oil." Peak oil is the hypothetical point in time when the global production of oil reaches its maximum rate, after which production will gradually decline. The reader should be aware that there are still vast amounts of oil left in the ground. While oil is a finite resource, the bigger concern is when it becomes scarce it becomes expensive, and that has a huge impact on economies and ultimately the quality of life. The upside of more expensive oil is that expensive renewable energy solutions like solar and wind become more cost competitive and therefore more likely to be adopted.

Table 2.4.1.2a Overview of oil energy for sustainable designers

Evaluation	Oil energy
Worldview	Industrial
Power (power generated per unit of energy)	High
Availability (24/7 availability to deliver electricity)	Yes
Portability (ability to transport energy source)	Yes
Decentralization (ability to generate energy on site)	No
Resources (remaining)	Finite
Pollution (pollution generated per unit of energy)	High
Short-term danger	High
Long-term danger	High

Source: Created and drawn by the authors.

2.4.1.3 Natural gas

Natural gas is considered the cleanest of the fossil fuels. It is unique in that it can be delivered by pipes directly to buildings for a wide variety of uses including heating and cooking. This is a huge advantage over oil and coal which relies on vehicles for transportation. Needless to say the infrastructure to deliver gas is complex, but the return in the form of convenience and energy reliability is unprecedented. Natural gas, like other fossil fuels, still comes at a heavy environmental cost. Carbon dioxide is emitted when burned, and in many circumstances, methane burn-off adds to the GHG footprint of the use of natural gas (Table 2.4.1.3a [next page]).

Hydraulic fracking started in the 1940s as a method for extracting natural gas. The process of fracking involves drilling a 1- to 2-mile well into the ground; then, fracturing fluid is pumped at high pressure into the well to fracture the surrounding rock and release the layer of oil or gas. Rising costs of oil and imported fuels, as well as concerns over energy security, have driven a recent boom of fracking in the U.S., raising many concerns. Although fracking is often promoted as a "clean energy," it has own unique set of challenges: excess fracking fluid, groundwater contamination, and decreased air quality (Investopedia 2018). There is also compelling evidence that fracking has triggered seismic activity in areas otherwise not prone to earthquakes. In Oklahoma, U.S., the rates of earthquakes are up to 100 times higher than normal levels (Livescience 2018).

Sustainable designers constantly seek the "best" energy solution to every project. Clearly, the long-term risks of global warming and climate change warrant every effort to avoid the use of natural gas; however, when it is used, the most efficient design is necessary. More importantly, because natural gas can be used at any time to generate electricity, it offers a backup in times when solar and wind are not producing energy. Until battery technology is developed, natural gas will continue to be a sought-after source of energy, even with all of its negative impacts and long-term threats.

Table 2.4.1.3a An overview of natural gas energy for sustainable designers

Evaluation	*Gas energy*
Worldview	Industrial
Power (power generated per unit of energy)	High
Availability (24/7 availability to deliver electricity)	Yes
Portability (ability to transport energy source)	Yes
Decentralization (ability to generate energy on site)	No
Resources (remaining)	Finite
Pollution (pollution generated per unit of energy)	High
Short-term danger	High
Long-term danger	High

Source: Created and drawn by the authors.

Additional resources

Facts about Fracking, www.livescience.com/34464-what-is-fracking.html

2.4.1.4 Nuclear power

Nuclear energy is also often promoted as a "clean" energy because the only by-products of direct energy production are water vapor and heat, so it doesn't have the same effects of global warming or air pollution that fossil fuels do. Nuclear power is an important energy innovation that provides huge amounts of power with minimal negative environmental impacts when compared to fossil fuels. However, the potential harm from nuclear power is significant. The extraction process of mining uranium can be a dangerous and unethical process. Nuclear power production comes with the possibility of nuclear accidents, which have decimated communities with consequences lasting multiple generations. The other danger with nuclear energy is the radioactive waste which has to be managed. This waste may take hundreds of thousands of years to reach safe levels of radioactivity (Amadeo 2018). Currently, radioactive waste is buried in cells designed to stop radiation from seeping into the surrounding environment. However, many question the likelihood that these cells will remain effective and continue to be maintained for a period of time longer than all of human history up to this point (Table 2.4.1.4a [next page]).

Many consider nuclear as the transitional energy source towards renewable energy because unlike wind or solar power, nuclear power is available anytime, anywhere, regardless of weather patterns. However, the disaster at Fukushima has caused concern and outrage because of the potential scale of death, destruction, and the extremely long recovery period which follows a nuclear leaking. It wasn't long after the accident that Germany decided to end its nuclear program permanently and switch to an increased reliance on renewable energy.

Table 2.4.1.4a An overview of nuclear energy for sustainable designers

Evaluation	*Nuclear energy*
Worldview	Information
Power (power generated per unit of energy)	High
Availability (24/7 availability to deliver electricity)	Yes
Portability (ability to transport energy source)	Yes
Decentralization (ability to generate energy on site)	No
Resources (remaining)	Finite
Pollution (pollution generated per unit of energy)	Low
Short-term danger	Very high
Long-term danger	Low

Source: Created and drawn by the authors.

2.4.2 Renewable energy

"Flows" are resources that can be used to produce energy without depleting the resource itself in the process. Solar, wind, and hydro are typical "flow" sources of energy. However, the cost of the technologies needed to harvest resources like solar and wind is perceived as expensive. The frame of reference in that perception lies primarily with the fact that the cost of fossil fuels is still relatively low, so the installation of a solar array is by comparison very high. However, a barrel of oil is burned once and is no longer available for use, and it emits high levels of pollution. Renewable sources of energy like solar are available each day for the life of an array – usually 25 years.

2.4.2.1 Hydro power

Hydroelectric power generation uses the flow of moving water to produce electricity. It's considered a flow because there is just as much water released into the natural system after the energy production process as before, but that does not mean that the water is unchanged, or that there are no impacts on the natural systems surrounding hydroelectric energy production. Historically, hydro energy meant converting the kinetic energy of moving water directly into kinetic energy to perform another task, such as using a water wheel to turn massive mill stones to grind grains into flour (Table 2.4.2.1a).

While dams can be small scale as well, we typically think of large hydroelectric dam projects designed to power entire communities. For this type of plant, a dam is built in an existing slow-moving river that floods the area above the dam to store the "potential energy" of the water in a reservoir. The water in the reservoir can then be released as needed converting the potential energy into kinetic energy, as it runs through the dam, which is then turned into electricity. These dams can have huge environmental impacts, mostly related to flooding. In a flat area, the damming of a river can flood thousands of square

Table 2.4.2.1a Overview of hydro energy for sustainable designers

Evaluation	Hydro energy
Worldview	Information
Power (power generated per unit of energy)	Medium
Availability (24/7 availability to deliver electricity)	Yes*
Portability (ability to transport energy source)	No
Decentralization (ability to generate energy on site)	No
Resources (remaining)	Infinite**
Pollution (pollution generated per unit of energy)	Low
Short-term danger	Low
Long-term danger	Low

*Except in the time of extreme drought.
**Prolonged drought is a threat.
Source: Created and drawn by the authors.

kilometers, destroying habitats and wildlife both in the flooded area and downstream, and also displacing entire communities. Although we think of hydropower as being "clean" in global warming terms, CO_2 is still produced in construction, operation, and demolition, and a significant amount of CO_2 and methane is released from the decomposing organic material that is destroyed by the flooding of the surrounding area. It's estimated that the life-cycle emission for a hydroelectric dam is 0.5 lbs/kWh, compared to 0.6–2.0 lbs for a natural gas plant or 1.4–3.6 lbs for coal-generated electricity.

2.4.2.2 Wind

Historically, windmills harnessed the kinetic power of wind to accomplish physical tasks like the milling of grain. Today, wind turbines are used to turn kinetic wind energy into electricity. Wind is widely available and relatively cheap to harness. Although wind doesn't blow all the time, there are areas where it blows consistently enough to be a viable source of energy, and it tends to reach higher speeds at night, when other renewable sources, like solar, are not available. This makes it an ideal component of a localized renewable energy grid (Table 2.4.2.2a [next page]).

The negative environmental impacts of wind energy are bats and migrating birds that can be killed by the turbines and the obvious aesthetic objections by neighbors. They can also be quite loud. Fewer birds die because of wind turbines than because of flying into building glazing, and recent applications of ecological research have significantly reduced bat and bird injuries. Offshore wind farms have actually been shown to increase local fish populations by acting as artificial reefs (Nuwer 2014).

Table 2.4.2.2a An overview of wind energy for sustainable designers

Evaluation	Wind energy
Worldview	Integration
Power (power generated per unit of energy)	Low
Availability (24/7 availability to deliver electricity)	No
Portability (ability to transport energy source)	No
Decentralization (ability to generate energy on site)	Depends*
Resources (remaining)	Infinite
Pollution (pollution generated per unit of energy)	Low
Short-term danger	Low
Long-term danger	Low

*Wind power can work in inhabited areas with high winds or high on skyscrapers.
Source: Created and drawn by the authors.

2.4.2.3 Solar

The sun is the ultimate energy source for all life on earth. Everything humans have accomplished is a result of solar energy converted into food energy by plants, and then converted into heat and kinetic energy by our metabolisms. Plants are the original solar energy convertors through photosynthesis (Table 2.4.2.3a [below]).

Photovoltaic (PV) Solar Panels convert light energy from the sunrays into electricity. This is a useful form of energy production because the sun's energy is available everywhere. However, the efficiency of electrical production depends on the technology we develop to harness it, and in most places, solar energy from PV panels is only available during the day, so we need batteries to capture and use it during the night. There are many ways to capture solar energy using PV, such as rooftop solar on individual homes. Some hazardous materials and rare metals are also used in the production of most PV cells.

Table 2.4.2.3a Overview of solar energy for sustainable designers

Evaluation	*Solar energy*
Worldview	Integration
Power (power generated per unit of energy)	Low
Availability (24/7 availability to deliver electricity)	No
Portability (ability to transport energy source)	No
Decentralization (ability to generate energy on site)	Yes
Resources (remaining)	Infinite
Pollution (pollution generated per unit of energy)	Low
Short-term danger	Low
Long-term danger	Low

Source: Created and drawn by the authors.

2.4.2.3.1 LIGHT

Light is defined as a form of electromagnetic radiation emitted by hot objects like lasers, bulbs, and the sun. Light contains photons which are minute packets of energy. When an object's atoms are heated up, it results in the production of photons. The electrons find excitement from the heat and results in earning extra energy. The energy is released in the form of a photon, and more photons come out as the substance gets hotter. For sustainable designers, "daylight" is an important ecosystem good that can be harvested through good design, thereby reducing energy costs and improving the quality of indoor light immensely. Light also provides critical ecosystem services such as basic functionality for humans (vision) as well as the intangibles of beauty and cultural meaning through the manipulation of light for a specific effect.

2.4.2.3.2 CONCENTRATED SOLAR THERMAL

Solar thermal is an emerging renewable energy source that uses mirrors to reflect sunlight to a tower that is filled with molten salt and is heated to over 150°C. This heat is used to create steam. The steam is used to spin a turbine to create electricity. This a very promising technology because it works well in deserts where there is a lot of space and sun. Birdlife is affected when they fly through the beams of sunlight aimed at the tower.

Additional resources

Solar Thermal Energy, https://en.wikipedia.org/wiki/Solar_thermal_energy

2.4.2.4 Geothermal energy

Geothermal power refers to using heat energy from deep within the earth which emerges as rising steam, and is used to rotate turbines and generate electrical energy. There are several environmental impacts associated with geothermal electricity. It produces 95% less CO_2 than coal-generated electricity. Geothermal electricity can only be produced in specific sites where lava exists in abundance. Iceland is the most well-known user of geothermal power. Wildlife and ecosystems are disturbed by the process of accessing and developing these sites, as well as by the massive drilling that takes place. Drilling is also a water-intensive process, and the waste "drilling fluid" is often deposited into nearby waterways with harmful effects. Although geothermal is considered a "flow," locally, the heat energy and hot water that's being converted to steam can be depleted through mass extraction, leading to changes in pressure and levels of underground water reservoirs, as well as changes in ground temperature and ground levels surrounding a site.

2.4.2.5 Biomass and biofuels

Biomass is a form of energy made from converting food waste, farm waste, and other waste forms into energy via an anaerobic reactor. Waste materials are burned, pressurized, and then used to spin a turbine, which in turn generates electricity. In addition to the electricity produced, heat is generated from the process making this systems a combined heat and power system (CHPS). The upside of this energy system is that it uses waste material. The downside of the system is that it requires massive amounts of waste materials to generate enough usable power to make the system worth the investment and creates emissions like any other form of combustion process.

Biofuels are made from organic matters, for example, ethanol from corn. This seems like a good renewable stock energy source; however, when plants are grown for energy, the land is no longer available to cultivate crops. Furthermore, fossil fuels are needed to help the corn to grow to make energy, a strange combination.

2.4.3 Conclusion

In conclusion, sustainable designers need to be able to sort out the different energy sources based on their advantages or disadvantages in order to make important decisions about how a product or building will be made, transported, and used. In Table 2.4.3a (next page), the comparison between nonrenewable and renewable energies is provided as a shortcut to comparing the different energy sources.

Table 2.4.3a Renewable energy as compared to nonrenewable energy

Nonrenewable energy	Versus	Renewable energy
Past	Vs.	Future
High power	Vs.	Low power
24/7 power	Vs.	Intermittent power
Portable transport	Vs.	Power loss over transmission lines
Centralized	Vs.	Decentralized and distributed
Finite resource	Vs.	Infinite resource
High pollution	Vs.	Low pollution
High short-term danger	Vs.	Low short-term danger
High long-term danger	Vs.	Low long-term danger

Source: Created and drawn by the authors.

2.5 LIFE – introduction

On top of this beautiful, strong, and amazing platform of elements, life as we know it exists. Plants and animals, including humans, are completely and utterly dependent on the elements of the earth to continue to survive. Diversity is critical to the success and resilience of any system, and biodiversity is essential to maintain the balance of our earth. In biology, biodiversity is a factor of two important qualities. The first is the diversity of organisms in the system: the number of types of organisms and the different functions they serve. The second is the diversity of relationships and interactions between those organisms. This means that a system with 1,000 different species is not diverse if there are no interactions between them, or if one of them is a predator which eats the other 999 with nothing in the system to check its growth.

A niche is a specific role that an organism serves in an ecosystem. When two organisms share a niche, they are competing for resources. In a healthy, biodiverse ecosystem, there are enough species filling each role that if one population fluctuates, the overall system remains resilient, but not so many that there is over-competition for scarce resources. In a healthy system, this process of differentiation into niches happens naturally.

Disturbance is an important part of a biodiverse system. Systems that reach their climactic state of existence and remain there are more vulnerable to collapse than systems that are continually disturbed and given the opportunity to regenerate. For example, regular forest fires are essential to the long-term health of a mature forest. There are ways in which human activity has tried to control small disturbances, leading to catastrophic disaster events which destroy the systems we've built. Flood levees and not allowing forest fires are just a few. There are also ways in which humans have created too much repetitive disturbance to systems without offering the ability for those systems to regenerate.

2.5.1 Flora and fauna

Plants are amazing, and they have developed mechanisms for surviving in earth's most extreme environments, including deserts and deep-sea steam vents. They are the bottom of the food chain, supporting all other life on earth. When they are threatened, an entire ecosystem can become dismantled. The International Union for Conservation of Nature (2018a, 2018b) estimates that 68% of plant species are threatened with extinction. This is largely due to the development of land and disruption of soil for buildings, infrastructure, and agriculture, but also due to indirect pollution, changes in weather patterns related to global warming, and introduction of invasive plant species which outcompete many other native species. In a recent study published in the journal *Nature* (Hughes et al. 2018), 50% of coral reefs in the world have died. In the summer of 2015, more than two billion corals inhabited the Great Barrier Reef. Half of them are now dead. Furthermore, coral reefs act as soft barriers to reduce the impact of storm surges. Without the reefs, trillions of sea creatures will perish, and the coastal ecosystems will cease to exist.

Animals (fauna) are also amazing. Their interactions with flora and the other elements of nature form the basis for a healthy ecosystem. They provide many direct ecosystems goods such as food and many ecosystems services such as pollination. Animal populations are in deep decline. According to the World Wildlife Fund (2016), "Populations of vertebrate animals—such as mammals, birds, and fish—have declined by 58% between 1970 and 2012. And we're seeing the largest drop in freshwater species: on average, there's been a whopping 81% decline in that time period." The impacts of humanity on the biodiversity are staggering. For sustainable designers, the ability to see these problems globally and take local action in design projects is a unique and critical skill set. The consideration of flora and fauna in design must take center stage if we are to begin to address the environmental collapse.

Additional resources

Living Planet Report, 2016 by the World Wildlife Fund, https://en.wikipedia.org/wiki/Solar_thermal_energy

2.5.2 Deforestation

Deforestation is another factor of climate change affecting flora, fauna, and humans. Deforestation is caused by human activities such as construction, development, mining, and clearing land for agriculture. Once an area is sufficiently deforested, the microclimate of that area changes, forcing the other flora and fauna that lived there to abandon it. CO_2 levels rise because the loss of trees and other vegetation, which lowers the amount of CO_2 that area can sequester. Every year, 46–48 thousand square miles of woodland are removed, equal to 48 football fields/minute, thereby making deforestation a significant threat to the climate and by default to us (Bennett 2017).

Forests serve as an important "sink" for carbon dioxide, absorbing it from the air and soil and storing it in biomass. When biomass is burned, that sink is destroyed, and hundreds of years' worth of carbon is released into the air in a matter of hour. It is estimated that 25% of the global GHG emissions are a result of deforestation and forest fires (Bennett 2017).

2.5.3 Habitat destruction

Habitat destruction is currently ranked as the primary cause of species extinction worldwide.

It is the result of excessive clearing of forests for crop cultivation and livestock, urbanization, increased pollution, introduction of invasive species, and climate change.

As the world population continues to increase, the pressure on farmers to increase yields will be intense, leading to more habitat destruction and more food insecurity and the rise of food prices around the world. The drivers for these problems are discussed in Chapter 3.

2.5.4 Species extinction

The Center for Biological Diversity (2018) estimates that 30–50% of species could go extinct before the middle of the 21st century. Most of this extinction is due to human behavior, habitat destruction, introduction of invasive species, or global warming. The species in the highest danger are amphibians that are extremely sensitive, and their loss signals that eco-system changes could affect other species. Birds, fish, mammals, invertebrates, plants, and reptiles are all being affected with the biggest causes being habitat destruction, pollution, and global warming. Due to the complicated interconnections of species with plants, larger ecosystems, and humans, the problems of attacking species extinction are hard to define.

2.5.5 Invasive species

Species imbalance in the plant and animal world is usually caused by the introduction of an invasive species that outcompetes local species or have no predators. Invasive plant species can flourish and take over indigenous species easily because local animal populations can-not use them for food. They grow unchecked and outcompete the surrounding species for sun and water. This also poses a problem for the animals because their food supply is cut down. Some plants have invasive or very shallow root systems that also destroy surrounding natural vegetation.

2.5.6 Food

While the study of biodiversity is an important part of environmental literacy, an explora-tion of food systems is also in order because of the tremendous negative environmental im-pacts from the industrialized production of food. In Chapter 1, the sequence of worldviews illustrated how humanity's attitude towards nature changed. The agricultural industry, as much as any other, has become an expression of the separation between humanity and the natural world. We saw earlier in the chapter the effects of fertilizers and waste on water quality.

2.5.6.1 Factory farming

The impact of monoculture farming has tremendous negative environmental impacts to the soil because it does not replenish the minerals. Fertilizers, pesticides, and chemicals are used to make up for this which make other changes to the soil and our food. Genetically modified organisms (GMOs) are planted to increase production without fully understanding the long-term consequences. Irrigation requires vast amounts of water to produce the crops needed to feed a growing population. In order to maximize profit, animals are treated in harsh conditions with a very low quality of life. Aside from the obvious ethical and medical threats, the impacts from factory farming contribute to deforestation, climate change and water pollution. Farming equipment needs fossil fuels to operate and plant, harvest, and care for the crops. Transport from large centralized farming requires also generates a high level of CO_2 emissions.

Large-scale meat-processing plants have similar negative environmental impacts and also come with significant ethical concerns over the treatment of animals. Yes, the animals even-tually die, but their quality of life is simply miserable. This is the dark underside of human-ity's desire for high-quality fruit, produce, and meat all year long. We would be enraged if the stores didn't carry lettuce or tomatoes in the winter, or if we could only get oranges a few months a year at a very high cost, or if meat prices skyrocketed. And yet, our dietary choices have huge environmental impacts that can't be ignored.

2.5.6.2 Bee Colony Collapse Disorder

Ecosystem services are usually perceived as water purification, or climate regulation, or soil enrichment, but the honey bee provides one of the most important ecosystem services. It is responsible for the pollination of nearly 25% of all our food. All of our fruits and nuts rely upon pollination (Sass 2011). "According to a Cornell study published in the May 22 issue of the journal Public Library of Science ONE, crops pollinated by honeybees and other insects contributed $29 billion to farm income in 2010" (Ramanujan 2012). More importantly than the loss of income to food producers is the food to humanity. With many things still being discovered about ecosystems, biodiversity and their interdependence, is the concern of the greater effect of Bee Colony Collapse Disorder that we haven't foreseen. Because of these concerns, the European Union banned a series of pesticides linked to Colony Collapse Disorder.

2.6 Climate change effects

The effects of both natural and human-induced climate change are profound. It is important to note that no single climatic event can be directly tied to human causes. Sustainable designers lose their credibility when they cry foul at every forest fire or hurricane. It is the long-term emergence of climatic patterns that are cause for concern. Even then, some long-term patterns are a natural result of being in an interglacial period.

2.6.1 Heat waves

The effects of climate change are felt primarily through higher temperatures. Among the many threats discussed in this chapter, heat itself is the cause of loss of life and illness among humans and animals. Frequent heat waves can damage or kill plants and animals that are not adapted, or cause certain invasive species to thrive killing more indigenous species.

Additional resources

Global Warming 101, www.nrdc.org/stories/global-warming-101#causes

2.6.2 Hurricanes

The link between global warming and increased hurricane activity is still inconclusive. The question as to whether the slight rises in the frequency and intensity of hurricanes are statistically significant is still under review. Despite the fact that several large hurricanes continue to hit landfall causing widespread damage and human suffering, the ability to pinpoint the cause as global warming is difficult because records in the early 20th century and prior to that are spotty and based on eyewitness accounts. There is a relationship between higher sea surface temperatures and increased intensity of storms. Added sea levels and the destruction of coastal wetlands are also the cause of increased destruction. It's predicted that the damage of storms will increase over time.

Additional resources

Climate Change Synthesis Report 2014, www.ipcc.ch/pdf/assessment-report/ar5/syr/
    SYR_AR5_FINAL_full.pdf
Global Warming and Hurricanes, www.gfdl.noaa.gov/global-warming-and-hurricanes/

2.6.3 Drought

A total of 2.7 billion people experience water scarcity each year, and potable water is becoming a globally scarce and valuable resource. Global warming is leading to increased
drought events by shifting precipitation patterns and increasing air temperatures. NASA
(2017) studies show that even in areas that are ecologically accustomed to drought, recovery
time between droughts is longer and often incomplete. This is leading to tree death which
reduces the system's ability to absorb carbon, provide shade, and store water. Together,
drought, famine, and desertification have been implicated in some of the major human ca-
tastrophes of the past, present, and future.

Additional resources

Causes of Drought: What's the Climate Connection, www.ucsusa.org/global-warming/
    science-and-impacts/impacts/causes-of-drought-climate-change-connection.html#.
    Wx5mCUgvwhc
Water Scarcity – Overview, www.worldwildlife.org/threats/water-scarcity

2.6.4 Flooding

While drought continues to threaten many areas of the world, climate change is also causing
extreme flood events because of increased precipitation events in some areas. The impact of
flooding on the design and construction industries is beginning to be felt in new zoning and
building codes that require structures to be raised up, or require new types of infrastructure
designed to address frequent flooding. Flooding and sea level rise are related because during
a storm event, a higher sea causes flooding further inland due to a higher base level of water
volume. This problem is especially pronounced during high tide.

Additional resources

Climate Change Indicators: River Flooding, www.epa.gov/climate-indicators/climate-
    change-indicators-river-flooding

2.6.5 Sea level rise

Projection models indicate that based on current warming trends, sea levels will rise up to
1.5 meters by the end of the century. While this may seem small, remember that many of the
globe's largest human settlements are situated along the coasts. Large portions of major cities and even entire islands are projected to be underwater, and much larger portions would
be susceptible to coastal flooding and storm surge events. Miami, Florida is already facing
the effects of seal level rise on a regular basis and has begun to raise roads to address the
problem. In Chapter 11, we will look at how Manhattan is addressing this issue.

Additional resources

Climate Change: Sea Level, NASA, https://climate.nasa.gov/vital-signs/sea-level/
National Climate Report – May 2018, www.ncdc.noaa.gov/sotc/national/2018/05/
   supplemental/page-1

2.6.6 Landslides

Landslides are incredibly destructive natural disasters that can happen without warning when large portions of loose rock or soil become destabilized and flow down a slope. As they move, they can destroy anything in their path, including forests, towns, and cities. This is due to the overdevelopment of steep-sloped sites, especially the removal of trees whose roots hold the soil in place. Climate change is making the situation worse because of the increase in the frequency of storm events and the melting of high-altitude snow and permafrost.

2.6.7 Permafrost melt

There is an even bigger problem associated with the melting of permafrost. Permafrost is soil that has been frozen for centuries in cold climates, under a layer of topsoil. It contains organic material that froze before decaying, so when it thaws, it releases centuries of stored CO_2 and methane into the atmosphere. Large regions of permafrost in the polar north and south are expected to become major sources of GHG emissions in the next 60 years. It is estimated that they will remain that way for over 300 years after (Rasmussen 2018).

2.6.8 Forest fires

Forest fires can occur naturally and are nature's way of maintaining a healthy mature forest. In the past, we believed that we were helping to protect the forests by preventing forest fires. But after years of doing this, foresters realized that forests were becoming more vulnerable and plant diversity was shrinking. Some nationals parks have adopted a policy of periodically burning sections of the forests in controlled ways to help eliminate extreme and destructive wildfires as well as increase soil nutrients. Climate change leads to hotter drier temperatures, which also makes these forest fires more dangerous and destructive.

Additional resources

Climate Change Indicators: Wildfires, www.epa.gov/climate-indicators/climate-change-
   indicators-wildfires#

2.6.9 Climate change impacts: humanity

As the impacts of climate change worsen, many people are forced to leave their homes because of climate disasters and dwindling resources. World Bank Report conservatively estimates that there will be 140+ million climate refugees by 2050, but it will likely be many more (Kirby 2018). These refugees will face many hardships and will face economic, infra-structural, and political stress. There is a growing body of literature that shows that climate insecurity can be linked to increased incidence of war throughout 12,000 years of human history. As our climate changes, so does our culture. Societies whose culture, behavior, and beliefs are already very tied to the patterns of nature are being forced to confront and respond to climate changes. To learn more about human settlement patterns and their effects, see Chapter 3.

Financial impacts

The financial impacts of climate change are staggering. National Oceanic and Atmospheric Administration's (NOAA)'s research indicates a dramatic shift in the amount, intensity, and costs of storms. While one individual storm cannot be tied directly to human-induced climate change, the pattern of destruction is unmistakable. 2016 and 2017 are the two highest years of financial damage on record in the U.S. with over 15 billion dollars of damage each year.

Additional resources

Billion Dollar Weather and Climate Disasters: Table of Events, www.ncdc.noaa.gov/billions/events/US/1980-2018

Conclusion

We are, indeed, in a crisis of epic proportions. Many decision makers in the world have yet to move past the Age of Industry, remaining stuck in a worldview dominated by the motivations of maximizing profit, unlimited growth, and infinite resources. Consumerism continues to drive the overuse of material goods and energy to satisfy the desire for human excess. Clearly, there are many significant barriers to the widespread change required to address the environmental threat we face.

Chapter 3 explores the reasons why humans continue to destroy their own environment or fail to take definitive action in the face of certain destruction. Chapters 4 and 5 introduce foundational frameworks and approaches to the problems with a discussion about sustainable values and integral sustainable design. Chapters 6 through 8 study the emerging solutions of bio-inspired design, resilience, and health and well-being centered design. Chapter 9 covers integrative design processes as a means to discover and implement bold solutions that rely upon empathy, transdisciplinary design, and rating systems. In Chapters 10–14, the opportunities of sustainable design are presented by scale, starting with the global scale, followed by the urban, district, site, buildings, interiors, and products scale. The book concludes with a look at change management as a means to equip the reader with some tools to go out and attack the problem of their choice.

The environmental crisis outlined in this chapter may seem daunting, overwhelming, and frightening, but the human species is remarkably intelligent, inventive, and resourceful. If enough people join in the effort to make changes, the problems we face are not only solvable but may just be the catalyst to reveal the empathetic, compassionate, and altruistic sides of ourselves. And if that happens, the future is indeed bright.

Additional resources

Years of Living Dangerously Film Series, https://facebook.com/YearsOfLiving/

References

Amadeo, K., 2018. *Nuclear Power in America, How It Works, Pros, Cons, and Its Impact*. The Balance. www.thebalance.com/nuclear-power-how-it-works-pros-cons-impact-3306336 edn.

Asia Pacific, 2014. *Beijing Marathoners Don Face Masks to Battle Toxic PM2.5 Smog*. Science Health. www.japantimes.co.jp/news/2014/10/19/asia-pacific/science-health-asia-pacific/beijing-marathoners-don-face-masks-to-battle-toxic-pm2-5-smog/#.W1kXTtJKiUk edn.

Baldé, C.P., Forti, V., Gray, V., Kuehr, R. and Stegmann, P.T., 2017. *Global E-Waste Monitor – 2017*. Bonn, Geneva and Vienna: United Nations University (UNU), International Telecommunication Union (ITU) and International Solid Waste Association (ISWA).

Bennett, L., 2017-last update, Deforestation and Climate Change [Homepage of Climate Institute], [Online]. Available: http://climate.org/deforestation-and-climate-change/ [July 29, 2018].

Carrington, D., 2017. *Plastic Fibres Found in Tap Water Around the World, Study Reveals*. Guardian News and Media Limited. www.theguardian.com/environment/2017/sep/06/plastic-fibres-found-tap-water-around-world-study-reveals edn.

Center for Biological Diversity, 2018-last update, The Extinction Crisis [Homepage of Center for Biological Diversity], [Online]. Available: www.biologicaldiversity.org/programs/biodiversity/elements_of_biodiversity/extinction_crisis/ [July 29, 2018].

Cho, R., 2018. *Losing Our Coral Reefs*. Earth Institute, Columbia University. https://blogs.ei.columbia.edu/2011/06/13/losing-our-coral-reefs/ edn.

Derocher, A.E., Lunn, N.J. and Stirling, I., 2018. *Polar Bears in a Warming Climate*. Oxford. https://academic.oup.com/icb/article/44/2/163/674253 edn.

EPA, January 10, 2018c-last update, Estimated Animal Agriculture Nitrogen and Phosphorus from Manure [Homepage of United States Environmental Protection Agency], [Online]. Available: www.epa.gov/nutrient-policy-data/estimated-animal-agriculture-nitrogen-and-phosphorus-manure [July 26, 2018].

EPA, June 20, 2018b-last update, Health and Environmental Effects of Particulate Matter (PM) [Homepage of US Environmental Protection Agency], [Online]. Available: www.epa.gov/ghgemissions/overview-greenhouse-gases [July 25, 2018].

EPA, April 11, 2018a-last update, Overview of Greenhouse Gases [Homepage of US Environmental Protection Agency], [Online]. Available: www.epa.gov/ghgemissions/overview-greenhouse-gases [February 13, 2018].

FAO, 2018-last update, Food Loss and Food Waste [Homepage of Food and Agriculture Organization of the United Nations], [Online]. Available: www.fao.org/food-loss-and-food-waste/en/ [July 29, 2018].

Geyer, R., Jambeck, J.R. and Law, K.L., 2017. *Production, Use, and Fate of All Plastics Ever Made*. American Association for the Advancement of Science. http://advances.sciencemag.org/content/3/7/e1700782.full edn.

Harmony, 2018-last update, The Facts about Textile Waste [Homepage of Harmony Enterprises], [Online]. Available: https://harmony1.com/textile-waste-infographic/ [July 29, 2018].

Hughes, T.P., Kerry, J.T., Baird, A.H., Connolly, S.R., Dietzel, A., Eakin, M.C., Heron, S.F., Hoey, A.S., Hoogenboom, M.O., Liu, G., McWilliam, M.J., Pears, R.J., Pratchett, M.S., Skirving, W.J., Stella, J.S. and Torda, G., 2018. *Global Warming Transforms Coral Reef Assemblages*. Springer Nature Limited. www.nature.com/articles/s41586-018-0041-2 edn.

Hurwitz, Z., 2011-last update, Mining Giant Joins Belo Monte Dam [Homepage of International Rivers], [Online]. Available: www.internationalrivers.org/blogs/258/mining-giant-joins-belo-monte-dam [July 29, 2018].

Investopedia, 2018. What Are the Effects of Fracking on the Environment? https://investopedia.com/ask/answers/011915/what-are-effects-fracking-environment.asp [July 29, 2018].

IUCN, 2018b-last update, International Union for Conservation of Nature Website [Homepage of International Union for Conservation of Nature], [Online]. Available: www.iucn.org/ [July 25, 2018].

IUCN, 2018a-last update, IUCN Redlist of Threatened Species [Homepage of International Union for Conservation of Nature and Natural Resources.], [Online]. Available: www.iucnredlist.org/ [July 29, 2018].

JWN Staff, 2016-last update, Infographic: 17 Things that can be Made from One Barrel of Oil [Homepage of JWN Glacier Media Inc], [Online]. Available: www.jwnenergy.com/article/2016/9/you-might-be-surprised-these-things-come-barrel-oil/ [July 29, 2018].

Kahn, B., 2017. *We Just Breached the 410 Parts per Million Threshold.* Climate Central. www.climate central.org/news/we-just-breached-the-410-parts-per-million-threshold-21372 edn.

Kirby, A., 2018. *World Bank Expects 140 Million Climate Refugees by 2050.* Zuade Kaufman. www.truth dig.com/articles/climate-refugees-may-reach-many-millions-by-2050/ edn.

Livescience, 2018. Oklahoma Suffers Its 2,724th Earthquake since 2010. www.livescience. com/62410-oklahoma-earthquakes-wastewater-injection.html [August 4, 2018].

Lowry, C.A., Hollis, J.H., de Vries, A., Pan, B., Brunet, L.R., Hunt, J.R.F., Paton, J.F.R., van Kampen, E., Knight, D.M., Evans, A.K., Rook, G.A.W. and Lightman, G.A.W., 2007. *Identification of an Immune-Responsive Mesolimbocortical Serotonergic System: Potential Role in Regulation of Emotional Behavior.* Elsevier. www.sciencedirect.com/science/article/pii/S0306452207001510?via%3Dihub edn.

Molles, M.C., 2015. *Molles, M. (1999) Ecology: Concepts and Applications.* 7th edn. Boston, MA: McGraw-Hill Education.

NASA, July 25, 2018-last update, The Relentless Rise of CO_2 [Homepage of NASA's Jet Propulsion Laboratory], [Online]. Available: https://climate.nasa.gov/climate_resources/24/graphic-the-relentless-rise-of-carbon-dioxide/ [July 25, 2018].

NASA, 2017-last update, Study Finds Drought Recoveries Taking Longer [Homepage of NASA Jet Propulsion Laboratories], [Online]. Available: https://climate.nasa.gov/news/2617/study-finds-drought-recoveries-taking-longer/ [July 29, 2018].

Nogrady, B., 2016. *Your Old Phone is Full of Untapped Precious Metals.* BBC. www.bbc.com/future/story/ 20161017-your-old-phone-is-full-of-precious-metals edn.

North Carolina Climate Office (NCCO), 2018-last update, Composition of the Atmosphere [Homepage of North Carolina Climate Office], [Online]. Available: https://climate.ncsu.edu/edu/Composition [July 25, 2018].

NRDC, 2018. *10 Threats from the Canadian Tar Sands Industry.* https://nrdc.org/stories/10-threats-canadian-tar-sands-industry.

Nuwer, R., 2014. *Offshore Wind Farms Offer Seals a Smorgasbord of Fish.* Smithsonian. www.smith sonianmag.com/smart-news/seals-have-figured-out-offshore-wind-farms-offer-smorgasbord-fish-180952117/ edn.

Oxford English Living Dictionary, 2018b-last update, Definition of Artificial [Homepage of Oxford Press], [Online]. Available: https://en.oxforddictionaries.com/definition/artificial [July 24, 2018].

Oxford English Living Dictionary, 2018a-last update, Definition of Nature [Homepage of Oxford Press], [Online]. Available: https://en.oxforddictionaries.com/definition/nature [July 24, 2018].

Princeton, 2014. *A More Potent Greenhouse Gas than Carbon Dioxide, Methane Emissions will Leap as Earth Warms.* Science Daily. www.sciencedaily.com/releases/2014/03/140327111724.htm edn.

Ramanujan, K., 2012. *Insect Pollinators Contribute $29 Billion to U.S. Farm Income.* Cornell University. http://news.cornell.edu/stories/2012/05/insect-pollinators-contribute-29b-us-farm-income edn.

Rasmussen, C., 2018-last update, Far Northern Permafrost may Unleash Carbon within Decades [Homepage of NASA], [Online]. Available: https://climate.nasa.gov/news/2691/far-northern-permafrost-may-unleash-carbon-within-decades/ [July 29, 2018].

Sass, J., 2011. *Why We Need Bees: Tiny Workers Put Food on our Tables.* Natural Resources Defense Council. www.nrdc.org/sites/default/files/bees.pdf edn.

Sweeny, G., 2015. *It's the Second Dirtiest Thing in the World—And You're Wearing It.* AlterNet Media. www.alternet.org/environment/its-second-dirtiest-thing-world-and-youre-wearing-it edn.

Ukela, 2017-last update, Sources of Water Pollution [Homepage of UK Environmental Law Association], [Online]. Available: www.environmentlaw.org.uk/rte.asp?id=90 [July 26, 2017].

US Fish and Wildlife, 2018, Nutrients: *Too Much of a Good Thing* [Homepage of US Fish and Wildlife Service], [Online]. Available: www.fws.gov/nc-es/edout/albenutrient.html [July 26, 2018].

White, C., 2015-last update, Resilience: Growing Topsoil [Homepage of Post Carbon], [Online]. Available: www.resilience.org/stories/2015-11-10/growing-topsoil/ [July 29, 2018].

World Waste Facts, 2018-last update, World Waste Facts – Tons of Waste Dumped [Homepage of The World Counts], [Online]. Available: www.theworldcounts.com/counters/shocking_environmental_facts_and_statistics/world_waste_facts [July 29, 2018].

World Wildlife Fund, 2016. *Living Planet Report 2016.* https://worldwildlife.org/pages/living-planet-report-2016 [August 4, 2018].

3 Motivations for sustainable design

3.0 Introduction

The full scope of environmental challenges facing society are covered in Chapter 2. From significant climate change impacts, to staggering loss of biodiversity to extreme pollution, the planet earth is in peril and it's time to take action. This chapter marks the beginning of the long search for solutions. We begin by examining the competing motivations of self-interest and empathy. By recognizing the most fundamental drivers of human behavior in ourselves and in our communities, we can begin to work out a grand narrative for sustainable design – a foundation upon which to develop the more recognizable aspects of sustainable design such as bio-inspired design, resilience and health and wellness. Only when we begin to see the underlying motivations of change, will we be able to envision a future society where the built environment works to regenerate the ecosystem, promote health and well-being, and provide equal access and opportunity (Figure 3.0a [below]).

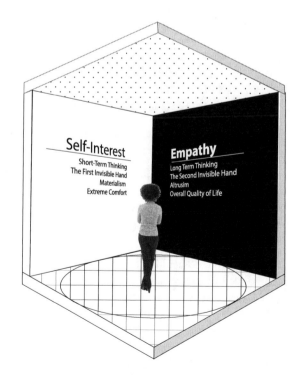

Figure 3.0a Self-interest versus empathy.
Source: Created and drawn by the authors.

3.1 Humanity, space, and time

In Chapter 1, we took a quick fly through of history to understand how our worldview shifted our relationship to the natural world, and that affected our decisions regarding the built environment and how we used the natural resources available to us. Here, we look at the motivations underneath each successive worldview.

As hunter-gatherers, we evolved from a connected society that was integrated with the natural world. As the planet entered an interglacial period called the Holocene, the Age of Agriculture began and we learned to *manage nature* via farming and animal husbandry. Humanity settled into villages and towns initiating a population explosion because of plentiful food, shelter, and the advent of medicine. With the Industrial Age, human population continued to rise and technological advancement made new levels of comfort possible and increased the availability of material possessions. To achieve technological advancement and wealth, we learned how to *dominate nature*, taking resources and transforming them into buildings and products. Many saw an increase in their quality of life, while many more worked in coal mines or factories with long hours, for low wages, in horrible or dangerous conditions. By the worldview shift to the Age of Information, human population continued to rise. Society was deeply focused on reaching higher levels of affluence, resulting in a consumerist culture that was dominated by white-collar, information-based employment. Rachel Carson published *Silent Spring*, profiling the long-term damage to the environment by pesticides. Society was becoming aware of the need to *save nature*, still demonstrating human power or dominance over the natural world. The environmental movement began in earnest leading to the creation of the concept of sustainable development. The continued struggles for civil rights during that time led to the inclusion of social equity as a component of sustainable design.

We are now at the doorstep of the next worldview shift to the Age of Integration. Humanity is beginning to see a *unified nature* where humans and the natural world are interrelated. Solutions and strategies are sought that establish a co-creative and coevolutionary processes, focused on synergies, efficiencies, and reduction. The science and design professions are seeking new ways of working *with nature*. The focus is on success for all species, human, plants, and animals, in an interdependent relationship – a return, in a sense, to the Age of the Hunter-Gatherers. The effects of previous worldviews have resulted in negative impacts we are seeing such as climate change and the associated rising temperatures, drought, sea level rise, and host of other life-threatening events. The Holocene climatic period is giving way to the Anthropocene, the first human-induced climatic period. Technology, previously used to increase wealth, needs to be redirected to restoring the balances of nature.

3.2 Humanity and nature

How did we get to this point? Why have we allowed ourselves to enter into such a warlike relationship with the natural world? After all, humans are an amazing species, self-aware, highly intelligent, empathetic, and remarkably adaptable. And therein lies the crux of the problem. Our ability to physically adapt to any ecological condition also comes with the ability to mentally adapt as well, to convince ourselves that everything will be "OK" – that someone else will solve these problems, or that the problems are not really that threatening.

$$I = P \times A \times T \left(I = PAT \right)$$

In the early 1970s, Paul Ehrlich, among others, established the equation: I = PAT. It describes the relationship between environmental impact (I), human population growth (P), affluence (A), and the use of technology (T) to deliver goods and services to the society.

3.2.1 Environment impact

Impact is the sum total of the effect upon the earth's ability to sustain itself as a self-regulating system. In the previous chapter, the environmental impacts of humankind were well documented, and the likelihood that the earth's resources can continue to support life as it currently exists, is seriously in question. *Beyond Limits* and other publications in the 1970s studied the earth and it's *carrying capacity* to support human life. They realized that humanity was, in fact, on a collision course with itself and would cause its own destruction. The I = PAT equation is an elegant approach to describing the problem and identifying solutions.

3.2.2 Population growth

A large number of problems that we face today are not due solely to human indifference or evil polluters, but simply due to the fact that there are so many of us. The population growth rate would be considered a wonderful thing if we thought that the earth could support us. The explosion of human population is a perfectly normal and natural occurrence, given the remarkable characteristics of the human brain, our opposable thumbs, our unique ability to adapt varying climates for settlement, and the eradication of predators. In the equation I = PAT, reducing the population rate is a logical place to attack environmental problems. China enacted a "one child" policy which helped to curb population growth and led to a wide range of positive and negative impacts. The United Nations sees education of women as a primary strategy to reducing population growth. Overall, the planet's population growth rate has slowed offering a glimmer of hope for an overburdened planet.

3.2.3 Affluence

The two most populated countries in the world, India and China, are growing in *affluence*, placing a greater burden on ecosystem goods and services. The desire for physical comfort, safety, and shelter is a perfectly natural motivation. The problem is the materialistic drive towards higher and higher levels of consumption and consumption being an indicator of success. If this continues unabated, the level of environmental impacts will continue to grow. Therefore, systemic change in the societal views that drive human behavior must change to reach some level of sustainability at the global scale. Sustainable designers cannot by themselves convince people to stop consuming. Something that sustainable design strategies can do is mitigate the negative effects of consumption on the environment.

3.2.4 Technology

Technology need not be a negative factor in the I = PAT equation. Currently, fossil fuel-based technologies and energy systems have a negative impact on the environment. When more people with more affluence demand more fossil fuels, the overall impact on the environment increases exponentially. The number for "T" becomes larger. Consider for a moment the positive effects renewable energy systems could have on the equation. "T" becomes smaller when renewable energy systems such as solar and wind are used. Technology can be a tool to reduce impacts and possibly even reverse them in some cases. For example, designing a refrigerator that uses 95% less energy than the same model did ten years ago is an example of how design can play a role – not in changing the culture, but at least mitigating the negative impacts of technology.

I = PAT becomes a much more sophisticated equation when the variable of green design and clean energy are added, which were all, technologies not yet developed or readily available in the 1970s. The world population is rising slower than previously anticipated, suggesting a positive change in the formula. The determining factor to whether positive change will continue to take place fundamentally lies in changing how societies view population growth, affluence, and technology, all key to the emerging integral worldview.

3.3 Motivations for sustainability

Humanity will face the following challenges within the next ten years: extreme weather events, natural disasters, and the failure of climate change mitigation. These challenges illuminate the stark reality that we are very far away from the type of sustainable future that is advocated for in this book. In fact, the slow movement towards addressing climate change has left us in the unenviable position of reacting to the effects of climate change rather than proactively seeking comprehensive solutions.

Maximization of profit or doing well

In Figure 3.3a (next page), each of the two fists represent fundamentally different motivations for human existence. On the left is the ever-present motivation of self-interest – a constant reminder, from millions of years of evolution, that survival is the prime directive. The desire to survive comes with an ever-growing demand to increase comfort and security. It is often said that money is the "root of all evil." The term "profit" expresses the accumulation of wealth, which brings financial security and enables us to buy what most social structures define as success, or the "good life." It's not money, wealth, or profit that is the "root of evil," it is being exclusively self-interested and short sighted, thinking only of financial profit regardless of the cost or harm to others and the environment.

Today, we see the legacy of self-interest gone astray: air pollution, water scarcity, loss of biodiversity, ocean acidification, and many more. Despite the clear and present threat to the quality of life for present and future generations, the environment remains under attack: Tar sands extraction in Alberta destroys millions of acres of pristine land, leaching toxic chemicals into wetlands; rainforests are burned daily, emitting millions

Self-Interest   **Empathy**

Figure 3.3a The two invisible hands of sustainability held in opposition.
Source: Created and drawn by the authors from stock photo.

of tons of carbon dioxide to make space for methane-emitting cattle ranches. The wide-spread use of factory farming leads to dangerous levels of phosphorus and nitrogen in the soil.

The temptation is to brand these activities as "evil." Yes, the quest for profit often leads to "evil deeds," but the survival instinct is embedded very deeply in our psyche it is akin to instinct. For thousands of years, societies have been focused on financial profit as the indicator of success. We as a culture have come to associate this type of activity with "rational thinking" or the "logical" expression of human purpose. Financial and material growth are seen as the measure of human progress for many people.

This logical, self-interested thinking is often used as justification for many human tragedies in history: the plundering of nature's resources, the colonization of indigenous countries, and the widespread use of slavery. This is the dark underbelly of the industrial worldview: The ends justify the means – especially when maximizing profit is concerned. Sadly, much of the world still holds this view today. And yet, the Industrial Age brought about the great advances in medicine, sanitation, energy and communications technology, and manufacturing that have greatly improved the quality of life and the health and well-being of generations of people. The past is a platform to discover and pursue a new, more positive relationships with nature, and with each other.

Maximization of goodness – doing good

While the motivation to maximize profit will always be there, a second invisible hand, shown in Figure 3.3a [above] on the right, represents a second fundamental human motivation – empathy. Empathy, like self-interest, is an inherent characteristic of humans. It draws from the ethical core of each person to motivate each of us towards the long-term, shared benefits of a healthy environment and equitable society.

Altruism, ethics, and "goodness" are the most apparent manifestations of empathy. Unlike self-interest, the benefits of empathy are not always measurable or physically seen. If a person is altruistic, they may derive happiness and satisfaction by completing a good deed. In the example of a design project, each team member derives satisfaction from collaborating because it creates a solution that benefits and improves everyone involved. This leads to "enlightened self-interest." Empathy is the primary human motivation that drives collaboration and cooperation in pursuit of the greater good and long-term happiness.

Figure 3.3b The two invisible hands working together to achieve sustainability.
Source: Created and drawn by the authors from stock photo.

In short, self-interest compels us to do *well*. Empathy compels us to do *good*. Like self-interest, empathy has its own underbelly of negativity in the form of overregulation and over-taxation which can stifle the optimistic entrepreneur.

Throughout history, the two motivations of doing well and doing good have been portrayed as being in opposition, with a black-and-white choice of profit or empathy or as "good" versus "evil." Extreme capitalism gone astray seeks to maximize profit at the expense of all else. The flip side is communism or socialism whose professed motivations are to meet the needs of the many but end up repressing individualism, entrepreneurship, invention, and creativity, all fundamental human traits necessary for productive and successful societies.

Admittedly, this is an oversimplification of two extremes, and there are many exceptions and nuances to this argument, but the broad, general descriptions highlight the difference between the two motivations.

We are all prejudiced towards our own positions, and interestingly, depending on which side one stands, "good" and "evil" have different meanings or actions associated with them. Those focused only on profits are against overreaching governments when they require an increasingly higher percentage of profits through higher taxes, increasing environmental regulations, and imposing laws like the minimum wage because these are all direct threats to the maximization of profit. Those focused only on empathy want greater government control and solutions to low wages, poor living and working conditions, environmental destruction, greed, corruption, and conspicuous consumption.

We need to ask: how can we find a third way? Can our inherent self-interest also support empathic goals of helping each other? Can helping each other generate profit? Can we find a model that removes the duality of good versus evil to build a more ambitious model of doing well *and* doing good? (see Figure 3.3b [above]).

The moment a new social model is understood, a leap in consciousness occurs. This happens at the personal level, and when enough people shift their thinking, entire societies change consciousness, leading to changes in the way they operate and a new worldview is born. We as a society are understanding that both profit and empathy are necessary. Profit is necessary to fund research, pay employees, provide healthcare benefits, give back to charitable organizations, or do pro bono work, and continue to support all the physical aspects of running a business that provide the goods and services necessary for our lives. The third way looks to create profit in responsible, sustainable, ways and uses it in a fair and empathic way (Figure 3.3b [previous page]). This requires engagement and creativity, but some major corporations and small businesses are finding this approach to be very effective. The benefits are of higher value than pure monetary gain because they include direct and indirect, short and long term, physical and economic and, affect and reach individuals and communities. These types of benefits also reach the deeper more permanent, emotional, spiritual, and moral aspects of our individual being and society as a whole.

Additional resources

World Economic Forum, www3.weforum.org/docs/WEF_GRR18_Report.pdf

3.3.1 The first invisible hand – self-interest

The core tenets of the Age of Enlightenment (1685–1815), reason, individualism, and skepticism, illustrate the focus on the quantifiable, measurable aspects of reality. This was a much-needed and welcome shift from the previous Agricultural Worldview of reality as mysterious, under the control of the gods and unpredictable. The Age of Industry was buttressed by the emerging sensibility that the world could be defined, measured, described, and eventually manipulated for human benefit. Most of humanity today benefits from the scientific approach of reality in the form of better medicine, better technology, increased comfort, longer life, and happiness – for some. However, there has been a price to pay for this approach.

Many philosophers during that period wrestled with self-interest as a primary motivator. Jeremy Bentham is probably the most famous one to address self-interest, breaking it down into the simple motivations of maximizing pleasure and avoidance of pain. In 1780, he published *An Introduction to the Principles of Morals and Legislation*, where he wrote:

> "Nature has placed mankind under the governance of two sovereign masters, pain and pleasure. It is for them alone to point out what we ought to do, as well as to determine what we shall do."

> (Bentham 1988)

Here, we see the absence of the motivation of empathy, the sole focus on self-interest as a "sovereign" duty or responsibility. A "sovereign" refers to the absolute leader, usually of a country such as a king or queen.

Many other thinkers, philosophers, and politicians of the time were exploring individualism, individual rights, and the role of self-interest in government, but Bentham's simplified view of human motivation expresses the roots of society's current dilemma today. Affluence, population growth, and the use of technology to either increase comfort or minimize pain build the triad of an unsustainable society as defined by the equation I = PAT discussed earlier in this chapter.

The movement towards individualism and reason meant the beginning of the end of religion as the primary driving force underneath a worldview. The Age of Agriculture's focus on the village or clan as the collective, and upon goodness as expressed through god and worship, the connection to the land, and the expectation of long-term benefits through the actions of the collective changed. In the Age of Industry, the worldview changed to prize the individual, whose behavior was rationalized by short term "progress" or gain. Corporate structures were developed for protecting businesses and individuals, the environment, and lower social groups were viewed as resources to be used to reach a profitable end.

Adam Smith – invisible hand

Adam Smith was the author of *The Theory of Moral Sentiments* in 1759 and *The Wealth of Nations* in 1776. *The Wealth of Nations* introduced the concepts of Free Market Economy, Supply and Demand, and Division of Labor, all very common concepts today, but radically different from the current business and economic climate of the time.

His view was that by allowing people to freely pursue their own economic interests, society as a whole would benefit. The merchant would want to produce the best possible product to inspire the consumer to choose their product. The consumer would have a better quality and larger selection. Supply and demand addresses the theory that the amount of available products will increase if there is a demand for them, and decrease if not. This also addressed the dependency of pricing on the demand relative to the supply. Division of Labor addressed breaking each task within production and increasing efficiency. This is the basis of assembly lines and industrial factories during the industrial age. Inherent in Smith's argument was the presence of the rule of law, moral codes, and holistic benefit to society. The principles outlined in *The Wealth of Nations* are still used today.

In *The Theory of Moral Sentiments*, written before *The Wealth of Nations*, Smith argued that pure selfishness was not the goal of his free market economy. It must include a conscious and clear moral capacity (Smith 2011). Included was consideration and sympathy for society at large, an example of empathy. Smith also recognized the dangers of a globalized society, arguing for the benefits of a local economy. Although Smith argued against mercantile restrictions and other intrusions into the free market, and a wealthy landowning man himself, the ideas he expressed demonstrated some early examples of "enlightened" self-interest.

Adam Smith described his principles this way: "It is not from the benevolence of the butcher, the brewer or the baker that we expect our dinner, but from their regard to their own interest." (Smith 2011) He believed that economic markets didn't need interference by governments or any other organizing body. He also believed that an "invisible hand" rewarded businesses that were most effective and responsible, balanced supply and demand, and set the prices of goods and services where they "should" be. On the opposite side, if consumers had sufficient information, they would choose the best products, and the producers of those products would be rewarded with sales. The producers that didn't make consumers happy would either change what they did or go out of business.

Smith was a deep thinker with sophisticated ideas about economics, morality, and the purpose of life. And yet, much of his valuable thinking was sadly misappropriated and misunderstood, and more importantly his "invisible hand" was misused. Other negative theories and ideas of the Age of Enlightenment have been attached to capitalism as well as influencing society. Darwin's phrase "Survival of the Fittest" often wrongly interrupted as "only the strong survive" offers even more ammunition to the false narrative that the free market is a natural expression of strength and power to be used for domination. And yet, the power of free market capitalism examined solely as a potential force for change – either positive or negative – has to be reckoned with and ultimately placed into broader service of a more empathetic set of basic operating parameters for a sustainable economy. In other words, capitalism in some form is needed to drive the sustainability movement.

The free market

Without profit, business owners and researchers cannot afford to explore and manufacture technologies or make the changes necessary to drive our present economic system toward sustainability. However, it comes with a set of severely damaging assumptions that continue to inhibit progress towards sustainability.

Some of the major assumptions of the free market include the following:

- Resources are infinite or substitutable.
- Long-term effects are discounted or considered "externalities" that the company does not have to worry about.
- Costs and benefits are internal.
- Infinite growth is the measure of success.

Infinite resources

The perception that resources would be considered "infinite" and that growth could be infinite makes sense since the size of the earth during the Industrial Revolution was still considered so massive that there was no reason to imagine a scarcity of resources or materials. This sense of abundance is expressed throughout capitalism and is a powerful motivator for profit seeking, but not an incentive to conserve resources – an important part of sustainability. This underscores the role of thinking across space and scale as a critical factor in sustainability. Today, we see the earth as much smaller and its resources as finite.

Substitutable resources

Resources that can be substituted to deliver the same function are *substitutable*. For example, copper wires for communication are substituted with fiber optics – essentially made of glass. If copper runs out or becomes too expensive, then fiber optics becomes a substitute. Photovoltaics are potentially a substitute for oil, but since oil is portable and solar power is not, a true substitution is not possible. However, battery technology is quickly evolving which will make solar and wind power more portable and more consistent in varying conditions.

Cost and benefits are internal

Keeping all decisions internal simplifies decision-making, but it ignores the input of stakeholders from outside the system who may be affected by the decisions. Stakeholders can be customers, suppliers, or nearby residents. Stakeholders are often the ones who feel the negative impacts of internally focused decisions that ruin wildlife areas or pollute.

Externalities

Pollution and now global warming are externalities to the widespread use of fossil fuels, beef farming (cows emit methane), and a host of other industries. Profit-driven thinking seeks to externalize these problems, leaving governments responsible for cleaning up the problems. This is a powerful way to maximize profits but a sure pathway to environmental destruction.

Short-term thinking

This assumption underscores the fixation with short-term, profit-motivated, decision-making processes. When the long term is not considered, complex decisions become much simpler and the ethical context of a single decision is simplified. Filtering out long-term externalities like climate change intensifies the short-term economic benefits. The net effect of this has yielded unprecedented wealth for some and a better quality of life for others. Still countless others have suffered for generations from the legacy of worker exploitation, colonialism, and slavery. Furthermore, the environment itself has also suffered greatly under this model.

Paul Hawkin's statement in his interview looking back on the publication of *The Ecology of Commerce* states: "This intransigency [slowness to change)] hasn't changed as much as one would have hoped, but is more about human nature than it is about commerce, the inability to subordinate short-term monetary gains to the long-term well-being of humanity" (Grist 2004). Time and again, the underlying motivations of humanity lie at the core of the long march towards a higher-level consciousness.

3.3.1.1 *Tragedy of the commons*

The tragedy of the commons is an economic concept originally developed by William Forster Lloyd in 1833 and became more widely known through the work of Garrett Hardin (1968). Hardin argued that "Individuals, acting independently and rationally according to each one's self-interest, behave contrary to the whole group's long-term best interests by depleting some common resource."

The "Commons" as in a common resource is meant to express any shared or open access to a resource such as air, water, or open space. The commons can also be labeled as ecosystem goods and services which include the climate, oceans, rivers, fish stocks, soil, and any other shared resource that are not privately owned or publicly shared.

The tragedy of the commons describes a problem when individuals attempt to gain the maximum benefit from a shared resource. As demand and competition for a shared resource rises, the supply of that resource over the long term is placed into jeopardy. As each individual continues to try and extract more value from that resource, for their own personal gain, the resource itself is exhausted or diminished and the entire group of people sharing that resource lose profit.

AN EXAMPLE OF THE TRAGEDY OF THE COMMONS

The example used most often for the tragedy of the commons is the issue of overgrazing a shared pasture. Quite simply, there are a maximum number of cows that can get enough food from the shared pasture to grow and remain healthy. When multiple farmers are sharing the pasture equally, everyone is receiving a consistent level of income – a good situation. However, if an individual farmer who seeks to increase profits, in the short term, adds more cows to the land, it decreases the resources available for all the other cows. Now, the pasture itself becomes a scarce resource because of overgrazing, and the farmers will compete to properly feed their animals, not thinking they could lose the pasture in the long run. As a result, everyone's cows are smaller and sicker, and are less profitable. Over the long term, the land is overgrazed, making it difficult to regenerate and be used again for grazing, thereby placing everyone's entire business in jeopardy.

The shared pasture is a metaphor for the entire planet earth. As more and more people populate the planet, and as more and more people try to extract value, the earth's ecosystem goods and services begin to break down. Clean water, clean air, energy, food and more all become scarcer, meaning diminished profit for investors and diminished quality of life for those that depend directly upon those resources for survival. In short, this is an unsustainable situation.

A RESPONSE TO THE TRAGEDY OF THE COMMONS

Governments are charged with establishing rules and guidelines to keep all the people and resources safe, to establish and maintain infrastructure, and to provide and protect the citizens. This also includes eliminating unfair advantages and monopolies, managing the commons, and providing safety nets for the disadvantaged. It is possible for economic growth to happen within limits set to protect the ecosystem and its resources. At the same time, these limits can also encourage intellectual and technological growth. Many countries have achieved this by blending free market and socialist concepts into hybrid models.

POSITIVE INCENTIVES

Incentives are a standard way to motivate companies to behave differently and invest in technologies that are beneficial to the long-term health of the commons. Tax incentives for the installation of solar arrays or rebates for electric cars are good examples. The government's role is to seek the long-term health of the planet, which affects the health of everyone. Positive incentives can be a way to change perceptions and spur change.

NEGATIVE INCENTIVES

Taxation, regulation, and penalties are another way to protect the commons by penalizing polluters with economic sanctions such as higher taxes and more stringent rules. It's argued that these tend to disincentivize investment by private business and encourage companies to relocate to other parts of the world where such penalties do not exist. Negative incentives can produce positive actions because change is made and the natural resources are protected. Regulations and ordinances are seen by some to be restrictive or not supporting individual freedoms, this is because the comprehensive, long-term effects of a choice are not always considered in free market thinking.

PRIVATIZATION

When a common resource is privatized, its care is assumed by someone who may be motivated by self-interest. It also means that restrictions for using this resource may be denied to the general population or a certain group. Different countries and different communities handle this question in different ways. The state of Hawaii has declared that all of its beaches are public property and that building is not allowed within a certain distance of the shore. There are also hundreds of access points with amenities so that all people have equal opportunity to share the commons. The Living Building Challenge, a rating system for sustainable projects, discussed in Chapter 13.9.3 does not allow land within gated communities to count towards their requirement for open space.

SUSTAINABLE DESIGN AND THE COMMONS

The commons that designers most often interact with are access to view, light, and air. These are not privately owned and therefore should be accessible to everyone. They are affected by the configuration and position of a building on its site. The streets of New York City in the early years of the skyscraper boom were getting darker and darker during the day because of the intrusion of tall buildings, reducing the common access to daylight. City leaders enacted a basic code requiring the design of tall buildings to step back from the edge of the property line in order to allow light to reach the streets. This seemingly simple regulation greatly reduced rentable square footage of high-rise buildings, but also greatly increased the amount of sunlight reaching the streets making life much more pleasant for everyone. This example underscores the important role of policy making in protecting access to common resources in the urban space, landscapes, buildings, and interior design.

Interior designers also deal with the commons. In a typical floor plan, the common benefits of light and air are reserved for those that work in the perimeter offices. Employees with lower levels of pay and status spend their days in office space with no access to fresh air, daylight, or views to the outdoors. Interestingly, studies have shown that employees working in such spaces are less productive than those with access to the light, air, and view (Seppala and Berlin 2017). Viewed strictly from the profitability side, loss of productivity equals loss of profit. Current trends in office design are reversing the placement of offices with open office space that allows light and air to enter deeper into the space and provides access to everyone. In this type of plan, offices are along interior walls facing the windows, with full glass walls that see out over other open spaces to the light. Building codes in European countries are now requiring all the employees have access to daylight and fresh air.

Additional resources

Sustainability and the 'invisible hand' of the market, https://aheadahead.wordpress.com/2010/02/21/sustainability-and-the-invisible-hand-of-the-market/

Why is the 'Invisible Hand' in the Middle of Smith's Works? https://fee.org/articles/why-is-the-invisible-hand-in-the-middle-of-smiths-works/

3.3.2 Cognitive empathy – the second invisible hand

While self-interest is a powerful driver of human behavior and has led to wealth, comfort, and security – for some, social structures and much of humanity are in a precarious place. Climate change and extreme weather are posing real and increasingly frequent threats, and widening extremes and stratification of wealth and resources is also threatening social cohesion. Empathy is presented here as the second invisible hand, not as a replacement for self-interest, but rather as a parallel fundamental human motivation. This second hand guides a whole host of human activities and market forces including, in part, the creation of sustainability. The mantra of *transcend and include* asks us to dwell in the ambiguous world of multiple right answers. We must build a new framework and dialogue for sustainability that is inclusive of multiple motivations. In other words, self-interest and empathy are both valid motivators.

Smith's invisible hand is a great descriptor of the unseen market forces and motivations that drive human behavior. Similarly, the second invisible hand activates market forces by tapping into an entirely different set of motivations. Empathy and altruism, two sides of the same coin, express a complex interaction between a person or groups of people and their morals, ethics, and beliefs. Altruism and empathy are woven into our DNA. They also drive the invisible forces behind our actions such as the creation of nonprofit organizations, voluntary donations by retailers to social/environmental causes, businesses choosing transparency and equity, citizens helping with disaster relief, or simply holding the door for a mother carrying a baby, all of which are acts of empathy and altruism. There are literally millions of examples of good deeds occurring every day.

Judy Wicks (2013), author of *Good Morning Beautiful Business*, wrote:

> If all people believed that life is interconnected, as the Eskimos and likely most other indigenous people do, then there would indeed be an invisible hand of enlightened self-interest guiding our decisions towards building an economy based on sharing, caring, and cooperation that would, in fact, serve the interests of society.

Society is still struggling to acknowledge and incorporate the second invisible hand thinking into mainstream for-profit companies. However, the tide is turning, and rest of this chapter will show how empathy lies at the heart of sustainability. Underneath still deeper, buried somewhere in the ethical core of society are empathetic drivers at work constantly evolving and shaping the world on behalf of all that is good.

Affective empathy

Empathy is a tricky word, often associated with a soft, fluffy image of a person helping a bystander. This is the first type of empathy called *affective empathy* or *emotional empathy*. It occurs when a person's display of emotion is mirrored within our own emotional state. For example, when seeing a person in distress and exhibiting fear in their movements and facial expressions, like crying, a mirror effect is happening to the onlooker, who "feels" for the person in distress and may sometimes find themselves crying. This is a normal biological reaction to a given situation, and we are hardwired to experience this for the obvious group benefit of shared survival. As much as we seek to be individual beings, our DNA is configured to be sensitive to group feelings and dynamics (Therrien 2018). In other words, we are hardwired to help our fellow human, especially if they are in our family and/or tribe. This is a perfectly natural and expected human trait. This human trait plays out in the moment, in the short term, and any assistance brought is a selfish attempt to relieve the anxiety generated by the other person.

Cognitive empathy

Cognitive empathy is defined as "… the capacity to understand another's perspective or mental state" (Lesley University 2018). Cognitive empathy is also referred to as "perspective taking," the tendency to spontaneously adopt others' psychological perspectives. The ability to assume the perspective of another individual is not something that happens automatically or instinctually. In fact, many of us are so locked into our worldview that it is very difficult to see things differently. But this is the essence of the second invisible hand – to be able to transcend your own selfish instincts and see the bigger picture, longer into the future.

Cognitive empathy is also a prerequisite to effective collaboration across disciplines and with project stakeholders – a critical aspect of the sustainable design process. The ability for an architect to work with an engineer is enhanced when each discipline not only respects the other but also is able to assume the viewpoint of that person. In other words, the architect can take the perspective of an engineer and look at a project through a completely different lens. This reveals all sorts of design opportunities and potential problems that were previously invisible.

Cognitive empathy from one person towards another is easy to understand. But what happens if empathy is expanded in the form of a circle around larger and larger groups of people. Examples of this are easy to see. A parent extends their empathy for the entire family, and the children become part of an empathic circle. The circle grows to include all extended family members and close friends, neighborhoods, larger communities, and even nations.

In short, biological evolution describes the extinction of some species while others thrive. Genes play a major role in this, and the evidence is mounting each year that we are as much biologically disposed to possess genes that produce altruism and empathy as we are to be self-interested. In other words, we are not only savage self-interested brutes but also capable of glorious acts of kindness and compassion at the drop of a hat. The truth, of course, is that we are both brutish, self-interested toddlers and wise, sage-like altruists. Sustainability, as we will see, tries to resolve these two motivations into a third driver of human behavior.

The baby lab

In a multi-year study at Yale, scientists discovered that babies as young as eight months are already discerning between right and wrong (Wallace 2014). Experiments conducted with toy bears playing out morality situations in front of babies were able to reveal that the babies exhibit a high level of ethical reasoning, without the benefit of socialization by the parents (Wallace 2014). At the same time, the babies also exhibited an innate sense of bias, a form of tribalism based on the most superficial differences between the toy bears. The roots of bigotry and racism are already present inside a baby upon arrival. Socialization by family and friends can overcome both the good and bad instincts within us. Sustainability is clearly rooted in an ethical foundation, but the self-interested person inside, the biased person, continues to exist and exert its forces in decision-making. We are beginning to see the complexity and impossibility of a simple good versus evil moral argument.

3.3.2.1 Empathy across space – globalization and private development

Being able to project cognitive empathy across space is a critical aspect of broadening the discussion around the role of empathy in sustainability. Try and imagine the faraway farmer dealing with drought, the faraway hunter watching rising seas overtaking their hunting grounds, or the far way fisherman catching fish with plastic in their bellies. Taking the perspective of citizens, we will never meet helps to "bring sustainability home." It becomes more real and more tangible. We begin not only to see the world as an entire interconnected system but also to connect to the people, animals, and ecosystems that are suffering the damage or consequences of bad decisions (Figure 3.3.2.1a [below]).

This is where globalization comes into play. From one point of view, it's easy and necessary to argue the negative impacts of globalization: Water pollution, loss of cultural heritage, and increased carbon emissions from transporting products are a few of the negative

Empathy Across Space

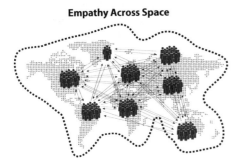

Figure 3.3.2.1a Empathy across space.
Source: Created and drawn by the authors.

expressions of globalization. However, when imagining globalization as a gigantic force for spreading altruism and empathy, the possibility of a global shift towards a sustainable future becomes more possible. Of course, the position of the authors as Americans affords us the luxury of making this argument, and we are aware of how the phenomenon of globalization can present painful experiences for many reading this book. Even still, what if the sheer power of globalization was turned into an overwhelming force for positive change?

For the sustainable designer, empathy across space plays out in very straightforward ways, especially through the specification of materials for a project. Many rating systems reward, or require, design teams to specify materials within a certain radius not just to reduce energy from shipping but also to stimulate the local economy. More to the point, specification of certain rainforest woods can either hurt or benefit a faraway culture. If a local company far away is sustainably harvesting hardwoods and making a profit and reinvesting that profit locally, it might be a good decision to specify that product regardless of the trade-off of travel distance. On the other hand, if the product is produced without regard to sustainability or to the local economy, the justification for the use of that product is unethical. Sustainable designers spend time weighing these variables during the decision-making process.

3.3.2.2 *Empathy across time*

Empathy across time is the most abstract but the most powerful expression of the empathic responses discussed so far. The concept of having concern for humans yet to be born requires a great leap in cognitive empathy. It is abstract. These unborn future generations are not necessarily part of your family, your country, or your religion, and yet, the expectation of sustainability is to have empathy for these people. In that sense, one must be able to transcend difference, denying our own tribalistic impulses to fear the "other." And assume the perspective of some future global citizen or someone who is completely unrelated to you. That is a tall order (Figure 3.3.2.2a [below]).

The Iroquois Nation developed the Seventh Generation Principle where every decision made by the group must consider its impacts seven generations into the future, approximately 150 years into the future. Clearly, the Iroquois have their roots in the hunter-gatherer culture, so the impulse to coexist with nature was already there, but placing the concepts in writing and using it as a basis for group decisions showed their commitment to empathy across time (Indigenous Corporate Training 2012). Each choice that we make today will affect people 150 years from now. For architects and designers, the site, building configuration, energy, and mechanical systems are elements of the building that are long lasting and hard to change. Future generations will pay the price for what is built today. If we are thoughtful about design projects, renewable energy systems can be added later, but if our

Empathy Across Time

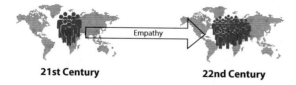

21st Century   **22nd Century**

Figure 3.3.2.2a Empathy across time.
Source: Created and drawn by the authors.

buildings are oriented in the wrong direction, that is impossible to fix. In Chapter 7, we will look at how to consider the future in the design process.

3.3.2.3 Empathy across difference

HUMANITY

Everyone has biases that pertain to their particular ethnic, cultural, socioeconomic, or professional background – not from some evil intent or purpose, but because that is the only reality that we know, and because humans have a deep connection to their "tribe" or group of belonging. Our family of origin, and all the experiences from childhood to the present have an effect on how we see the world and people's actions. Because of these ingrained biases, that most of the time we are not aware of, it can be hard to truly see an issue from another's perspective. But that is exactly what is needed if we are to practice cognitive empathy, participate in collaborative and transdisciplinary design, and create truly sustainable designs that holistically meet the needs of society over the long term. The good news is that as hard as the process is in the beginning, the more you participate in seeing and understanding different perspectives, the easier it is, and the more you want to do it. Reading about or traveling to different countries, or cultures, or even areas within your own city or region can also help to experience other ways of life and perspectives, making it easier to see life from a different perspective.

TRANSDISCIPLINARY THINKING – IMPACT ON DESIGN

Integrative or transdisciplinary design is ultimately about setting aside one's professional biases and using empathy to view the project from another's position. So, the engineer will look at the project through the architect's eyes, and the interior designer will look at the project through the builder's eyes. The process then emerges where conversations about ideas are focused on solving problems, rather than on guarding territory or advocating for your profession's best interest. Sustainability projects are complex with multiple layers of goals, so a coordinated approach involving active participation from all professions and stakeholders is needed to be successful.

3.3.2.4 Empathy for nature

We have always had affective empathy for animals. Who can forget the polar bear struggling on the floating ice in the warming arctic wilderness? Cognitive empathy is also increasing as we begin to understand the essential role that animals play in the ecosystem. One could make the argument that this is a simply an example of "enlightened self-interest" as we, of course, understand that we need the food chain to stay intact in order to support our ability to persist as a species. However, the argument here is that we are forming a deeper kinship with the animal kingdom as well as with nature itself.

We can begin to see nature as a stakeholder in the design process. The rise of land bridges over major interstate highways is an example of empathy across species. These bridges allow animals to cross major highways without being in traffic lanes. This, in turn, saves human lives by reducing collisions with animals. Examples of this can be found in the sources section below. Ultimately, empathy is a critical building block for the platform of sustainable design. In the last section of this chapter, the unification of self-interest and cognitive empathy will be explored as a structure for sustainability.

Additional resources

Wildlife Crossings, https://en.wikipedia.org/wiki/Wildlife_crossing

3.4 Sustainable development

Transcend and include is a way of thinking that does not seek dualities, dichotomies, dialectics, or contrasting forces that need to be resolved but employs the "both, and" approach where we seek to find new ways to combine and even leverage the two opposing forces to create a new, more powerful solution. Sustainability is just that – a transcendence from the old ways of right versus wrong, and good versus evil. It seeks a higher path of reconciliation but also of unification – not in sameness but of inclusion. The goal is a meta-framework of values that reinforce holistic system.

3.4.1 Our common future

In 1983, the United Nations General Assembly acknowledged that the human environment and natural resources were deteriorating at an alarming rate. They responded by forming the World Commission of Environment and Development (WCED) and appointed former Norwegian Prime Minister, Gro Harlem Brundtland, as chairperson because of her background in public health and the sciences. Four years later, the WCED released a report, *Our Common Future*, where they formally coined the term "Sustainable Development", stating, "Sustainable development is development that meets the needs of the present without compromising the ability of future generations to meet their own needs" (The Brundtland Commission 1987). In many ways, the formation of this Commission and the release of *Our Common Future* (1987) indicated a paradigm shift: the realization that issues of the environment and development affected the whole globe, and that these issues required global-scale as well as local-scale solutions. The passages in *Our Common Future* reinforce the points made so far in this chapter. The desire for economic growth is considered a required part of a long-term environmental improvement expressed, in their statement:

> Environment and development are not separate challenges; they are inexorably linked.

Notice that the health of the environment and the health of our economic system in the form of development are linked in a direct relationship. This simple statement brings into being an entirely new way of looking at the basic motivations of humanity. Instead of pitting self-interest against empathy in the typical battle, the statement expresses a synthesis of the two as interdependent.

The environment can be thought of as the commons, because it includes all the resources and services needed for humanity to survive. Here, "development" is understood as private initiatives based on self-interested persons to make a profit. Here, the motivations of self-interest and empathy for the commons are both addressed and given equal value. The beginnings of a new way to think about human activity expanded further below:

> Development cannot subsist upon a deteriorating environmental resource base; the environment cannot be protected when growth leaves out of account the costs of environmental destruction.

> (The Brundtland Commission 1987)

Environmental destruction, so often a result of development, cannot be treated separately by fragmented institutions and policies. They are linked in a complex system of cause and effect; therefore, the document engages the social side of sustainable development:

> ... environmental stresses and patterns of economic development are linked one to another. Thus, agricultural policies may lie at the root of land, water, and forest degradation. Energy policies are associated with the global greenhouse effect, with acidification, and with deforestation for fuelwood in many developing nations. These stresses all threaten economic development. **Thus, economics and ecology must be completely integrated in decision making and lawmaking processes**.
> (The Brundtland Commission 1987, emphasis by the author, not in original)

We also see in the selected passages the rise of social equity as a key factor in sustainability, not just for the obvious empathetic concerns but for more practical reasons as well. The Brundtland Commission realized that people in poverty were unlikely to be able to focus on the longer goals of sustainability when their day-to-day survival is in jeopardy. As a result, the goal of alleviating poverty is pursued from a "rational" point of view, not one of sympathy.

> Third, environmental and economic problems are linked to many social and political factors ... It could be argued that the distribution of power and influence within society lies at the heart of most environment and development challenges. Hence new approaches must involve programs of social development, particularly to improve the position of women in society, to protect vulnerable groups, and to promote local participation in decision making.
> (The Brundtland Commission 1987)

In this quote, the social side of sustainable development is expressed in both process and final product. The consideration of "vulnerable" groups and women is an example of using empathy across difference to find common ground and create a truly holistic movement. The most recognizable quote from *Our Common Future* is repeated below:

> Sustainable development is development that meets the needs of the present without compromising the ability of future generations to meet their own needs.
> (The Brundtland Commission 1987)

This often-quoted statement reflects empathy across time, where decisions made now need to include the needs of future generations. This was a new, powerful, and radical position for the mainstream, short-term thinking of the time. *Our Common Future* is such a profound document because it acknowledges the need for financial performance, protecting the ecosystem, and improving life of all stakeholders. It also understands that all these aspects of life are intertwined and allows the freedom to do well and to do good simultaneously.

Sustainability: the great resolution

The joining of the two invisible hands reflects a unique time in history, a pivot away from dichotomies, dualities, and biased thinking. It marks the beginning of a new worldview is based on the merging and leveraging of self-interest and empathy. The pathway won't be

Table 3.3.1a The two invisible hands unified to achieve sustainability

Sustainability		
Self-interest	+	Empathy
The first invisible hand	+	The second invisible hand
Materialistic	+	Altruistic
Short term	+	Long term
Direct benefits	+	Indirect benefits
Human	+	Humane
Progress	+	Progressive

Source: Created and drawn by the authors.

straightforward, but the impulse to unify is enough to change the basic operating paradigm of society and open the door to a more evolved sensibility about human potential.

The resolution of the basic motivations of self-interest and empathy has led to the ground-breaking development of sustainability, which, in turn, has led to the most remarkable design revolution since modernism – sustainable design. The effect of sustainability since the 1980s has been staggering, with hundreds of thousands, if not millions of people working to apply sustainable development principles in the formation of new companies, nonprofits, architecture, fashion, and the government sector. In Table 3.3.1a [above], notice the '+' symbol in between each seemingly competing motivation or thought process.

In Chapter 4, the simple collection of sustainability values: People, Profit, Planet, and Place, a shorthand of the Quadruple Bottom Line, drives holistic thinking into the mainstream. Chapter 5 will cover meta-frameworks for sustainable design leading to integral thinking and design approaches. The sustainable design movement is evolving and branching out into new approaches including bio-inspired design (Chapter 6), resilience (Chapter 7), and health + well-being (Chapter 8). New integrated design processes and methods (covered in Chapter 9) have emerged to change the way design practice is delivered leading to better, more holistically design projects. In Chapters 10 through 14, sustainability will be applied at every scale from global sustainable design, to eco-cities, to living buildings, to sustainable products.

In the end, the sustainable design movement relies on empathetic partnerships between disparate groups of people and also with nature itself, a coevolutionary relationship that builds the foundation for a sustainable future. Chapter 15 will look at the future of sustainable design and the steps needed to accelerate the movement to increase the scope and speed of impact.

Additional resources

Our Common Future, www.un-documents.net/our-common-future.pdf

References

Bentham, J., 1988. *The Principles of Morals and Legislation: Great Books in Philosophy Series*. Buffalo, NY: Prometheus Books.

Grist, 2004. *A Wide-Ranging Interview with Environmental Visionary Paul Hawken*. Grist Magazine. https://grist.org/article/pauling/edn.

Hardin, G., 1968. *The Tragedy of the Commons*. American Association for the Advancement of Science. http://science.sciencemag.org/content/162/3859/1243edn.

Indigenous Corporate Training, 2012-last update, What is the Seventh Generation Principle [Homepage of Indigenous Corporate Training, Inc.], [Online]. Available: www.ictinc.ca/blog/seventh-generation-principle [July 28, 2018].

Lesley University, 2018-last update, The Psychology of Emotional and Cognitive Empathy [Homepage of Lesley University], [Online]. Available: https://lesley.edu/article/the-psychology-of-emotional-and-cognitive-empathy [July 28, 2018].

Seppala, E. and Berlin J., 2017. *Why You Should Tell Your Team to Take a Break and Go Outside*. Harvard Business Review. https://hbr.org/2017/06/why-you-should-tell-your-team-to-take-a-break-and-go-outside edn.

Smith, A., 2011. *The Theory of Moral Sentiments*. Germany: Gutenberg Publishers.

The Brundtland Commission, 1987. *Our Common Future: Report of the World Commission on Environment and Development*. Oslo: World Commission on Environment and Development.

Therrien, A., 2018. *Genes have a Role in Empathy, Study Says*. BBC News. www.bbc.com/news/health-43343807 edn.

Wallace, K., 2014. *What Your Baby Knows Might Freak You Out*. CNN. www.cnn.com/2014/02/13/living/what-babies-know-anderson-cooper-parents/index.html edn.

Wicks, J., 2013. *Good Morning, Beautiful Business: The Unexpected Journey of an Activist Entrepreneur and Local-Economy Pioneer*. 1st edn. Chelsea, Vermont: Chelsea Green Publishing.

4 Sustainability values

4.0 Introduction

In the previous chapter, sustainable development was recognized as the beginning of a great reconciliation between the human motivations of self-interest and empathy. The Brundtland Commission of the United Nations in its publication of *Our Common Future* established a new platform for human activity – a new way of thinking about the world – the basis for a new worldview. Holding this new worldview together is a set of sustainability values: social equity, ecological regeneration, and economic prosperity.

4.1 Historical context of sustainability values

In Chapter 1, the brief history of humanity illustrated how the core values of society are constantly evolving as a response to the key drivers of climate, new energy sources and new communication technologies. The Ages of the Hunter-Gatherer, Agriculture,

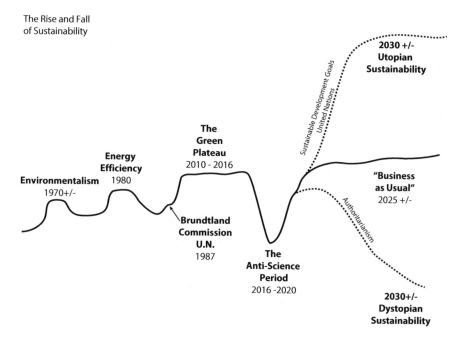

Figure 4.0a The rise and fall of sustainability.
Source: Created and drawn by the authors.

Industry, and Information each reflect a set of environmental, technical, philosophical, and climatic drivers that dramatically shifted the way humanity viewed and related to the natural world. At the writing of this book, we are in the transition period between worldviews. There are those fighting to continue in the industrial worldview, trying to save the dying coal industry, and desperately holding on to old power structures that favor privileged groups at the expense of the others. Climate change remains a "hoax" for this group of people who insist on ignoring the obvious changes to the planet all around them. The value of profit, at all costs, is the mantra. Others remain fixed in the Age of Information. Cognitive empathy runs deep in this group who fight for the cause of social equity and the environment. The next worldview is struggling to emerge. Integration and transcendence reflect another pathway for society, a departure from the duality between good versus evil, humans versus the environment, a third way forward that requires significant leaps in consciousness. This worldview acknowledges the interdependence and connection of humans to the earth and seeks to restore the ecosystems that support human life and create a balanced way for existence to move forward. This worldview promotes the understanding that by creating social, financial and ecological equity we all benefit and prosper. Sustainability is the framework that is organizing the activities of small groups of people all over the world seeking a better way. The values of Profit, Planet, and People are brought forward from the previous worldviews and recombined into new and important economic models designed to raise the quality of life for all people, regenerate the planet, and secure a healthy and resilient future. This chapter lays out the core values needed to reach that future.

4.2 Sustainability values and nature

Sustainability is most associated with environmental concerns and with good reason. As discussed in Chapter 2, the state of the planet's ecological health remains in danger. While the climate has always been changing and will always change by its own accord, the added impacts of humans have shifted the primary patterns of climatic systems leading to changes in weather, temperature, and even ocean currents. The changing state of nature, even as influenced by our own actions and inaction, is not in itself a bad thing. After all, humans are very good at adapting to change, and a two degree Celsius rise in global temperatures and some changing weather patterns do not sound like a problem. But in Chapter 2, we learned just how extreme the damage already is and will continue to be. Current changes in climate have caused entire coastlines to flood and have increased the frequency of severe drought and the intensity of hurricanes. Already the quality of life for millions of people has been affected and every year we fail to take action, we run the risk of millions more losing their lives. The transition to a new worldview will be driven in large part by the necessity to survive the changes to the planet.

4.3 Sustainability values and motivations for sustainable design

At the center of the problem, *and the solution*, lies humanity. We have the knowledge and power to attack environmental problems, thereby preserving the quality of life we now know into the foreseeable future. Up to this point, our efforts have either been too slow, not grand enough, or in many cases nonexistent. In some countries, like the United States, the ongoing pattern of climate denial comes into full force, slowing efforts by others to address the areas we do have control of. In Chapter 3, we explored the impacts of self-interest in a free-market economy, what cognitive empathy is, why it is important, and how it can be combined with self-interest to create a new third way.

The green design movement has made great strides in moving society up and onto what could be called a green plateau (see Figure 4.0a), providing vistas to more ambitious and more sustainable plateaus off in the future. Without the green design movement, the next step towards sustainability and eventually towards regenerating the environment could not have been made. Green building rating systems, such as LEED®, have played a pivotal role in the success to date. However, now is not the time to rest or pat ourselves on the back. There is a much work to be done to reach regenerative sustainability.

4.4 Sustainability values

Sustainability is a relatively new concept that is fundamentally changing the way we think and act in our daily lives and how we run our companies, governments, and nonprofit organizations. It does this because it changes the goal of success from maximizing personal profit in the short term to maximizing societal prosperity in the long term. It changes the equation from local or personal focus to considering others in the future or in other geographical locations. We are seeing a change from the attitude of needing to "save nature," to coexisting, coevolving, and even seeing nature as a model suitable for imitation. We are realizing that our survival is intertwined with other species and the ecosystem. As a society, we are building cognitive empathy, the ability to make decisions that not only benefit us now but also others far into the future. We are also beginning to understand that the earth will be fine with climate change, it's just that humans may not be able to survive.

Despite the desire of many to pursue sustainable design, there is so much confusion, disagreement, and frustration around how the term is used and what it actually means or entails. Defining core values to support and direct the design process is imperative to moving toward a sustainable future. Having a set of values organized by a straightforward framework helps to establish a consistent understanding of what true, restorative sustainable design is. Then, by default, it clarifies our thinking and goals, which then direct and clarify the most effective strategies in each specific situation and ultimately our measure of success. We will first examine the values of People, Profit and Planet, and then add the idea of Place. They are convenient shorthand terms for broad areas of concern in sustainability. However, they belie the complexity of the subject which requires a deeper understanding from those who wish to practice deeper holistic sustainable design.

Triple Bottom Line

The Triple Bottom Line (TBL) is an accounting framework applied to sustainability by John Elkington in 1994 in his book *Cannibals with Forks*. It has three parts, social, environmental, and financial, that are commonly referred to as the three Ps or People, Planet, and Profit. Many organizations, governments, and corporations have adopted the TBL framework to evaluate their performance and guide decision-making towards a sustainable future. The discussion in Chapter 3 regarding self-interest, profit, and empathy is the basis for TBL, and it reflects the core values used to develop *Our Common Future*, the seminal document created by the UN's Brundtland Commission.

The term "bottom line" is an accounting term for the final total of a balance sheet or underlying outcome of an analysis. By applying this principle to sustainable design, the conversation was enlarged from strictly energy generation, use of resources, or energy conservation, to include a broader range of "bottom lines." The TBL seeks to consider and maximize value in each of the categories to create a decision framework.

Graphically, the TBL is represented in many different ways, each depicting its own organization and hierarchy. Figure 4.4a (below) depicts a non-holistic and segmented view of each element where actual integration, or actual sustainability, only occurs at the intersection of the three circles, – a rare occurrence.

Figure 4.4b (below) shows a more integrated and hierarchical model of nested systems, where the "environment" is presented as the outer limit or the container for human activities of the economy and of society. This is a good representation because without the environment we would not be able to survive or function. The representation with nested systems also references the interrelationship of all three elements, but it does not represent any interaction, permeability, or overlap between the different values. In Figure 4.4c (next page) (Elkington 1999) expresses the TBL as a series of "tectonic plates" that shift depending on the importance a company, a person, or a government places on each value. When the "plates" are aligned, it shows that equal consideration is assigned to each value and holistic decisions are being made.

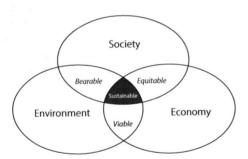

Figure 4.4a Triple Bottom Line diagram.
Source: Wikipedia, redrawn by the authors. By original: Johann Dréo (talk contribs) translation: Pro bug catcher (talk contribs) – own work Inspired from Developpement durable.jpg. Translated from Developpement durable.svg, CC BY-SA 3.0, https://commons.wikimedia.org/w/index.php?curid=1587372.

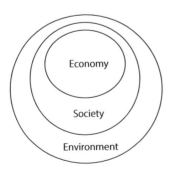

Figure 4.4b Nested Triple Bottom Line diagram.
Source: Wikipedia, redrawn by the authors. By KTucker – Own work, CC BY-SA 3.0, https://commons.wikimedia.org/w/index.php?curid=17030898.

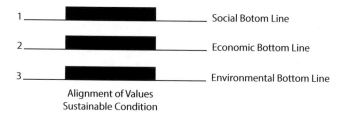

1 _____ ████████ _____ Social Botom Line

2 _____ ████████ _____ Economic Bottom Line

3 _____ ████████ _____ Environmental Bottom Line

Alignment of Values
Sustainable Condition

Figure 4.4c Tectonic plates of Triple Bottom Line sustainability by John Elkington.
Source: R. Fleming (2011).

Additional resources

Cannibals with Forks, John Elkington
Ecology of Commerce, Paul Hawken

4.4.1 Profit

The core value of profit is a natural expression of the self-interested human. Obtaining everything one needs to survive, be comfortable, and be financially secure is a necessary and natural motivator. It leads to the creation of businesses and other ventures that create jobs, drive innovation, create deep levels of satisfaction and purpose along with generating wealth. But the danger is, when profit is the only driver of decision-making, the values of environment and equity are left alone – resulting in a company being a destroyer of the environment and bad corporate citizen. When this happens, a "shear zone" is developed. This shear zone creates all kinds of short- and long-term problems – especially with the social and environmental "plates." Elkington's diagram, Figure 4.4.1a (next page) is a visual description of a shear zone. The misalignment of the tectonic plates is a common occurrence in many organizations that are not pursuing sustainability intentionally.

The Deepwater Horizon Oil Spill in 2009, in the Gulf of Mexico, is an example of non-TBL thinking. When the pressure to increase the production of oil became more important than protecting human lives and the environment, a shear zone in TBL thinking was created. The result was an explosion and leakage of the offshore oil well led to the loss of 11 lives, heavy environmental damage to the Gulf of Mexico, and billions of dollars in penalties. From an accounting perspective, all three bottom lines suffered huge losses.

4.4.1.1 Economic prosperity

Prosperity offers a more robust and compelling descriptor for the profit motivation. Prosperity is defined as: the condition of being successful or thriving, especially in the economic sense. Synonyms include ease, plenty, comfort, security, and well-being. The principle of well-being is woven through all the values in the TBL – social well-being, environmental well-being including physical health, and economic well-being or prosperity – and it includes a longer-term view of profit in the form of security, both for an individual and for an entire society.

FROM PROFIT TO PROSPERITY

Since well-being is at the heart of prosperity, economic goals are not an end unto themselves but are a vehicle to obtain critical social, environmental, and even experiential goals (design).

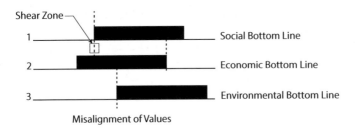

Figure 4.4.1a Misalignment of values in Triple Bottom Line sustainability by John Elkington.
Source: R. Fleming (2011).

Prosperity has a fuller, deeper connotation than merely financial gain, and it speaks to the quality of life for all *stakeholders* including the natural environment. The term "stakeholder" means anyone with an interest in or that will be affected by a project or a decision. It is a conscious change from the term "shareholders," which is anyone with a financial interest in a business or project. Prosperity encompasses thriving, happiness, health, pleasure, laughter, reduced stress, loving and being challenged by your job, doing well, or well-being. We all want these things for ourselves, and with cognitive empathy, we can imagine and want them for all other populations of the world. This type of future requires a transformation in our thinking, decisions, and actions.

THE VIRTUAL CYCLE FOR BUSINESS PROSPERITY

Most of the world still operates out of a short-term, profit-oriented mindset. Return on investment (ROI) is the main vehicle to evaluate the economic value and business success. The inclusion of sustainability values into the decision-making process asks for a triple ROI., or a financial, social, and environmental return. For example, an investment in indoor air quality means that workers will have less sick days and be more attentive and productive at the workplace. That translates into increased financial return, and social equity with better working conditions. Transparency and accountability mean that all workers will be more invested in their jobs, take ownership, be more productive, and look out for the company's best interest – all translating again to financial profit. Investment in the experiential aspects of company by making beautiful work spaces, or allowing outdoor work areas, or having flexible work hours, and creating a cohesive culture with social outings will lead to worker satisfaction. Worker satisfaction greatly influences retention rates, reduces sick leave, and increases the ability to recruit highly talented employees and motivates all employees. All these things directly impact the financial bottom line of a business but have been mostly ignored because their connection is not immediately visible when analyzing the short-term cost of construction or improvement cost.

RETURN ON INVESTMENT

The use of ROI does not come close to accounting for the complexity of achieving prosperity which includes the social or ecological considerations, and yet, this is one of the most common methods used in business to determine whether an investment is viable. There are companies around the world that use more sophisticated techniques to evaluate economic opportunities such as internal rate of return or even net-present value. These models which are too complex to describe here offer more sophisticated methods to account for the social and environmental values in decision-making. Many companies, countries, and organizations around the world have incorporated the TBL into their core processes, and the result is transformational and leading to long-term prosperity.

Additional resources

Do Green Buildings Cost More?, Desiree Hanford, www.facilitiesnet.com/green/article/Do-Green-Buildings-Cost-More-Facilities-Management-Green-Feature--8954

Green Buildings and Productivity, Miller et al., www.josre.org/wp-content/uploads/2012/09/Green_Buildings_Productivity-JOSRE_v1-41.pdf

Wood as a Restorative Material in Healthcare Environments, FP Innovations, www.woodworks.org/wp-content/uploads/Wood-Restorative-Material-Healthcare-Environments.pdf

4.4.1.2 Prosperity: circular business models

Prosperity cannot sustain itself if we continue to follow the traditional business models. In a typically linear supply chain, materials are removed from the earth, placed into service, and then discarded into landfills. This linear system is based on core values of the industrial worldview where our natural resources were considered infinite, and infinite growth was the achievable and only goal. This is an "open-loop" system leading to vast amounts of waste, extensive pollution, and lack of energy efficiency. The Circular Economy is designed to close these loops and make each link in the supply chain more efficient. Some core aspects of this approach include the following (Weetman 2017):

- Extend the life of materials and products over multiple 'use cycles' where possible ;
- Use a 'waste = food' approach to help recover materials, reuse them where possible and ensure if not that what is returned to earth is not toxic and has the ability to decompose or be broken down
- Retain the embedded embodied energy, water, and other process inputs in the product and the material for as long as possible;
- Use systems-thinking approaches in designing solutions and decision making;
- Regenerate or at least conserve nature and living systems;
- Push for policies, taxes, and market mechanisms that encourage product stewardship, for example, 'polluter pays' regulations.

4.4.1.3 Prosperity: sharing economy and collaborative consumption

A primary driver of the worldview shift to the Age of Integration is a change in communications technology. Communication technologies like social media are transforming the way business is conducted. The Sharing Economy, based on renting idle resources, relies on distributed networks like social media to link up all the different parts of the process via easy-to-use apps. Airbnb, Bikeshare, Zipcar, and so many more are helping to greatly reduce the amount of material consumption and emissions in the world, laying the groundwork for farther reaching sustainability goals.

Trust and empathy are important underlying principle for a sharing economy and collaborative consumption. They will be explored further below. This also illustrates a shift from materialism and consumerism that have driven society since the industrial age. Empathetic connections have been extended to other trading partners resulting in the change of many economic and social structures.

4.4.1.4 Prosperity: B Corp

BENEFIT CORPORATIONS

The most respected form of sustainable business organization is the Benefit Corporation (B Corp). A B Corp is a company that meets a third-party-verified set of requirements to assess businesses developed by the nonprofit B labs organization. There are standards to measure a company's social and environmental performance, as well as levels of transparency and accountability built into the business structure. B Corps employ a "Declaration of Interdependence" viewing profit as a vehicle for "maximizing goodness" as shown below (from the Blab Website).

B CORP DECLARATION OF INTERDEPENDENCE

We envision a global economy that uses business as a force for good. The economy is comprised of a new type of corporation – the B Corporation – which is purpose-driven and creates benefit for all stakeholders, not just shareholders. As B Corporations and leaders of this emerging economy, we believe that

- We must be the change we seek in the world;
- All business ought to be conducted as if people and place mattered;
- Through their products, practices, and profits, businesses should aspire to do no harm and benefit all;
- To do so requires that we act with the understanding that we are each dependent on one another and thus responsible for each other and for future generations.

Notice that B Corp uses the term "purpose-driven" which is interesting considering that the pursuit of profit is a well-established purpose. In this context, the word *purpose* is used to describe social and environmental aims, a chance to do more than provide economic gain. Another interesting shift is the shift from shareholder to stakeholder is clearly established – a powerful point of view that forms the foundation of sustainability. Also notice the term "place" is used as a descriptor for something that should "matter." This refers to the possibility for a fourth 'P' or Place, as a sustainability value. The last principle reinforces the idea of future generations, their consideration and the present generation's obligation to consider and be accountable for their choices. B Corporations are transformative, serving as an empathetic model in the profit-driven business world.

Additional resources

B Lab website: www.bcorporation.net/what-are-b-corps/the-b-corp-declaration
  Balle: https://bealocalist.org

4.4.2 Planet: environmental regeneration

RAZORS EDGE: FROM MITIGATION TO REGENERATION

For most green projects, the environmental goals are typically focused on reducing energy use or basic environmental impacts. This approach of doing "less bad" is certainly better than nothing, but it does not solve significant environmental problems. The need for transformational thinking is more important than ever. Figure 4.4.2a, from the Natural Step, illustrates the looming environmental challenges and possible solutions to those problems. The earth can no longer go on meeting the increasing needs of humanity in a balanced way. It is being shown that the earth has exceeded its carrying capacity, and its systems will move into decline. Some argue that we have already arrived at this point, while others, who still hold on to an outdated industrial worldview, believe that the planet's resources are infinite or that substitutes for dwindling materials can be found. Sustainability values need to be ambitious to help reshape societal priorities and move beyond merely lessening damage to ultimately regenerate the ecosystems that have been compromised by our decisions in the past. Renewable energy and technological discoveries, although not a direct substitute for oil, have arrived at just the right time to offer a new energy pathway for society.

4.4.2.1 Ecological regeneration

Creating a *regenerative condition*, whereby the demand for resources is reduced and the supply of renewable ecosystem services are increased, is the ultimate expression of sustainable design. This is reflected by the two inverted curves on the right side of the diagram in Figure 4.4.2a (below). The reason for focusing on "regeneration" is that if we only reduce our consumption, the earth's resources will be depleted because the carrying load is still greater than the

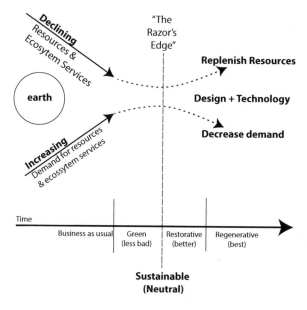

Figure 4.4.2a The natural step funnel.
Source: R. Fleming (2013).

resources available. Regeneration does not mean simply looking for another finite resource like oil, as a substitute, but rather finding energy sources that are self-renewing, or technology that increases production of existing resources or uncovers additional types of supplies. This approach works with existing natural systems and allows the earth to follow its own regenerative pathway. This can include, but is not limited to, generating more energy on-site, using landscape systems to store and clean stormwater and clean the air, configuring the building to increase daylighting and natural ventilation, specifying renewable, biodegradable and locally available materials, and restoring structurally sound historic or industrial sites for new purpose. There are also rating systems that help establish goals and strategies. These rating systems will be covered in Chapter 13.

4.4.3 People: social equity

Why are people and their social status, ethnicity, employment conditions, medical treatment or access to quality education included in the sustainability equation? After all, most people see sustainability as an environmental imperative requiring the reduction of energy consumption to "save the planet." When we add the social dimension level of complexity increases and so does the probability of achievement. Upon closer inspection, though, social equity is a critical aspect of true sustainability and is interlinked with the ability to pursue sustainability initiatives at the personal or community level.

From social equity as an "issue" to an "opportunity"

For so long, social equity and diversity have been perceived by groups in power as "issues" to be dealt with or as a problem. It is considered a "problem" if diversity is not present in an organization. It is a problem if people are paid differently, or treated with less value, or not given opportunities simply because of their ethnicity, social standing, gender, marital status, where they live, or anything other non-performance related aspect of who they are. It is a problem if social equity is a problem if people in the organization are fighting for better pay and better treatment. There are companies who have chosen to see these "problems" as an "opportunity" to discover new ways of operating and creating competitive advantages. People in positions of leadership have the choice to see this as a "problem" or to use it as "opportunity" to discover new avenues of competitive advantage. This impacts every profession, socioeconomic level, culture, and ethnic background. The design community is not immune. In 1968, in his keynote address to American Institute of Architects (AIA) convention, Whitney Young, a civil rights activist, called architects to task for their indifference towards the civil rights movement.

> ... you are not a profession that has distinguished itself by your social and civic contributions to the cause of civil rights, and I am sure this has not come to you as any shock. You are most distinguished by your thunderous silence and your complete irrelevance ... [thinking] that you are only giving the people what they want.
>
> (Young 1968)

Young's point was that architects, should be doing more with their leadership and influence; when they create the built environment, by encouraging women and minorities within their profession and by being politically active with social injustice. Today, the number of women and minorities that make up the profession is still shockingly low when compared to other professions, and it is a well-known fact that women still make less than their male counterparts who work in the same positions. Sustainability and the Integral Worldview posits that when one person prospers we all prosper, and that until we all prosper true sustainability will never be reached.

4.4.3.1 Social equity in the built environment

In the abstract, it would be nice to believe that everyone is equal and also treated equally, but common wisdom suggests otherwise. In the U.S., women make 81 cents for every dollar of their male counterparts for doing the same work (Sheth et al. 2018). This is an inequitable situation that leads to low morale, low production, high tension, and often the loss of talented employees. If one of the core values of sustainability is prosperity, it is hard to see how these types of social arrangements can exist in any organization claiming to be "sustainable." Social equity involves the conscious desire and demonstrable practices that afford a meaningful voice in the organization to underrepresented or traditionally oppressed segments of society. It also includes a more holistic decision-making process that includes stakeholders and the consideration of the long-term effects on those groups.

Intentionality towards social equity requires self-awareness, empathy, and a constant engagement with the issues and opportunities. Sectors of society are changing and are starting to use social equity as a determining factor in who they purchase from or even who they choose to work for. The loss of profit and good employees as a result of social inequality can be reversed by incorporating social equity as a part of a self-interested business model.

DIVERSITY AND INCLUSION

The definition of diversity implies variety or a mixture. And in our current social situation, it is most often used when referring to racial or ethnic backgrounds, gender, sexual orientation, religious beliefs, educational level, or even marital status. Diversity has become a very popular term, and it is used by corporations and organizations to demonstrate some aspects of social equity, without really defining what they are. To reach more authentic levels of diversity and inclusion, deeper structural changes need to be made. Merely increasing the "variety" of backgrounds or minority groups represented is not enough. The test for organizational diversity is how those people are treated. Is everyone included and given a voice of equal value? Are there barriers, or bias, or conscious or unconscious discrimination present? These questions are not easy to answer.

PRIVILEGE

Defining privilege is pretty simple: Those in power or who are in a more favored demographics have access to a wide array of conditions or "privileges" that improve their quality of life. Depending on which society one lives in and which worldview is in place, privilege can come with gender, ethnicity, or economic standing. In a sustainable equitable model, privilege is acknowledged as a reality, and those in privilege are compelled by the second invisible hand of empathy to use that privilege on behalf of building a more equitable society – which, in turn, helps lead to overall sustainability.

SOCIAL AND ENVIRONMENTAL JUSTICE

Social justice addresses poverty, gender inequity, racism, and any other form of discrimination, seeking to move them further into the mainstream discourse. It can take a more confrontational approach such as public demonstrations or be reflected in our design directives and educational efforts. Understanding and including social equity issues in the design directives at every stage of the design process and at every scale can also effectively promote change and awareness. It is an important part of sustainability to call attention to discrimination and oppression when we see it and work toward equitable change and restoration. Environmental justice addresses the continuous pattern of locating polluting and toxic industrial processes near low-income communities. Those communities struggle to advocate on their own behalf either because of the cost involved in opposing large corporations or knowledge of their rights. Neighborhoods with higher socioeconomic levels often have the necessary political power and connections to block "undesirable" projects from being built in their neighborhoods. Environmental justice is a lens that can be used by designers to better understand the impacts of their decisions upon the occupants of local communities.

POVERTY ALLEVIATION AS A DRIVER OF SUSTAINABLE DEVELOPMENT

The argument for a future without poverty is not solely based on altruism. It is in part a rational approach to pursuing a sustainable future that argues: If people are trapped in poverty, they tend to be consumed with meeting short-term needs such as obtaining food and shelter, and therefore, do not have the capacity to pursue longer-term sustainable actions. If the majority of the world lives in poverty, focused solely on survival, resources will continue to be exhausted without regard to long term survival. Many people fear that poverty alleviation is equated to "wealth redistribution" and is often blocked. Research, education, training, and greater inclusion are required to find holistic long-term solutions to reduce poverty and allow whole populations the ability to focus on long-term sustainable actions.

LIVING WAGE AND FAIR TRADE

Most sustainability advocates focus on the living wage as the pathway to poverty alleviation. Some cities in the U.S., such as Seattle, have already established a living wage requirement for all companies. This could transform the economic approach of businesses. The living wage concept and fair trade have a lot in common. Fair trade is designed to make sure that workers employed within the various steps of the supply chain are treated fairly and receive a fair wage. Businesses are increasingly integrating fair trade practices into their normalized business plans.

FOOD ACCESS AND HEALTH EQUITY

Access to healthy food and quality healthcare have been a constant struggle for many people. As an expression of social equity and poverty alleviation, work is being done to provide for healthier food and better healthcare in lower-income neighborhoods. This allows people without dependable transportation to have better access, save time in emergency situations, and be more likely to get preventative care. This correlates directly with health equity. Better access to health care and high-quality food improves the quality of life for those in lower economic neighborhoods. These are critical pieces to alleviating poverty, which leads to a better chance of pursuing overall sustainability.

All of these topics are connected by empathy, a critical component of a society that is able to provide equal opportunity for anyone to be successful regardless of their background. For designers, the use of social equity as a core value is critical because, without it, a project often benefits the privileged at the expense of others. Sustainable designers, then, should be advocates for all stakeholders in the design process.

4.4.3.2 Transparency

The need for transparency in decision-making is critical to sustainability because stakeholders are affected by the decisions made by an organization. Humans have a fundamental need for equity and justice, and when it's missing, tensions form and resentment builds. Allowing input, being open, or at least informing people of what is going on goes a long way in creating cooperation, buy in, and loyalty. When designing projects, at all scales, transparency can play a critical role in making the process more equitable, effective, and efficient. For sustainability projects which typically require more and different types of maintenance and also have ambitious energy-saving goals, making all employees a part of the process is key for the long-term effectiveness of a sustainable project. In Chapter 9, the specific processes needed to deliver transparency are discussed in detail.

4.4.3.3 Accountability

Transparency and accountability are linked. Once the door is open to transparent processes, accountability will automatically follow as there is an expectation that the ideas and strategies discussed during a decision-making process will actually be carried out. Care must be taken to frame brainstorming discussions as a process to generate a large base of ideas, allowing all stakeholders a voice, yet developing the understanding that just because a suggestion was made, further study and research will be needed before implementation. This allows openness and freedom in exploration while also allaying fears over possible unwanted solutions. Building and zoning codes exist as another level of accountability for designers, developers and clients. Being accountable to all areas of the TBL or any other sustainable building rating system the designer or client chooses should go without saying.

CORPORATE SOCIAL RESPONSIBILITY REPORTING

Corporate social responsibility (CSR) is where transparency and accountability are expressed by an organization. This type of reporting is the best example of how transparency and accountability are finding their way more directly into the mainstream. Most major corporations now have sustainability departments and report on the social and environmental bottom lines of the company. While not all CSRs are created equal, the simple act of reporting on progress or lack of progress is a critical step in the overall evolution of an organization towards achieving sustainability.

4.4.4 Place: the fourth bottom line of sustainability

While the TBL has served to expand society's understanding of sustainability over the past 25 years, Elkington's tectonic plates illustrate how we should be ordering priorities and the resulting changes to sustainability. We also have rating systems that are great guidelines for reducing consumption, increasing efficiency, and even the best materials to select. But is this enough? What about beauty, history, culture, and design? We have made tremendous progress in understanding and rating energy performance in the years since the Brundtland Commission released its report, but to be truly sustainable, we have to address all the needs of humanity.

As a reminder, the sentence that best summarizes the Brundtland Commission's report is:

> Sustainable development is development that meets the **needs** of the present without compromising the ability of future generations to meet their own **needs**.
>
> (WCED 1987)

Beauty, aesthetics, "Place," history, and cultural connections are all needs for human existence. The built environment does more than simply protect us from the elements. Fred Gage states in the forward to Brain Landscapes: The Coexistence of Neuroscience and Architecture that "I contend that architectural design can change our brains and behavior" (Eberhard 2009, p. XIV). Architects and designers have known this since the beginning of the profession, but they had no means to quantify it as we now do. Vitruvius, considered the father of architecture also saw this and expressed it this way when talking about the three most important qualities of architecture; "Firmness, Commodity, and Delight" (Mallgrave 2006, p. 6).

Psychology, philosophy, environmental psychology, evidence-based design, and even the bio-inspired design approaches all show a common thread of the visual, aesthetic, beauty, cultural connection, and historical reference being, not just desired, but essential needs for healthy, whole human existence. All of these fields of study, including neuroscience, confirm and support that our physical environment shapes are actions, physical health, productivity, cogitative ability and even our treatment of others.

This conflict between performance and efficiency, verses beauty, culture, and place has caused resistance and hesitation by many to pursue sustainable design. In Chapter 3, we included the seemingly conflicting motives of profit and empathy in sustainable design. To be fully inclusive and meet all the "needs" of people expressed in the Brundtland Commission's state, "Place should be in addition to People, Planet, Profit."

Maslow's hierarchy of needs

Maslow's hierarchy of needs was developed by psychologist Abraham Maslow in 1943 when he started studying human motivation. Briefly, Maslow's theory was based on the idea that human beings are motivated by a hierarchy of needs. The more basic needs, like food, water, and shelter, must be more or less met before the higher ones, like creativity, are pursued. He also saw that the order of these needs was flexible and based on external circumstances or individual differences. This also means that we can fluctuate between levels depending on circumstances or be at different levels in different areas of our lives. He also observed that most behavior is multi-motivated, meaning that it could be caused by more than one need in more than one category or level (McLeod 2018). Figure 4.4.4a (next page) uses Maslow's pyramid as a basis for creating built environments to fully meet the underlying needs of humanity.

Ultimately, our goal is to "transcend" or reach the top of the pyramid where self-actualization is reached. True sustainable design should meet the needs at all three of Maslow's levels.

Expanded Maslow's hierarchy of needs

During the 1960s and 1970s, Maslow's traditional five levels of need were expanded to either a seven- or eight-level model. This was done to elaborate and expand the scope of Self-actualization needs, originally level five the pyramid to include (McLeod 2018):

1 Cognitive needs (new level five) – knowledge and understanding, curiosity, exploration, need for meaning, and predictability.
2 Aesthetic needs (new level six) – appreciation and search for beauty, balance, form, etc.
3 Self-actualization needs (new level seven) – realizing personal potential, self-fulfillment, seeking personal growth and peak experiences
4 Transcendence needs (new level eight if included) – motivated by values that go beyond the personal self (e.g., mystical, aesthetic experiences, service to others, the pursuit of science, religious faith, and knowledge).

Figure 4.4.4b (next page) also indicates what are considered deficiency needs, those motivated by a lack, and growth needs, those motivated by the desire to grow. Short-term, long-term, empathic, and self-interested needs are also shown to help relate Maslow's hierarchy to earlier discussions on the motivations of sustainable design.

Notice how the deficiency needs are aligned with self-interested short-term needs, and the higher-level needs are associated with empathy and long-term thinking. This is the crux of sustainable design – to consider it our responsibility meet all of the needs today and address the long-term so future generations will still be able to meet theirs.

For easy comparison, both Figures 4.4.4a and 4.4.4b in the chart on the next page are summarized below.

Level 1: Basic needs

Maslow's first level of needs is the most important level. They represent immediate and life-threatening *deficiency needs*, meaning that the lack of these motivates us to find them. These are the bare minimum for survival. For sustainable design, this means creating structures to protect us against the elements of nature, and other physical harm. This level of needs correlates to short-term needs and self–interest (Table 4.4.4a).

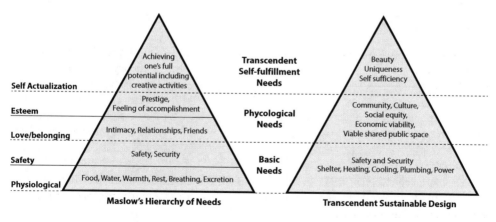

Figure 4.4.4a Maslow's hierarchy of needs from a sustainable design perspective.
Source: R. Fleming, design education for a sustainable design (2013).

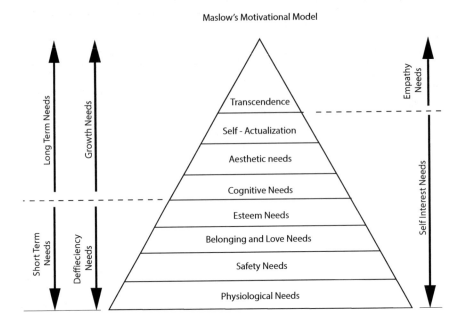

Figure 4.4.4b Maslow's expanded motivational model – altered by the authors.
Source: Drawn by the authors.

Table 4.4.4a Maslow's hierarchy of needs: level 1 – basic needs

Level 1 – basic needs	
Maslow's intrinsic physical needs	*Sustainable built environment needs*
• Food, water, air	• Provide shelter and physical safety for the current generation
• Shelter, warmth + physical safety	• Stop degradation of ecology to benefit future generations
• Sense of security + protection	• Preserve energy and resources for future generations
• Resources free from pollution + toxins	
Considered deficiency needs	
• Motivation comes from level 1 needs not being met	
• Motivation gone when needs are met (hopelessness)	
• Short-term focus	

Source: Created and drawn by the authors.

Engineers help to meet these needs by delivering basic necessities such as clean water and systems to provide temperature regulation of our homes and workplaces. Designers and engineers play a critical role in this and all design professionals are excellent at meeting these physical needs. Some examples of this are basic mechanical systems, HVAC, building codes, ordinances, zoning, and the metrics for rating systems and sustainable strategies.

Table 4.4.4b Maslow's hierarchy of needs: level 2 – psychological needs

Level 2 – psychological needs	
Maslow's intrinsic psychological needs	*Sustainable built environment needs*
• Intimate relationships • Friends • Self-esteem, prestige • Accomplishment	• Strengthen community • Create viable shared space • Public interaction and connection • Community rituals, icons, festivals • Environmental psychology
Considered deficiency needs • Motivation comes from these not being met • Motivation gone when needs are met • Short-term focus	

Source: Created and drawn by the authors.

Level 2: Psychological needs

This level addresses subjective aspects of human experience at the scale of buildings and communities. These needs are important to a person's overall development, especially if they want to reach the top of the pyramid to self-actualization and help others in a transcendent way. Family and community are critical factors in the development of these levels. Design goals would include planning communities and public spaces which foster interaction, connection, pride and investment in the community and each other (Table 4.4.4b).

Community and public spaces within individual buildings encourage community and interaction. Urban and community scale strategies include creating engaging and safe public spaces that encourage interaction. These types of spaces provide areas to meet, gather, and participate in community activities like festivals, graduation ceremonies and experience a sense of community. The built environment, when well-designed and well-constructed, adds an individual's self-esteem and collective pride. Sustainable and resilient buildings design increases longevity and allows future generations to benefit by the cultural heritage.

Level 3: Self-actualization

This level is comprised of cognitive needs, aesthetic needs, and self-actualization. These are growth needs, not motivated by a lack but by the desire to grow and expand, to become better. Cognitive needs include the thirst for knowledge and lifelong learning and working to reach their personal highest potential. We are hardwired to constantly learn new things. Aesthetic needs include a wide range of experiences that involve spaces, places, objects and all the senses. Internal experiences including music, art, writing, reading, and poetry are also part of the aesthetic experience. All these increase the desire to know and understand a broad range of ideas and principles (Table 4.4.4c [next page]).

The design goals for this level of need are addressed by striving for beauty and meaning at the deepest levels. Beauty in its myriad of forms provides an essential element to the development of individuals in society. Maslow uses terms such as uniqueness, unity, aliveness, perfection, completion, justice, order, simplicity, richness, effortlessness, playfulness, self-sufficiency, and meaningfulness in design. Maslow's description of higher-level needs should be the most powerful and profound drivers of sustainable design.

Table 4.4.4c Maslow's hierarchy of needs: level 3 – self-actualization needs

Level 3 – self-actualization needs	
Maslow's intrinsic self-actualization needs	*Sustainable built environment needs*
• Self-actualization • Self-acceptance • Cognitive needs • Aesthetic needs • Creativity • Help others	• Beauty • Uniqueness • Self-sufficiency • Holistic design strategies • Integrity • Great design aesthetic
Considered deficiency needs • Motivated by internal desire to grow • Satisfaction can lead to transcendence • Long-term needs • Self-interested needs	

Source: Created and drawn by the authors.

Sustainability should be addressing the deeper, less tangible and more ambiguous dimensions of design and the human psyche. If we pull out Maslow's descriptors and briefly elaborate, we start to get a fuller picture of what sustainable design could be.

Truth = transparency | accountability | visibility | honesty of materials | form
Beauty = delight | attraction | color | texture | composition | ornamentation
Wholeness = holistic | non-fragmented | resolution | unification
Aliveness = excitement | visual + physical movement | responsive | bio-inspired
Justice = social equity | integrity | motivations | strategies | design quality
Richness = deeper meaning | materials | integral complexity | breadth | quality
Uniqueness = surprise | individuality | beauty | connection |
Order = completeness | regulated | tradition | configuration | deeper patterns
Simplicity = form | materials |
Goodness = right | true | holistic |
Self-Sufficiency = autonomous | resilience | adaption | net-zero | strength | purpose

Level 4: Transcendence needs

Transcendence is at the peak of the pyramid, and it goes beyond the needs of the individual to focus on others. Internal motivation seeks to make a difference and serve without expecting a return. Empathy and compassion are natural products that extend to humanity, other species and for nature itself. In *The Farther Reaches of Human Nature*, Maslow expresses it like this:

> Transcendence refers to the very highest and most inclusive or holistic levels of human consciousness, behaving and relating, as ends rather than means, to oneself, to significant others, to human beings in general, to other species, to nature, and to the cosmos.
> (Maslow et al. 1971, p. 269)

Transcendence is what allows people to truly make a difference, to give back, and to contribute to the greater good. It's the place where cognitive empathy is activated, and people are

Table 4.4.4d Maslow's hierarchy of needs: level 4 – transcendence self-fulfillment needs

Level 4 – transcendence self-fulfillment needs	
Maslow's transcendence needs	*Sustainable built environment needs*
• Know and understand self • Find meaning • Reach out to others • Make a difference • Selfless service	• Meaning in design • Empathy across time and space • Social equity • Health + well-being for future generations
Considered growth needs	
• Internal motivation to make a difference • Empathy • Long-term focus • Hopefulness	

Source: Created and drawn by the authors.

motivated to make long-term change. In Chapter 15, we cover this in ways to effect lasting change (Table 4.4.4d [above]).

Quadruple bottom line sustainability

Adding the value of Place to the triad of People, Profit, and Planet creates a more holistic set of core values for sustainable design and expands the traditional role of the designer. It acknowledges the importance of beauty and connection in the built environment to shape personal lives and the community.

> Although we fundamentally shape our surroundings, ultimately place exists independently of human life, in turn shaping us.
>
> (Trigg 2012, p. 2)

Place-making as a core sustainability value

Beauty, aesthetics, culture, and history, represented by "Place," need to be an integral part of sustainable design. As we briefly examined above, the built environment has the power to touch human experience and change individuals and society. The addition of Place does not diminish the other areas originally part of TBL. Designers must constantly strive to include all of the values at all times. The creation of Place will be explored though the rest of the book, but most closely in Chapters 5 and 8.

Included strategies vs integrated design

Without a comprehensive understanding of all the values of sustainable design and their long-term importance, it can be easy to think that including a few strategies makes design "sustainable." While "doing less bad" is better than nothing, it is not the goal of true sustainable design, and it will not effect any great change. A statement expressing the design

Figure 4.4.4c Great design as a subset of sustainable design.
Source: Created and drawn by the authors.

communities' hesitation to follow true sustainable design is: "Sustainability is just part of good design." This statement indicates that sustainable strategies can simply be added on at the end of the traditional design process – this is not a holistic approach. If we want to create a new model, we need to change from:

> Sustainability is **just** part of Great Design
> to
> Great Design is an **integral part** of Sustainability

With the Quadruple Bottom Line, all "sustainability values" are equally valued and considered at the beginning of the design process and integral part of sustainable design. This leads to a more comprehensive and inclusive approach to sustainable design (Figure 4.4.4c [above]).

In practice, the term "sustainable design" is used to differentiate between superficial and casual greening and a more comprehensive design applying all the sustainable values. Moving to sustainable design is not easy, but it happens when the designer realizes that aesthetics are not the sole focus of a project, and uses cognitive empathy to understand all the stakeholder needs.

Conclusion

Sustainable design needs to address long-term social and environmental impacts of design and work to restore the damage that has been done to the ecosystems and society. By adding "Place" as the fourth bottom line to People, Planet, and Profit framework, sustainable efforts can be more comprehensive. Chapter 5 will introduce a meta-framework to effectively integrate these sustainable values into an actionable system for holistic sustainable design. The remaining chapters look at effective design strategies and how they are applied at various scales.

References

Eberhard, J.P., 2009. *Brain Landscape: The Coexistence of Neuroscience and Architecture*. Oxford and New York, NY: Oxford University Press.

Elkington, J., 1999. *Cannibals with Forks: Triple Bottom Line of 21st Century Business*. Gabriola Island, BC: Capstone.

Mallgrave, H.F., 2006. *Architectural Theory*. Malden, MA: Blackwell Pub.

Maslow, A.H. and Maslow, B.G., 1993. *The Farther Reaches of Human Nature*. Arkana: Penguin.

McLeod, S., 2018. *Maslow's Hierarchy of Needs*. Simply Psychology. www.simplypsychology.org/maslow.html edn.

Sheth, S., Gal, S. and Gould, S., 2018. *6 Charts Show How Much More Men Make Than Women*. Insider Inc. www.businessinsider.com/gender-wage-pay-gap-charts-2017-3 edn.

Trigg, D., 2012. *The Memory of Place: A Phenomenology of the Uncanny*. Athens, OH: Ohio University Press.

Weetman, C., 2017. *A Circular Economy Handbook for Business and Supply Chains: Repair, Remake, Redesign, Rethink*. London and New York, NY: Kogan Page.

World Commission on Environment and Development, 1987. *Our Common Future*. 1st edn. Oxford: Oxford University Press.

Young, W.M.J., 1968. Keynote Address at the 1968 AIA Convention in Portland, Oregon, *AIA Convention 1968*, American Institute of Architects.

5 Integral sustainable design

5.0 Introduction

In Chapter 4, the core values of sustainability – People, Planet, Prosperity, and Place – were explored in depth to uncover a holistic and comprehensive foundation for sustainable design. In this chapter, we will explore integral sustainable design, a meta-framework that encompasses and multiplies each of the four sustainability values into a new and powerful set of principles for design. In Figure 5.0a (below), the four major principles of sustainable design are organized as four perspectives which will be explained throughout this chapter. "Form follows function" is a much used phrase in the design world, and it defines the systems of a design. "Form follows wordview" references design as an expression of worldview shifts, covered in Chapter 1. "Form follows experience" ties into the intuitive side of design, along with the pursuit of beauty and biophilic experiences. "Form follows performance" speaks to the drive for efficiency in natural and technical systems.

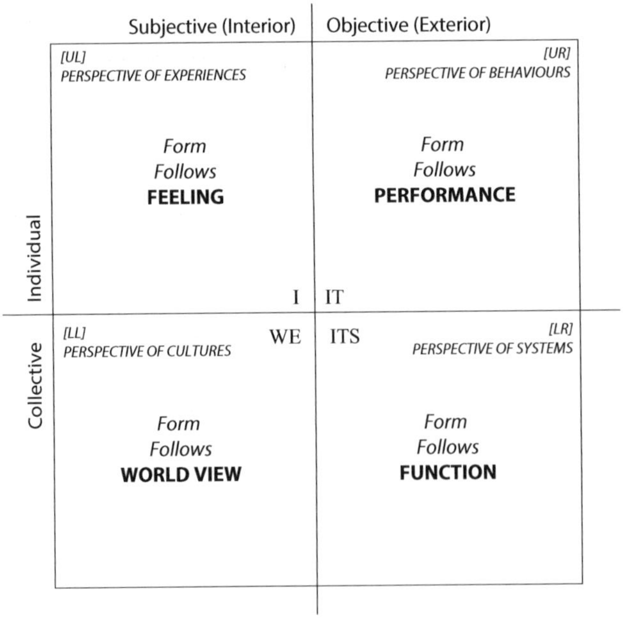

	Subjective (Interior)	Objective (Exterior)
Individual	*[UL]* *PERSPECTIVE OF EXPERIENCES* *Form* *Follows* **FEELING**	*[UR]* *PERSPECTIVE OF BEHAVIOURS* *Form* *Follows* **PERFORMANCE**
	I	IT
Collective	*[LL]* WE *PERSPECTIVE OF CULTURES* *Form* *Follows* **WORLD VIEW**	ITS *[LR]* *PERSPECTIVE OF SYSTEMS* *Form* *Follows* **FUNCTION**

Figure 5.0a Integral perspectives on major movements.
Source: Created and drawn by the authors.

The most widely accepted definition of sustainable design was developed by Jason McLennan, the founder of the Living Building Challenge:

> **Sustainable design** is the philosophy of designing physical objects, the built environment, and services to comply with the principles of social, economic, and ecological sustainability.

> —McLennan 2004

Interestingly, McLennan defines sustainable design as a philosophy and not as a practice. Sustainable design is actively applying sustainable values to design challenges at all scales in all areas of human existence. By including empathy across time and scale, depth is added to sustainable design. The authors propose an updated definition:

> **Sustainable design** is a holistic methodology for designing physical objects, the built environment, and social policy with the goal of integrating the values of beauty, social equity, economic prosperity, and ecological regeneration over the long term and across scales.

In this definition, the sustainable values covered in Chapter 4 are clearly represented. The importance of thinking across time and scale is added to acknowledge that all of our choices have long-term ramifications that should be considered during the design phase.

In this chapter, we will introduce new meta-frameworks for sustainable design that offer a holistic approach. We will also briefly introduce the reader to a wide range of subtopics within the frameworks to tease out some of the basic approaches and concepts needed to pursue sustainable design at a high level. Despite the best efforts of the authors, communicating the depth and complexity of sustainable design, time and space allows for only a brief overview of the major topics which is meant to spur further exploration and study. This chapter, like the other chapters of the book, begins its examination of sustainable design by looking first at the history to get to the heart of sustainable design.

5.1 Sustainable design over time

Applying the lens of history to sustainable design may seem like a non-rigorous approach, but it does open interesting viewpoints of past design movements. Design philosophies and approaches have changed over time as a response to changing views of nature, increased communications, new forms of energy, and a changing climate.

In the **Age of the Hunter-Gatherer**, sustainable design was automatic, a pragmatic approach to the simple need to survive. Local materials, proper sun orientation, and site selection were all examples of working with what was available to create comfort. The early cliff dwellers of the Southwest U.S. chose locations that provided shade in the summer, direct heat gain in the winter, and, of course, protection from intruders.

In the **Age of Agriculture**, in the 4th century BCE, the worldview shifted as the planet became warmer, surplus food was now a reliable energy source, and communications exploded through the use of writing. Most importantly, the climate was not only warmer but also predictable, meaning that humans stood a better chance at survival and had a reliable food supply. Many foundational texts were initiated during this period including the Tao Te Ching, which was developed as a means to understand the relationship between humans, nature, and spirituality. It offered a philosophy focused on how to build a positive relationship between humankind and the natural world. It wasn't long after that Feng Shui was developed as a comprehensive design system for the organization of spatial relationships, materials, and circulation patterns with the cardinal directions. The manipulation of Qi (spatial energy) through the arrangement of walls and placement of objects represents a clear goal of understanding and guiding the subjective aspects of design. The combination of the tangible and intangible aspects of design constitutes the first example of an integral approach to design. Around the same time period in India, Vastu Shastra was developed. Like Feng Shui, it puts into place a systematic view of design that sought alignment between buildings and cosmic energies mainly in the form of air, water, earth, fire, and space. Integration with nature is a key aspect of Vastu Shastra. These early forms of integrated design are still used today and often used in combination with contemporary sustainable design practices. They predate sustainable design rating systems by thousands of years.

By the **Late Age of Industry**, the prevailing worldview saw nature as an unlimited resource at the disposal of human production. As a result, ecosystems suffered harm from extraction and pollution and caused a real threat to human health and comfort. In the city of Pittsburgh, PA, if men went out to eat lunch, they had to bring another shirt to change into because the soot in the air was so heavy their shirt would be dirty when they came back from lunch. An undercurrent of environmental awareness paralleled the technological explosion of the industrial revolution. The Romanticists followed by transcendentalists were two groups that believed in the "inherent goodness" of nature. They were influenced by eastern philosophy and offered a subjective, non-rationale opposing force to the harsh technological, rationale forces of *progress* and mass production. This foreshadows the contemporary frameworks for integral sustainable design that will be presented in this chapter.

By the time the Modern Movement of architecture reached its zenith in the early 20th century, the quiet, yet persistent influence of nature-based thinking found its way into the work of the Frank Lloyd Wright, Le Corbusier, and Alvar Aalto. Each explored ways to address the natural world in new and important design methods. All of them agreed that the dark, stuffy, "room-based" architecture of the past had to give way to open, airy, bright spaces. Aalto explored this in diverse and varied ways, especially in his Paimio Sanatorium, built in 1933. Le Corbusier's five points of architecture included roof terraces and long horizontally banded windows – both offering new and imaginative ways to connect buildings to nature. But it was Wright, heavily influenced by Unitarianism (which is related to transcendentalism), who found the deepest connections in his "organic" architecture. Wright's work expressed the strengths of objective thinking through his ingenuitive structural systems and the indoor spaces that blurred the line between interior and exterior. Instead of walls with windows, Wright employed the use of continuous screen walls to directly bring nature into buildings, and he used fractal based geometry, local materials, and dematerialized building corners. Wright's work still stands today as a form of proto-sustainable design, especially as it's presented in this book – the union of objective and subjective approaches.

By the 1960s, in the **Age of Information**, ecology was gaining ground as a new science aimed at uncovering the complex interdependent relationships of ecosystems, partly as a means to study the origins, interactions, and effects of pollution. Ian McHarg's *Design with Nature*, published in 1970, laid the groundwork for an ecological understanding of design, with a focus on the reintegration with the natural world. His use of research and pre-GIS technology to define the macroscale of a project indicated a change in design process where the complexity of a local and regional ecology would be understood before intervening with a built project. McHarg and Ray and Charles Eames were some of the first designers to truly grasp the importance of thinking about design across scales. Furthermore, influences from eastern cultures continued to impact western thinking, especially hippie counterculture. Once again, the importance of so-called subjective thinking, although presented as "alternative" by the mainstream, continued to weave spirituality and appreciation of nature into the hearts and minds of designers.

In 1987, the concept of sustainable development was launched by the Brundtland Commission of the United Nations, and six years later in 1993, William McDonough delivered *A Centennial Sermon, Design, Ecology and the Making of Things*. This firmly placed the responsibility to pursue sustainability on the backs of designers. President Clinton launched the Greening of the White House Initiative, placing green design in the forefront of the American public and design thinking. As a result, the United States Green Building Council (USGBC) was founded as a multidisciplinary organization. The race towards a sustainable future was now well on its way.

5.2 Sustainable design and the natural world

By the year 2000, the deplorable state of our environment, the widespread destruction of our most precious ecosystems, and the harsh reality of finite energy sources were becoming all-too-familiar problems. Climate change was now a serious topic which designers, engineers, and architects were taking notice of. Green design had finally arrived in full force and is generally accepted today as an expectation of state-of-the-art design projects. However, the line between actual "greening" and "green marketing" is very fine, causing cynicism among some regarding the authenticity of corporate or government actions under the banner of sustainability. Green building rating systems like BREEAM and LEED transformed the design industry because the metrics for green building projects were well defined, scientifically based, quantifiable, and more importantly came out of a consensus multidisciplinary process. The Passive House Rating System was also in its infancy in Germany. It would not be long before a plethora of rating systems would emerge including Green Rating for Integrated Habitat Assessment (GRIHA) in India and eventually the formation of the Living Building Challenge.

5.3 Sustainable design and humanity

In the year 2000, William McDonough and partners published *The Hannover Principles* for the World Fair in Hannover, Germany. This set of *guiding principles* for design was not only important for its words but for the belief that design can stem from principles rather than concepts. Therefore, *The Hannover Principles* is the logical extension and expression of the Centennial Sermon (McDonough 1993) delivered and published in 1993. The principles are listed below.

The Bill of Rights:

1   Insist on the right of humanity and nature to coexist in a healthy, supportive, diverse, and sustainable conditions.
2   Recognize interdependence. The elements of human design interact with and depend on the natural world, with broad and diverse implications at every scale. Expand design considerations to recognizing even distant effects.

3   Respect relationships between spirit and matter. Consider all aspects of human settlement including community, dwelling, industry, and trade in terms of the existing and evolving connections between spiritual and material consciousness.

4   Accept responsibility for the consequences of design decisions upon human well-being, the viability of natural systems, and their right to coexist.

5   Create safe objects of long-term value. Do not burden future generations with requirements for maintenance or vigilant administration of potential danger due to the careless creations of products, processes, or standards.

6   Eliminate the concept of waste. Evaluate and optimize the full life cycle of products and processes, to approach the state of natural systems in which there is no waste.

7   Rely on natural energy flows. Human designs should, like the living world, derive their creative forces from perpetual solar income. Incorporating this energy efficiently and safely for responsible use.

8   Understand the limitations of design. No human creation lasts forever, and design does not solve all problems. Those who create and plan should practice humility in the face of nature. Treat nature as a model and mentor, not an inconvenience to be evaded or controlled.

9   Seek constant improvement by the sharing of knowledge. Encourage direct and open communication between colleagues, patrons, manufacturers, and users to link long-term sustainable considerations with ethical responsibility, and reestablish the integral relationship between natural processes and human activity.

Of special note is the clear expression of empathy for nature in the first principle. The second point reestablishes humanity's position in the natural world as an equal member and lays the groundwork for current thinking on coevolution today. The rest of the principles are very powerful concepts, and the reader will find them woven throughout this book.

5.4 Sustainability values

In Chapter 4, we studied the Quadruple Bottom Line, a framework that captures all the important values of sustainability: People, Profit, Planet, and Place. Decision-making within this framework gives all the values equal weight. We saw how the power of these values is moving society to new levels of collective consciousness – a greater awareness of ecology, a reconsideration of material happiness, a preference for experience over stuff, and a greater sense of equity. The goal of sustainable design is to consider these values in the design of products, built environments, and policies.

5.5 Integral sustainable design

Integral theory was developed and popularized by Ken Wilber, who presents himself as a philosopher, mystic, transpersonal psychologist and prolific author. He released his first book on the subject, *The Spectrum of Consciousness*, in 1977. Wilber looked at history, psychology, and Eastern Mysticism as he sought to combine the knowledge from these disparate fields into one comprehensive philosophy. Wilber's theory was that all understanding and perception could be categorized into four broad categories or quadrants. It is based on the understanding that human perception is deeper than physical sight or quantifiable factual evidence. We are multifaceted humans that understand, perceive, and act on more than just factual data. We are shaped by the past, which, in turn, influences the present.

Studies in psychology, phenomenology, and environmental psychology all confirm that people's reactions to the built environment go far deeper than cognition and rational thought.

Integral theory is not about homogeneity or erasing differences. Wilber describes it as follows:

> Integral: the word means to integrate, to bring together, to join, to link, to embrace. Not in the sense of uniformity, and not in the sense of ironing out all of the wonderful differences, colours, zigs and zags of a rainbow-hues humanity, but in the sense of unity-in-diversity, shared commonalities along with wonderful differences.
>
> (Wilber 2000)

In order to place into practice the quote above, and to define a holistic and comprehensive view of reality, Wilber proposed the four quadrants below as a means to capture and organize the various points of view held by humans. It should be noted that the four quadrants are one facet of many that comprise integral theory.

Notice that columns are identified as "subjective (Interior)" and "objective (Exterior)" which together offer a comprehensive view of reality. This is a revolutionary idea because it unites and equates the tangible and intangible aspects of reality – a perfect blend for applying the sustainability values as discussed in Chapter 4. The rows are labeled "Individual" and "Collective." This is quite an important distinction. Sometimes an individual expresses their personal emotions, but often times their views are shaped by cultural forces such as religion, education, and family values. Each quadrant reflects a fundamentally different approach to reality as defined by the rows and columns. Later in the chapter, a detailed explanation of each quadrant will be covered.

Another fundamental principle in integral theory is viewing all entities as holons. "Holon" is the term and concept that Wilber (1977) borrowed from the writings of Arthur Koestler. A holon is "… an identifiable part of a system that has a unique identity yet is made up of subordinate parts and in turn is part of a larger whole" (www.integralworld.net/edwards13.html). A simple example is the human body. A person is considered a whole or its own entity, yet the human body is made up of many parts and organs. An organ, like your heart, is considered a unique and whole entity, but is composed of four chambers, tissue, and other things. And yet you as a person are also a small part of the group of people that you associate with, or the whole population of the world. This is one of the underlying principles of systems thinking and can help us understand the interconnectedness of all things. It also reinforces the practice of thinking across scales as an essential approach to sustainable design.

Because everything is a separate and a whole unit, each "holon" has an interior and exterior perspective. This is another way to define the two columns in the integral theory framework shown in Figure 5.5a. Meaning we can look at or perceive everything as if looking out from the entity or subjectively, or from being on the outside looking towards the entity, objectively. Wilbur also argues that all four of these quadrants comprise every perception and level of understanding possible. The concept of holons should remind the reader of Chapter 1 when thinking across different scales was covered. For example, rooms are nested within a building which is nested within a site which is nested within a district which is nested within a city and so on. The two diagrams in Figure 5.5a illustrate the difference between looking out into the world as a phenomenological approach and looking at an object as an analytical approach.

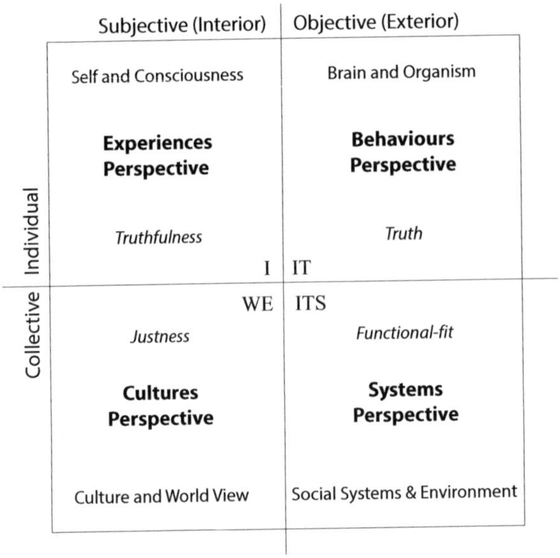

	Subjective (Interior)	Objective (Exterior)

Self and Consciousness | Brain and Organism

Experiences Perspective | **Behaviours Perspective**

Truthfulness | *Truth*

I | IT

WE | ITS

Justness | *Functional-fit*

Cultures Perspective | **Systems Perspective**

Culture and World View | Social Systems & Environment

Individual / *Collective*

Figure 5.5a Ken Wilber's integral theory.
Source: R. Fleming (2013).

A very new way of thinking

The four perspectives (quadrants) of integral theory are a very clear reminder that we should be thinking about all aspects when designing objects or creating policy. Maybe a better way to explain the relationship between the quadrants and how they should be incorporated is to think of them as lenses that we can "look" through to get a multifaceted vision of any aspect of reality. Each quadrant reflects an important point of view on its own which often influences, enhances, or conflicts with the other quadrants.

Figure 5.5b shows the mind's eye perceiving the multi-perspective reality of the four quadrants through four lenses simultaneously. Cognitive empathy is the means by which to accomplish this. By assuming the perspective of different "people" who view the world through one of the lenses and by honoring their viewpoint, each perspective can exist as "truth." Figure 5.5c (next page) is a more realistic way of thinking about the four perspectives where each perspective is layered on top of the other with the first (closest) perspective providing most of the filtering of reality and the others adding details, nuance, and conflicting information.

Another way to visualize integral theory is by studying a Celtic knot (Figure 5.5d [next page]). It is always flowing back and forth in a never-ending relationship. Each quadrant affects and is related to the others. For example, the systems within a building affect how the building performs, and how a space performs affects the experience or perception of that space, and also the culture or relationship between people, which goes back to how the people in the organization perform.

Whatever approach one chooses to use to help them visualize reality through the four quadrants is fine. Like a language, integral theory requires practice and a discipline of the mind to take full advantage of the framework's potential. The important thing to remember

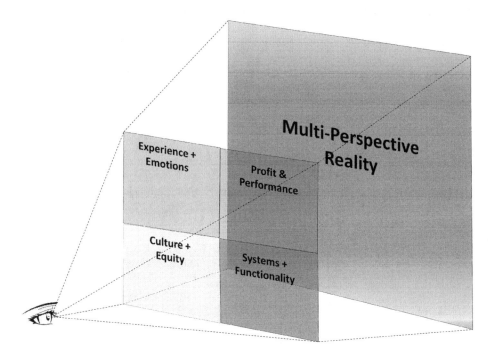

Figure 5.5b The four lenses of integral theory viewed simultaneously.
Source: Created and drawn by the authors.

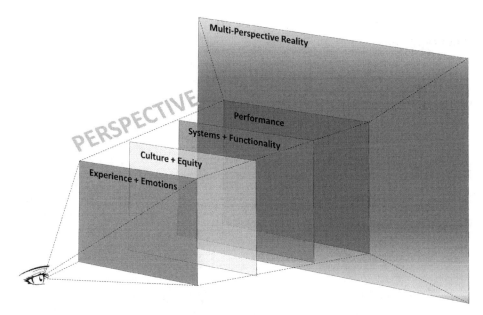

Figure 5.5c The four lenses of integral theory viewed in succession.
Source: Created and drawn by the authors.

Figure 5.5d Celtic knot illustrating the interactive and interdependent nature of integral sustainable design.

Source: Created and drawn by the authors.

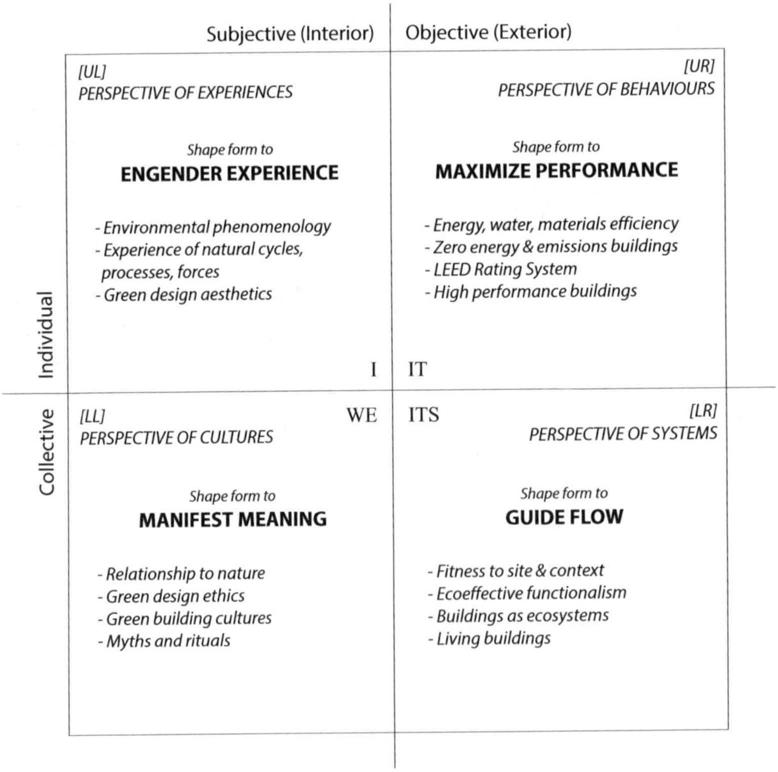

	Subjective (Interior)	Objective (Exterior)
Individual	*[UL]* *PERSPECTIVE OF EXPERIENCES* *Shape form to* **ENGENDER EXPERIENCE** - *Environmental phenomenology* - *Experience of natural cycles, processes, forces* - *Green design aesthetics* I	*[UR]* *PERSPECTIVE OF BEHAVIOURS* *Shape form to* **MAXIMIZE PERFORMANCE** - *Energy, water, materials efficiency* - *Zero energy & emissions buildings* - *LEED Rating System* - *High performance buildings* IT
Collective	WE *[LL]* *PERSPECTIVE OF CULTURES* *Shape form to* **MANIFEST MEANING** - *Relationship to nature* - *Green design ethics* - *Green building cultures* - *Myths and rituals*	ITS *[LR]* *PERSPECTIVE OF SYSTEMS* *Shape form to* **GUIDE FLOW** - *Fitness to site & context* - *Ecoeffective functionalism* - *Buildings as ecosystems* - *Living buildings*

Figure 5.5e Integral sustainable design by Mark DeKay.

Source: M. Dekay (2011) and redrawn by the authors.

is that the quadrants should always be influencing how we think and approach our decisions. It sounds complicated and overwhelming at first, but with a little experience, you will find yourself doing it automatically.

We all come with our own prejudices or favorite quadrant approach. Collaboration, open mindedness, and inclusion are the best strategies for moving forward to pursue integral sustainable design. The challenge is to use empathy to value all perspectives. Most importantly, we need to realize that by allowing additional influences and professions into the equation, our contribution is not compromised, but made better. It gives us a broader information base – a platform to reach higher levels of integration, which only makes us stronger and more capable of meeting the next challenge. For example, the engineer, who focuses on the systems of a project, is now made more aware of the values of the interior designer, who is focused on the experience of space. They can connect and work towards the goal of a superior project more easily and take the new broader information with them to the next project they work on.

It may seem redundant or fragmented to go through analyzing a particular problem or opportunity through each quadrant with its different perspective, but in the end, the designer emerges with a greater understanding and more holistic design directives. By combining the four complete perspectives into a holistic meta-perspective, we can begin to pursue the task of integration (sustainable design) in a more comprehensive and holistic manner. Our designs will resonate deeper with clients and the general public, be more energy efficient, support cultural connections, and become an integral part of the ecosystem, therefore becoming truly sustainable and restorative.

Criticisms of integral theory

Many within the architectural profession and some academic circles have been hesitant to embrace Wilber's thinking, perhaps because its application is so recent. Other objections regarding integral theory revolve around Wilber's ties to Eastern Mysticism and New Age religion. The fact that Wilber is selling a lifestyle may also camouflage the underlying principles and cause resistance. Others are put off by the quadrants in a grid formation, saying that the very goal of unity is lost due to fragmented graphic system.

All that said, MBA programs are starting to apply this model to their thinking and teaching methods because of the approach to multiple levels of perception and understanding it fosters. The American Institute of Architects (AIA) (2018) has profiled and endorsed Mark DeKay's book using integral theory as a "... powerful conceptual framework." Peter Buchanan (2012) in *The Big Rethink: Toward a More Complete Architecture* says, "... **vastly inclusive yet disciplined** ... giving equal value to the subjective and objective while also grounded in empirical evidence ... **providing a conceptual framework** that **stimulates new insights** by highlighting neglected areas of investigation." While it may be difficult for people to move beyond binary thinking at first, design programs at Jefferson and Chatham Universities are finding great value in using integral theory as a framework for comprehensive sustainable design.

Integral sustainable design

Mark DeKay is a professor at the University of Tennessee and the author of several books on sustainable design and high-performance architecture. DeKay realized during collaborative work in his professional practice that each segment of the profession came to the table with an approach that focused on their discipline. While it is very important that the structural engineer pays close attention to the structure of a building, it can also cause tension if they are not open to others' points of view for the project. The same could be true of architects, who are fixated on form and expression, sometimes ignoring the functional aspects of the design. DeKay also noticed that using a collaborative process where each profession had equal value and importance in the design process, the designs turned out to be better – more holistic, beautiful, and met the client's needs better.

DeKay was the first person to adapt Wilber's integral theory and apply it to sustainable architecture. In his book, *Integral Sustainable Design*, he started with Vitruvius's three tenants of architecture, "Firmness, Commodity, and Delight" (Mallgrave 2006, p. 6), or strength, usefulness, and beauty. These are considered the foundations of the architectural profession and speak to the multiple facets of purpose and perception.

DeKay's grid classifies the four quadrants as follows:

- "Maximize Performance" or the factual basis of a thing or how we objectively view a thing from the exterior, evaluating what we can factually see or measure, and/or how it actually performs;
- "Guide Flow" which looks objectively at the way systems interact or their interdependence and is one of the base principles of systems thinking;
- "Manifest Meaning" or how we ascribe meaning as a culture, ethnic, or social group or how we subjectively experience things as part of a group;
- "Engender Experience" or the things that we subjectively experience consciously and subconsciously like beauty, or it can be thought of us looking out at the world.

Integral sustainable design (Figure 5.5e [above]) recognizes that all these varying aspects of human existence and that the subjective, sometimes invisible aspects are just as important as the objective things we can measure. In any profession, these are valid and important areas of consideration. Integral sustainable design is not looking to do away with the past tenets of architecture, or any other profession, nor the beauty and poetics of design, but rather it attempts to broaden its scope to include efficient use and preservation of the earth's resources and promote the understanding "… that the rights of one species are linked to the rights of others and none should suffer remote tyranny" (McDonough 1993, p. 8).

Empathy, vision, and embedded values

Each of the sustainability values that we looked at in Chapter 4 have its own set of opportunities and challenges. Even more difficult is the requirement to translate those values into built form, products, and policies via design principles.

In Figure 5.5f (next page), the shift from values to principles is described by changes in phrases that are *actionable* in design. For example, the value of *Place-Making* is broadened to *Creating Deep Experiences*. This opens the design principle to be applicable across scale, discipline, and different contexts. *Economic Viability* is now expressed as *Maximize Performance* which expands the goal beyond financial to include energy and resource efficiency, spatial adjacencies, or user interaction. *Social Equity* now becomes *Embed Equity and Culture*, and finally *Ecological Regeneration* becomes Think in *Systems*.

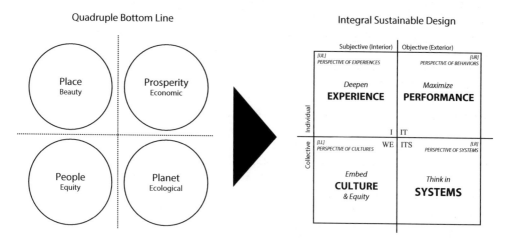

Figure 5.5f Sustainability values linked to integral sustainable design principles.
Source: Created and drawn by the authors.

5.5.1 Performance quadrant – integral sustainable design

The performance quadrant shown above is concerned with measurable, quantifiable aspects that influence the building's performance. The area of focus and research that is needed to design successfully in this quadrant is understanding site placement, typical climate data such as wind patterns and speeds, seasonal and diurnal temperature changes, relative humidity, the microclimates, and topography for each site and a detailed understanding of how the building will be used by its occupants to optimize user performance. These are all physical characteristics of a given site (Figure 5.5.1a [below]).

This quadrant also focuses on the performance of mechanical systems, renewable energy resources like wind power, solar collection, and geothermal heat pump systems. Any engineering-based technologies or calculations would also be included in this quadrant. Examples of these are solar array outputs, overhang calculations, 'R' values for walls, 'U' values for windows, and any other metric or performance values of an object or material. Any embodied energy calculations, waste reduction or recycling rates, and most of the rating systems fall into this category as well. The performance of a building is crucial to integral sustainable design, but is only one part of a holistic solution.

Figure 5.5.1a Performance quadrant.
Source: Created and drawn by the authors.

PERFORMANCE KEYWORDS: Datapoints | Measurable | Facts | Energy Performance | Resources | Efficiencies | Eco-efficiency | Dematerialization

Questions for the design process from a performance perspective are as follows:

- What is the best site placement based on use and local climatic information?
- How to maximize performance of passive and active systems to reduce mechanical requirements?
- What is the best way to optimize occupant performance?

Relevance: From a sustainability point of view, measures of efficiency and pollution are understood from this perspective. The performance of an automobile is measurable. The amount of carbon emitted from a car is measurable. The weight and shape of the car is defined by exact measurements. This is one lens by which we can understand the world – but not the only one. As an example the chart in Figure 5.5.1b (below) presents an evaluation of solar panels. The performance perspective is shown in black. The items listed in the other quadrants are all true and real, but not part of the performance quadrant. The overall quality of the photovoltaic (PV) panels is a function of all the quadrants.

5.5.1.1 Sustainable design principle: design to performance standards

One of the most fundamental principles of sustainable design is the use of performance targets to drive creativity, innovation, and measure accomplishment. Think about it like a workout plan or a diet. Humans need goals to help them push themselves beyond where they thought possible. The same is true for sustainable design projects, which almost always

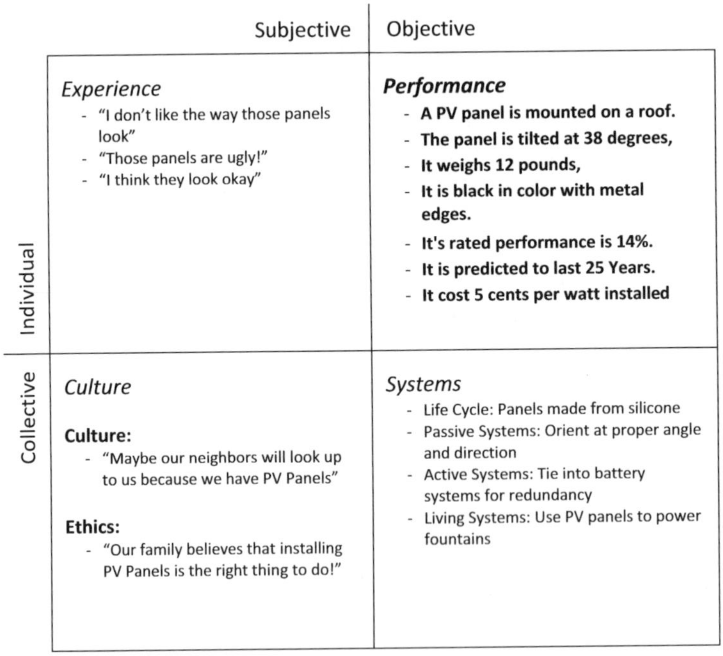

	Subjective	Objective
Individual	*Experience* - "I don't like the way those panels look" - "Those panels are ugly!" - "I think they look okay"	**Performance** - **A PV panel is mounted on a roof.** - **The panel is tilted at 38 degrees,** - **It weighs 12 pounds,** - **It is black in color with metal edges.** - **It's rated performance is 14%.** - **It is predicted to last 25 Years.** - **It cost 5 cents per watt installed**
Collective	*Culture* **Culture:** - "Maybe our neighbors will look up to us because we have PV Panels" **Ethics:** - "Our family believes that installing PV Panels is the right thing to do!"	*Systems* - Life Cycle: Panels made from silicone - Passive Systems: Orient at proper angle and direction - Active Systems: Tie into battery systems for redundancy - Living Systems: Use PV panels to power fountains

Figure 5.5.1b The performance perspective in context.
Source: Created and drawn by the authors.

have associated performance metrics, especially the use of rating systems. EUI or energy use intensity is a common measurement system used by professionals to communicate how efficient they are, specifically in comparison to other buildings of the same kind or a baseline. Baselines are critical because they are the benchmark, or bar, that the project is measured against. For example, the LEED Green Building Rating System uses ASHRAE 90.1 as a baseline. Points are awarded depending on how far above the baseline the project reaches. This is predicted by energy modeling, a key virtual technique that is central to the sustainable design process. Benchmarks are similar to baselines in that they provide a point of reference for a design project.

Standards are also set by governments to drive efficiency. They employ long-term thinking with research, data and past experience and then create standards and codes to protect human and animal life and the commons: air, water, energy, and ecological systems from contamination, depletion or destruction. The efficiency improvements in refrigerators correlate with state and national standards for the reduction in emission of greenhouse gases. This pattern is true of cars, homes, airplanes, buildings, and appliances.

Figure 5.5.1c The Bullitt Center's overhanging roof maximizes the performance of the building.
Source: Wikipedia By Joe Mabel, CC BY-SA 3.0, https://commons.wikimedia.org/w/index.php?curid=30809089.

The development of standards happens in the public sector through policy creation. Many of the benchmarks in sustainable design occurs in this way – not in direct design innovation, but in developing the policies that drive innovation. Financial performance is often based on short-term profit, and as a result, there is tension between the need to meet ever-increasing environmental performance standards and maintaining financial growth or profit.

On a typical sustainable design project using the LEED rating system, the client, design team, and stakeholders use the "goal-setting process" to collaboratively set the performance goals, including which points will be obtained and which will not. The performance goals form the ethical backbone of the project (culture perspective) and can often impact the aesthetics of a project (experience). The Bullitt Center (Figure 5.5.1c [above]), designed by Muller Hull Architects shown above, has an extraordinary roof which was needed to increase the capacity of the PV panels to reach net zero performance goals. The addition of a visible main stair was included and designed to encourage residents to walk between floors instead using of the elevator, which reduced energy and increased the level of fitness for the occupants.

5.5.1.2 Design for eco-efficiency

The net-positive energy goal of the Bullitt Center project is 105% of the total energy use of the building. This is a combination of a metric (net-positive) and the principle of eco-efficiency. This led the design team to seek every possible innovation in reducing the overall energy and resources used. The solar panel arrays project far beyond the building's façade, which is atypical. This had the dual benefit of increasing the solar collection area while providing greater shading, which reduced the air-conditioning load.

Eco-efficiency or eco-innovation is one of the most recognizable aspects of sustainable design. The desire to reduce environmental impacts and increase energy efficiency is shared by most people regardless of whether they are practicing sustainable design or not. For example, refrigerators are over 75% more efficient than they were just 45 years ago, while at the same time getting larger (Rosenfeld and Poskanzer 2009). The primary driver of eco-efficiency design is compliance with performance standards set by the federal government. Similar improvements have been made with buildings, cars, planes, and televisions. Notice the role of performance standards in driving efficiency.

5.5.1.3 Dematerialization

Dematerialization is the principle of using technology and engineering to create synergies, or advanced processes that allow structures or tasks to be completed with reduced resources. In economics, dematerialization refers to the absolute or relative reduction in the quantity of materials required to serve economic functions in society (Rosenberg 1982). Put simply, dematerialization means doing more with less. Dematerialization is a form of eco-efficiency. It can be applied to business practices, buildings, products, and even record keeping. Miniaturization is another term used for this process. An obvious example is the switch from large cathode ray tube (CRT) computer monitors to flat-screen monitors. Not only did they reduce energy use, but the new monitors emit almost no heat meaning less air conditioning load. This is an example of synergy where one technology helps to make another technology more effective.

Dematerialization is similar to ephemeralization, or doing more with less, as proposed by Buckminster Fuller. Fuller worked across multiple fields with the focus of "making the world work for 100% of humanity" (BFI 2018). His design principles were based on natural forms and structures which are inherently efficient – the maximum benefit for the least amount of materials and structure. Biomimicry, which will be explored in Chapter 6, is the latest evolution of this approach. Looking to nature's multibillion year history of innovation is a logical place to seek sustainable design solutions. Fuller is best known for inventing the geodesic dome, a structure that created the maximum volume of space with the minimum of material.

Dematerialization challenges all professions and all people to be the most efficient, seeking to advance design, engineering and technological knowledge. It also speaks to the societal changes that are the result of the increased connectivity and changes that need to happen regarding consumerism. It looks to focus on quality of life for all living things instead of acquiring stuff.

ROBOTICS, AUTOMATION, AND SUSTAINABILITY

The rise of robotics and automation is another form of dematerialization. Just as technology is used to reduce the required physical resources while still delivering value and performance, the benefits of robotics and automation dematerialize time – from a human perspective. Ethical questions aside, the ability for machines to assume traditionally human tasks is transforming the way society sees its relationship with time. The use of prefabricated architecture was promoted by Kieran Timberlake in the book *Refabricating Architecture* which outlined a transformation in philosophy based on observations of the design and manufacturing processes for automobiles and airplanes. They applied the principles of eco-efficiency to construction of the Loblolly House in 2006, a sustainable house using off-site construction. It started with a complete examination of the site, extensive drawings, and computer modeling to assure all the preassembled modules could easily be joined on site with only hand tools. The house was raised on piers to minimize ecological impact and used all-natural materials. The Loblolly house initiated a wide spread movement of architects and builders exploring ways to dematerialize the design and construction process, which could lead in theory, to higher-quality construction at the same or lower cost.

THE ROLE OF VIRTUAL SPACE AND TIME IN DEMATERIALIZATION

The mastery of virtual space and time is the ultimate vehicle for dematerialization. The ever-shrinking computer chip coupled with ever-expanding digital storage capabilities via the internet means that the physical laws of traditional space and time are not as restrictive. Space is no longer only measured by square meters, but by disk storage space. Time in the professional world is measured by the speed of the internet and the power of computers to process information. The positive environmental effects of the continual computer revolution are astounding. The energy saved because of a virtual world which reduces travel, physical material usage, and freeing up of time to pursue other activities is significant and makes digital dematerialization very attractive. The negative aspects of the digital revolution are the threat to human relationships, the never ending work week, and our relationship with nature as we stayed glued to our screens. Furthermore, the proliferation of digital devices means that literally trillions of devices are causing an increasing demand for rare earth metals and placing a burden on the mining industries and ecosystems at extraction points, not to mention the recycling and landfill implications associated with constant upgrading.

ADDITIONAL RESOURCES

Refabricating Architecture, https://amazon.com/Refabricating-Architecture-Manufacturing-
    Methodologies-Construction/dp/007143321X/ref=sr_1_1?ie=UTF8&qid=1529369338&
    sr=8-1&keywords=timberlake+architecture
Loblolly House by Kieran Timberlake, https://kierantimberlake.com/pages/view/20

5.5.2 Systems quadrant – integral sustainable design

The systems quadrant (Figure 5.5.2a [below]) seeks to discover the deeper patterns and the reciprocal interconnectedness of all things. It is concerned with evaluating all the possible systems and
connections between building, site, ecosystem, and the larger surrounding community. The focus is creating all things to coexist with the natural world and be a restorative force. Systems
thinking acknowledges that all decisions have a ripple effect as they move to larger-scale systems
like the ecosystem or inward to smaller-scale systems like human health. The systems quadrant
looks at things not as a group of separate entities functioning in a shared space, but an intimately,
interconnected "web of life" with each action being met with a corresponding reaction.

SYSTEM KEYWORDS: Reciprocal Interdependence | Relationships | Deeper Patterns |
Flows |Process | Dynamic Balance | Holon | Networks | Restorative | Synergies | Func-
tionality | Regeneration | Life-Cycle Systems | Passive Systems | Active Systems | Living
Systems | Human Systems

The communities, structures, and products designed with the directives from this quadrant are
always changing, always responding to external stimuli, and always pointing towards something
else. For example, a building is designed to correlate to the site and the people, not the other way
around. An excellent example is the renovation of Arizona State University's Campus Renovation. It was a comprehensive renovation that included the existing site, the buildings, and several
new features and building additions. What resulted is not only beautiful and functional but ecologically restorative and sensitive to the local climate and the people using the campus.

    When working from a systems perspective, there is an almost mystic quality to design de-
cisions, because they rely on interpreting the deeper patterns within the data associated with
each unique design project, allowing components to "speak." This is a very different way to
approach design. As DeKay says, "If we indeed desire to create an integral architecture we
need an integral path to get there" (2011, p. 24). A systems view of a building is crucial to
integral sustainable design, but is only one part of a holistic solution.

Figure 5.5.2a Systems quadrant.
Source: Created and drawn by the authors.

Questions for the design process from a systems perspective are as follows:

- How can human design interventions restore natural systems?
- What patterns should be used for making effective design decisions?
- How can architecture tie into the larger systems of the site and the smaller systems of individual rooms?

ADDITIONAL RESOURCES

ASU Polytechnic Academic District Renovations, www.aiatopten.org/node/34
Looking at the Toyota Prius from a systems thinking perspective, the systems quadrant is shown in black. The items listed in the other quadrants are all true and real, but not part of the systems quadrant. The overall analysis of the car is incomplete without the other perspectives (Figure 5.5.2b [below]).

WHAT IS SYSTEMS THINKING?

Systems thinking is a remarkably powerful, complex, and effective approach for designing sustainable products and projects. It's not uncommon for designers to develop a form or concept and then later integrate systems into the design to make it more efficient or less damaging. Systems thinking is a critical underlying force of sustainable design. The achievement of high-performance goals and metrics, as described in the previous section, is the direct result of the design of systems. System thinking can and should be considered at the very beginning

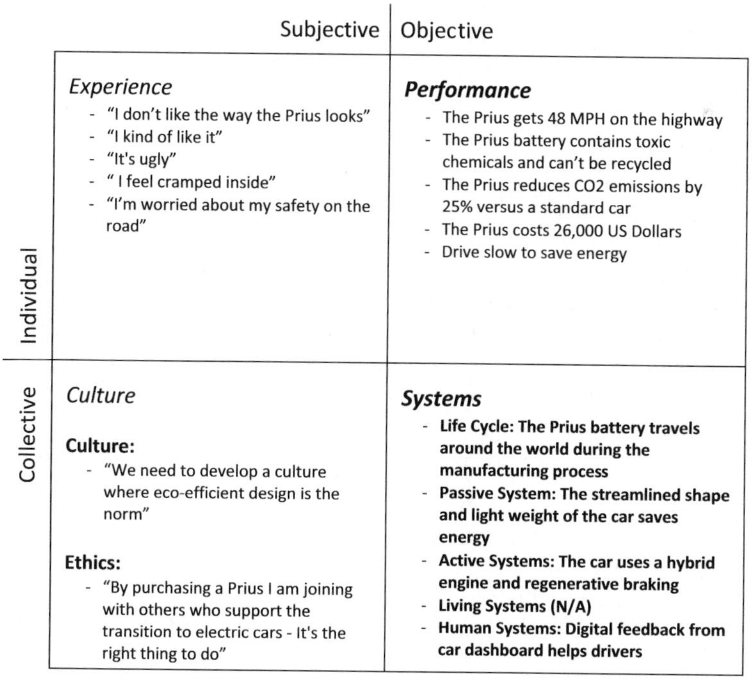

	Subjective	Objective
Individual	*Experience* - "I don't like the way the Prius looks" - "I kind of like it" - "It's ugly" - " I feel cramped inside" - "I'm worried about my safety on the road"	*Performance* - The Prius gets 48 MPH on the highway - The Prius battery contains toxic chemicals and can't be recycled - The Prius reduces CO2 emissions by 25% versus a standard car - The Prius costs 26,000 US Dollars - Drive slow to save energy
Collective	*Culture* **Culture:** - "We need to develop a culture where eco-efficient design is the norm" **Ethics:** - "By purchasing a Prius I am joining with others who support the transition to electric cars - It's the right thing to do"	*Systems* - **Life Cycle: The Prius battery travels around the world during the manufacturing process** - **Passive System: The streamlined shape and light weight of the car saves energy** - **Active Systems: The car uses a hybrid engine and regenerative braking** - **Living Systems (N/A)** - **Human Systems: Digital feedback from car dashboard helps drivers**

Figure 5.5.2b Integral sustainable design analysis of a Prius hybrid car.
Source: Created and drawn by the authors.

of a design process and can serve as a primary driver for design decisions. At the most basic level, systems thinking includes a series of core tools or styles to foster broader thinking.

The list below offers an illustrated view of the basic approaches or types of systems thinking along with the traditional type of analysis.

"Interconnectedness" mimics nature's ecological organization which is interlocked across scale and species.

"Circular" or cyclical systems means that the system has closed loops with no waste – a key aspect of a sustainable life cycle for a product.

"Emergent" implies collaboration across discipline to discover new ideas and solution for sustainability that could not be discovered working with a single discipline (1 + 1 = 3).

"Holistic" systems take into account all the variables and perspectives of given problem, consider them valid and try to account for them to eliminate externalities.

"Synthesis" systems imply a set of systems integrated together to create a high-functioning efficient outcome.

"Relational" systems are inclusive requiring a look at all things all the time.

The spirit of systems thinking is the direct response to the reductionist worldview of the industrial revolution which saw objects and buildings as disconnected and isolated from their context and each other. In the design world, we use the term context to refer to the need to make sure projects are interconnected with the local and global ecosystems and to the other human-made systems.

Sustainable design is part of the emerging worldview of integration, a time when the interconnection of humanity and nature forms a new deeply woven system that benefits both – a mutually beneficial system. System thinking is a discipline unto itself complete with its own well-codified theories, processes, and tools. For this book, we will limit the discussion to the systems that most impact the design, production, and use of products and spaces. What follows is the shortest most straightforward breakdown of systems for sustainable design: life cycle, passive, active, living, and human systems.

5.5.2.1 Life cycle systems

Life cycle assessment (LCA) is one of the most underrated systems when it comes to sustainable design. We tend to focus on the energy efficiency of a technology or building, but the embodied energy associated with the extraction, manufacture, transport, use, and end-of-life disposal is an important consideration for sustainable designers. By thinking about the energy used in the full life cycle of a product or project, we can discover many ways to save energy. One of the most obvious examples is the use of local materials in building projects. This achieves a reduction in energy and pollution from the transportation involved in each step of the manufacturing process. This also helps to stimulate the local economy.

5.5.2.2 Passive systems

Passive systems typically have no moving parts but are still able to generate a desired effect by using shape, form, position, and/or color to manipulate light, air, heat gain, and water for a desired affect or to meet a performance goal. For example, a shading device on a building interacts with the movement of the sun to alter the direction of light, thereby helping to cool the building. It's the interactions and relationships that make this a systems approach.

(The amount of energy saved by reducing cooling loads would be in the performance quadrant.) The Omega Center, built in New York State, and designed by BNIM, is one of the first Living Building Challenge-certified projects in the world and is a great example of the use of passive systems to reach ambitious performance goals. The project can be viewed from the links in the Resources section below.

ADDITIONAL RESOURCES

The Omega Center by BNIM, https://bnim.com/project/omega-center-sustainable-living

5.5.2.3 Active systems

Active systems are most often recognized as technological systems that serve human needs but require some form of energy or mechanical means to operate. Usually, fossil fuels or electricity are the power that drives these systems. Each system can be designed for the maximum efficiency and in the best cases be developed to create synergies with the other technologies or systems. As an example, LED lights which use much less electricity for the same amount of light also give off less heat than typical bulbs, thereby reducing the need for cooling, reducing the size of the cooling unit, and decreasing energy use, saving money, and reducing damage to the environment. Heating, ventilation, and cooling systems (HVAC) are active systems, along with PV panels, automated shading, motion sensors, and geo-exchange systems.

5.5.2.4 Living systems

Ecology is a tremendously complex system with trillions of interactions occurring all the time at all different scales. These systems have embedded feedback loops that help to inform the systems so that they can adjust to changing conditions such as seasons or temperature. That is why they are referred to as living systems.

   At the scale of a city, district, or site, ecological restoration is a major goal in the design of sustainable systems. Here, landscape architects, urban designers, and others are not only using living systems such as trees, wetlands, and meadows for the design of spaces and places, but they also become embedded in the larger ecosystems leading to a more robust, resilient overall ecology. At the building and interiors scale, living walls, green roofs, and living machines are examples of using living systems as part of an overall sustainable design approach. The design process, research, site inventory, and analysis phases are critical in order to learn about the specific ecological systems for a project. This knowledge can establish design directives to push towards an ecologically restorative outcome.

5.5.2.5 Human systems

All of the systems listed above are maintained or impacted by human systems. Ecological systems are placed in jeopardy when human economic systems encourage the widespread use of natural resources without consideration of the environmental impacts. The maintenance of our streets, public buildings, and public transport all rely on human systems to operate at a functional level. When designing a building, systems can be used to impact the behavior or performance of an occupant. For example, training employees to turn off lights and computers will lead to higher levels of energy performance. On the flip side, the use of ventilation from an operable window, a passive system, requires the interaction of humans to know when to open and close the window. At the bigger scale, humans develop all sorts of policies and procedures to reach higher levels of performance. For example, a citywide policy requiring solar or green roofs is an example of a human system designed to improve energy or environmental improvements.

HYBRID SYSTEMS

Clearly, there is an important relationship between life cycle, living, passive, active, and human systems. The better job we do in designing energy-efficient passive systems, the easier it will be for the active systems to meet performance goals. Hybrid systems are a combination of active, passive, and living systems. An example would be a building that uses the passive systems of larger windows, shading devices, better glass, and better-insulated wall assemblies to create a higher-performing building envelope. While these design choices add to the first cost of the project, the design team can now specify smaller mechanical systems or less lights to meet the needs of a project, thereby saving money for years on energy costs.

INTERACTION BETWEEN THE SYSTEMS QUADRANT AND OTHER QUADRANTS

The skill that a sustainable designer brings to the table is the ability to understand all the systems and how they interact with all the other quadrants. Systems are designed to reach high performance standards, but if the culture of the building doesn't value natural light, or has functional needs that demand high light levels, the electric light may still be turned on even though the daylighting systems are adequate. The systems designed to save energy don't work because of the occupants. When a designer takes a systems approach to the same project they understand the organization's culture and take into consideration the perceptual needs and past habits of the occupants designing to overcome these barriers. In another example, a university building, with a mission to be more ecologically sustainable, will use living systems such as an eco-machine to clean the water with the active participation of the occupants because their performance goals align with the culture. We can do the best job in the world designing systems to meet all the performance requirements, but if we are not aware of the cultural and experiential expectations, and include them in our considerations when designing, our systems will not be as efficient as possible.

Another area where the sustainable designer's expertise comes into play is understanding the difference between project types with similar energy efficiency goals. For example, a hospital that is required to have a high rate of air exchange because of sterilization issues will need a far different ventilation system than a vacation house. The first will need an active system with heat exchange, while the second can rely more on the passive systems for heating, cooling, and ventilation.

Comfort level of the occupants, along with their culture, is another crucial area for a sustainable designer to understand. If the users of a building have a high expectation for very specific temperatures and humidity levels, the mechanical systems will work harder to meet needs. If, however, there is a culture where occupants are willing to tolerate a wider array of temperature swings inside a building and use the operable windows, performance increases because there is a lower demand on the mechanical systems. Each project comes with a different cultural context, environmental conditions, use, client, and budget. Systems thinking takes into account all the specific conditions and their interactions.

5.5.3 *Cultures quadrant – integral sustainable design*

The cultures quadrant (Figure 5.5.3a [below]) is concerned with designing to create a personal bond with nature, historic culture, and the surrounding community. The cultural quadrant seeks to understand the history, vernacular architectural forms, community development patterns, and shared intellect or social practices that are the basis of a cultural background, of the project's site as well as the client. Promoting social equity and a sense of "Place" are large components of this quadrant. Ethnic or religious rituals, myths, or religious beliefs would also be categorized in this quadrant.

"Place" is how people bond or connect with the built environment. Dylan Trigg expresses it this way,

> As bodily subjects, we necessarily have a relationship with the places that surround us ... Over time, those places define and structure our sense of self ... Place is at the heart not only of who we are, but also of the culture in which we find ourselves.
>
> (2012, p. 1)

Creating place has been shown to be extremely important in people's willingness to care for buildings and even each other.

CULTURAL KEYWORDS: Cultural Forms | Rituals | History | Place | Equity | Connection to Diurnal (Daily) + Seasonal Cycles | Purpose | Interaction with Nature | Ethnic + Geographic Norms

Connecting with culture does not necessarily mean the replication of the existing forms. There are times when connection can mean complimenting or even creating a new aesthetic that connects people with the present aesthetic and current lifestyle. A mixture of styles and time periods can enhance the understanding and give a broader scope of an area's cultural heritage and changes that have occurred over time.

The importance of connecting with the built environment is well summarized by Stipe (2003, p. XIII): "... we seek to preserve our heritage because our historic resources are all that physically link us to our past ... [they] must be preserved if we are to recognize who we are, how we became so ..." Creating cultural connections and social equity are imperative to integral sustainable design, but are only a part of a holistic solution.

Questions for the design process from a culture perspective are as follows:

- How can social equity play a meaningful role in the design process?
- How can cultural connections in places go beyond replicating the local context?
- How can we design so nature is sensed as a cyclical force, fostering deeper connections to nature?

Figure 5.5.3a Cultures quadrant.
Source: Created and drawn by the authors.

Relationship to nature

Meaning in architecture is increasingly about connecting to nature and each other. The age of hermetically sealed boxes with nature being viewed from behind inoperable windows is giving way to a multisensory experience. We now understand that being aware of the sun's passage, the scent and feel of a warm breeze, or the fragrance of flowers planted outside a window is not only pleasant, but vitally important for productivity and health. This connection generates a sense of place, a sense of belonging, and an internal satisfaction that life has meaning and purpose. Being aware of the seasons, and daily cycles connects us to the process of life and balances neurological rhythms. Creating opportunities for engagement with nature that are multisensory and diverse is critical to integral sustainable design.

5.5.3.1 The principle of social responsibility

The goal to manifest meaning in architecture has taken a number of interesting and important directions. First and foremost, we are beginning to see the first evidence of social equity becoming a core value in the design of buildings. The question as to who designs buildings, who uses buildings, and how those buildings reinforce or destroy local communities is now becoming a fundamental design principle for more and more design firms. Clients are increasingly seeking design firms whose values align with theirs and the diversity of the end users.

The engagement of stakeholders early in the design process is an indication that real change is taking place. We will learn more in Chapter 9 on integrated design processes. It is sometimes called participatory design and specifically seeks input from the surrounding community, occupants, and anyone that will be touched by the project. Equity is one of the core values of sustainability. The culture lens allows us to examine a design project from the perspective of fairness, inclusiveness, and equity. These sustainable values were covered in Chapter 4.

Equity plays out all scales. At the regional scale, environmental justice is an issue when a polluting power plant is proposed to be built in a low-income area. The rights to fresh air, clean water, and to be healthy is jeopardized simply because of their economic standing. Or a large forest is removed to make way for a housing development. Here, the intrinsic rights of nature as well as the community that used the forest for hiking are compromised. Alternately, if a historic building is restored and now has a new function that serves the neighborhood, there is a signal of respect that is sent to the local neighborhood and part of their past is there as a constant reminder of where they came from.

Social equity for designers deals with issues and opportunities of access and inclusion. William McDonough has noted "… that the rights of one species are linked to the rights of others, and none should suffer remote tyranny" (1993, p. 8). An excellent example of social responsibility in design is the Neue Staatsgalerie Art Museum designed by James Stirling in Stuttgart, Germany. The museum allows public access between two main streets in the city with a combination of ramps and gradual steps that circulate through an enclosed sculpture garden on the museum's roof. Along the ramps and stairs are multiple large glass sections of the museum that allow the public to view the art as they pass by or gather in the plazas on the roof. The design serves as a connection between two parts of the city, but it also exposes the citizens to art as part of their daily routine and creates a culturally rich experience. It also allows access for those who are not able to pay the museum entry fee (Figure 5.5.3b [next page]).

Figure 5.5.3b Neue Staatsgalerie Pedestrian route through the external rotunda.
Source: Wikipedia By Veuveclicquot m - Own work, CC BY-SA 3.0, https://en.wikipedia.org/wiki/
File:Neue_Staatsgalerie_Stuttgart.jpg#/media/File:Neue_Staatsgalerie_Stuttgart.jpg.

Within buildings, the rights to light, fresh air, and a view are often negated by putting the executive offices at the perimeter. In all of these examples, the battle between self-interest and empathy is clearly expressed. Sustainable design seeks to address all of these conflicts holistically by looking for new solutions that leverage benefits over the long term.

Ethics is extremely important to consider in the culture lens. The ethical foundation of any project starts with the initial question of whether a building or product should be made in the first place. If so, where and how? From there, the long road of questions about values and beliefs is traveled until the completion of the project.

The culture and equity lens can be applied to any systems like the 2013 Savar building collapse in Bangladesh. It was originally designed for shops and offices, but ended up housing a garment factory on the top four floors instead. Even after structural problems were discovered, workers were ordered to return to work or lose a month's salary. The building collapsed with approximately 3,100 garment workers inside, more than 1,100 people died as a result. This also led to ethical questions regarding overseas garment manufacturing. Products with complicated life cycle systems require examination with the cultural lens. The accident generated a range of policy decisions (human systems) to attack the problem such as greater oversight over manufacturing processes by using a better auditing systems as well as increased collaboration with local partners.

A positive example is the Low Income Housing Project in Quinta Monroy Housing, Mexico. Architect Tatiana Bilbo interviewed the tenants and found the underlying desire for dignity, a sense of completion, spaciousness, and the traditional form of pitched roofs were the things most important to the owners. Bilbo went on to design the housing units to have a large completed central core. This core housed mechanical and plumbing spaces in such a way that each family could expand as required, with the types of spaces to best serve their individual needs. The houses all used recycled materials, and were designed to capitalize on passive systems to reduce energy use and the volumetric aspects that provide the perception of spaciousness. The project achieved engagement and investment from the community and the individual owners. Each family now has the ability to meet their varied spatial needs. There is individuality of form creating interest and pride of ownership because of the flexibility that came from looking deeper and desiring to create equity (Goldhagen 2017, p. 278).

5.3.3.2 The principle of shared values

Sustainable design projects are difficult to realize because stakeholders, design professionals, and clients come to a project with differing values and expectations. In addition, different people operate out of different worldviews, some still using an industrial mindset, others an age of information approach, and still others trying to use an integral approach. The conflicting goals of differing worldviews, or past, or no experience with collaborative design can make a project challenging or even a cause for resistance.

A set of collectively defined shared values should be a normative first step in sustainable design projects. Other clients defer to the LEED Rating System, Living Building Challenge or Passive House to guide their efforts and serve as a proxy for shared values. In some cases, the rating systems are seen as much a nuisance than a means to the end of realizing an ethical project. In Chapter 9, the details of how shared values are placed into practice as well as some other key shared strategies such as transdisciplinary design, design charrettes, and integrated project delivery are discussed.

5.5.3.3 The principle of cultural vitality and heritage

Design projects that are lively, dynamic, authentic, and engaging rely on the design principle of vitality for success. This can take place at many scales and include activities, cultural events, restaurants, shops, art and music venues, as well as materials and finishes, and architectural configuration and urban planning.

Architectural form, materiality, or ornamentation that is unique to specific geographic area is called vernacular architecture. By including these elements at the urban, district, building, or interiors scale, we can create a sense of connection to a specific place. We can acknowledge the cultures and traditions of the past, and we can create new ones that speak of the future.

Historic preservation and adaptive reuse are two common ways to include cultural forms at the urban and district scales. A classic example is old factories, sometimes on the waterfront, being used for apartments with retail and restaurants on the first floor. The form speaks to the past, and the use speaks to the present. Many times, a whole new culture is created in an area of the city that was deserted and decaying.

The National Building Museum in Washington, D.C., an old pension building from the civil war, was repurposed to house The National Building Museum. The historically certified building is full of interesting ornamentation, giant marble columns, beautifully daylit spaces, and "stairways" for horses. Installing exhibits of contemporary architecture creates a dynamic juxtaposition of worldviews in the space and highlights the effectiveness of passive systems.

It is the sustainable designers' responsibility during research to identify any potential negative aspects in a specific culture such as any societal taboos, past atrocities, inequities, or oppressions. These need to be understood, respected, and carefully avoided at every scale of a design project.

5.5.4 Experience quadrant – integral sustainable design

The experience quadrant (Figure 5.5.4a [below]) is concerned with designing for the subjective experiences of individual in space. To design well in this quadrant requires an understanding of health, behavioral, and perceptual or psychological changes associated with architectural design or volumetric configuration required to create connection with nature, light, form, aesthetics, and the meaning of color. Many of the areas in this quadrant are being researched by environmental psychologists and the field of environmental design.

EXPERIENCE KEYWORDS: Aesthetics | Beauty | Delight | Personal Perception | Place | Phenomenology | Psychological Reaction | Subconscious Behavior | Transcendence | Qualitative | Multisensory | Emotional Responses

The experience quadrant acknowledges that personal perception is a valid and important feature in sustainable design. Beauty is an issue currently being addressed within sustainable design by David Orr and Lance Hosey who talk about the importance of meeting the

Figure 5.5.4a Experience quadrant.
Source: Created and drawn by the authors.

Figure 5.5.4b Salk Institute by Louis Kahn.
Source: Wikipedia By TheNose (www.flickr.com/photos/tatler/339218853/) [CC BY-SA 2.0 (https://creativecommons.org/licenses/by-sa/2.0)], via Wikimedia Commons. https://commons.wikimedia.org/wiki/File:Salk_Institute.jpg.

inherent need in humans for beauty. They argue that beauty is needed to advance sustainable design in prominence, and so that buildings will be desired and cared for. Aesthetics can also be used to direct our attention to the deeper significance or purpose of an individual organization, or of life in general. Louis Kahn achieved this with his design of courtyard at the Salk Institute (Figure 5.5.4b [previous page]). Dr. Jonas Salk and his institute conducted a groundbreaking research that led to the first inactive polio vaccine. Kahn captures this exquisitely in the center courtyard design, where the stream of water and architectural form visually lead to the horizon and sea beyond, symbolizing future hope.

All of the research spaces share that similar view of the horizon, reinforcing the visionary mission of the institute and their research. Beauty can serve a profound function, and understanding its critical role is a large part of sustainable design, but it is only a portion of restorative, integral sustainable design (Figures 5.5.4c and 5.5.4d).

Questions for the design process from an experience perspective are as follows:

- How to create relationships with nature, and natural cycles?
- What is the role of structures in the shaping of human experience, memory, and actions?
- What type of experience do we want the occupant to be feeling?

Notice in the analysis Figure 5.5.4d (next page) that the experiential aspects of the project are expressed in very subjective terms which cannot be proved or disproved.

Lance Hosey (2012) in his book, *The Shape of Green*, ties into integral theory very well in his triad of principles: "shape for efficiency, shape for pleasure and shape for place." Pleasure correlates well to the experience quadrant (UL), while "place" also falls into culture quadrant (LL) as well as aesthetics, and finally shape for efficiency would fit in the performance quadrant (UR).

Figure 5.5.4c The Bilbao Museum.
Source: Wikipedia by PA – Own work, CC BY-SA 4.0, https://commons.wikimedia.org/w/index.php?curid=45169992. Web Site https://en.wikipedia.org/wiki/Guggenheim_Museum_Bilbao#/media/File:Bilbao_-_Guggenheim_aurore.jpg.

Experience	Performance
"I love the Bilbao, It's beautiful" **"I love the way the metal skin reflects the sky and water"** **"The irregular forms are organized in a casual way that makes the building pleasing"** **"I feel so free in the interior spaces because they are so high"** **"I feel a sense of wonder, and surprise when I'm inside"** **"The form reminds me of the mountains"**	The Bilbao costs 200 million dollars, we could have built 10 art museums all around Spain The building is energy inefficient The Project has brought millions of Euros in spending and investment to City of Bilbao
Culture "The museum is a great expression of national pride and investment in the arts" Ethics: "It is unethical that they spent so much money on this museum when people are starving in the world" "People are dying mining titanium"	System **Life Cycle:** The building has a huge life-cycle footprint **Passive System:** The shape of the building offers no benefit to performance **Active Systems:** The mechanical system is very large to heat and cool the structure **Living Systems:** The water in front of the building offers an opportunity for an ecosystem

Figure 5.5.4d Integral sustainable design analysis of the Bilbao Museum.
Source: Created and drawn by the authors.

5.5.4.1 Multisensory design

We need to consider engaging all of the senses when we design. Sight is surely our most dominant sense, and we rely on it heavily in many areas. But if you were to close your eyes, all of a sudden you would be overwhelmed by the sounds, smells, and how things felt. Environments that really speak to us and are sought after incorporate and activate all our senses. That is why gardens are so beloved; beyond the obvious visual beauty, there are the scent of flowers, the sound of running water, and cool breezes passing by from the shade of trees. In some gardens, we can even pick fresh fruits or vegetables and engage our sense of taste.

Juhani Pallasmaa (2005) in his book, *The Eyes of the Skin*, talks about ocular centrism, meaning architecture or design that focuses primarily on the visual. He says that this can push us into isolation, devaluing space and even the other senses. Richness in the experience quadrant is achieved by the diversity in sensory opportunities and dimensions. Industrial, fashion, textile, and product designers whose work dwells mainly at the human scale are typically more aware of the possible multisensory experiences, where the sense of touch is critical to experiential success.

PLACE

"Although we fundamentally shape our surroundings, ultimately place exists independently of human life, in turn shaping us" (Trigg 2012, p. 2).

"Place" can be an obscure concept that you may have heard about or not. It is a physical location, but it is also much more than that. The physical world is where we live our lives, meaning everything that happens to us, good or bad, is at a physical point on the earth. Christian Norberg-Schulz in *Genus Loci* talks about human's need to orient themselves within and identify himself with an environment, and it needs to be meaningful and have a distinct character to add quality to existence. The concept of place spans the fields of psychology, philosophy, and phenomenology, which is the study of experience and how it affects cognition and sense of self. Architects from the beginning of time have talked about and designed for the unseen effects of space and architectural configuration, but they could never quantitatively define it.

Dylan Trigg (2012), author of *The Memory of Place: A Phenomenology of the Uncanny*, argues,

> As bodily subjects, we necessarily have a relationship with the places that surround us ... Over time, those places define and structure our sense of self ... Place is at the heart not only of who we are, but also of the culture in which we find ourselves.

The observations of Trigg are being supported with current work and research within the fields of environmental psychology and evidence based design. Both of these fields study the effects of the built environment on the occupants. The amount of supporting data of the measurable changes in health, behavior, and learning capability directly related to the built environment has increased in recent years and will be covered more fully in Chapter 8.

The Academy of Neuroscience for Architecture started in 2003 by the AIA focuses on exploring the built environment and the neurological changes that occur in human brains as a result of experiencing architecture. Fred Gage states in his forward to *Brain Landscapes: The Co-Existence of Neuroscience and Architecture* that "I contend that architectural design can change our brains and behavior. The structures in the environment ... affect our brains and our brains affect our behavior" (Eberhard 2009, p. XIV). This statement corresponds with Norberg-Schulz when he talks about Genius Loci, or "spirit of place," and the concrete reality it represents in human existence, and designers' sacred responsibility in creating it. The environments being designed today are shaping how people and societies live and function in the future. This is a great responsibility, and a great honor!

5.5.4.2 Psychological responses

Psychological responses at a deeper level can be conscious or subconscious. They can be related to memories or experiences in the past that shape the way we perceive space now. They can also be related to the colors or materials used in the built environment and the corresponding responses. An example is the colors of walls and lighting and how they affect our hunger levels. Or lighting and how it can affect our retail behavior. These areas are also studied by environmental psychologists and the field of evidence-based design, and can even overlap with concepts of place.

Certain architectural and volumetric configuration also produces reactions. Higher ceilings inspire activity, awareness, and creativity or openness. Spaces that have lower ceilings tend to make us more settled. When we have protection from behind and above, this gives a feeling of protection, safety, and more of an ability to be introspective, which is also known as the theory of prospect and refuge.

These are all valuable tools that can be used to create enriching experiences and sense of place and extremely important in designing to meet the experiential needs and desires of the occupant and client. We will be exploring this more in Chapter 6, with biophilia, and Chapter 8.

EXPERIENCE OVER TIME

Passage of time is another aspect of experience that we normally don't think about. When we see the patina of copper, the marks of use on wood, or the ancient ruin, we have a sense of time passing, and life, and history. It gives us a sense of connection to our life and our memories which are a large part of who we are as individuals. Think of your favorite shirt, it may be well worn, yet you constantly return to it and every time you put it on you remember things associated with it.

The creation of memories through design is one of its most important objectives. Usually, memories are shaped by our interaction with other people and categorized by the place they occurred. Some spaces are memorable unto themselves, and images are carried forward in memory as an image. This might explain why our visual sense is the most favored by many design professions.

EXPERIENCE ACROSS SCALE

For centuries, different proportioning systems, sometimes based on nature or mathematics, have been used to choreograph the experiences by people in space. The sense of scale and proportion in design is critical and traditionally was strictly regulated by formulas devised during the classic Greek and Roman periods of architecture. Le Corbusier established a proportioning system based on the human body, Le Modulor, and was developed to create visual unity through out a building. The "Ken" of Japanese architecture is another example of a standardized unit that was used to determine pillar locations and room sizes. The Golden Ration or the Fibonacci series are other proportional standards that have been used in art and architecture for centuries.

Our reactions to scale are very deep and instinctual. Scale has the ability to make us feel welcome or isolated, invited in or excluded. In a cathedral, scale reflects the relationship of humans to a celestial being. Scale and the interactions it produces visually, or with physical space, can be invigorating, pleasurable, or uncomfortable, even when we are unable to articulate the exact reasons. Some would argue that the source of this attraction would be from cultural expectations or norms. Others would cite the source as being the inherent human desire for beauty and innate spatial understanding, and that is the key to understanding this quadrant. Divorcing cultural expectations of beauty versus pure beauty is a challenge. For example, stained glass windows in a church can be beautiful because they tell a meaningful story about the origins of a religion (culture), or because of the different colors of light can be enjoyed for their own sake (experience).

Understanding scale when designing urban environments is crucial to creating diverse and holistic relationships between the buildings, natural open spaces, and the human occupants. The Japanese Ken or Le Corbusier's Le Modulor are examples of proportioning systems that help to unify spatial experiences.

EXPERIENCE AND THE OTHER QUADRANTS

Styles, aesthetic attitudes, and tastes are all part of the experience quadrant and are constantly shifting with each worldview, but the need for beauty and aesthetic connection or joy doesn't. Aesthetics can affect productivity, healing rates, and even sales figures in retail environments, all these quantifiable measurements would be in the performance quadrant.

Architecture itself serves intrinsic needs with its beauty, or its preservation of history or its pointing to the future…"We require from buildings two kinds of goodness: first, the doing their practical duty well: then that they be graceful and pleasing in doing it."

("The Stones of Venice", p. 77, John Ruskin)

Conclusion

The meta-frameworks covered in this chapter are like any new knowledge or new language. They may be difficult at first, but extremely valuable with a little practice. Throughout the rest of the chapters, the integral framework will be used to make sure that the information covered is comprehensive and holistic. Health and well-being are affected by all the quadrants of integral theory. Research and science are now confirming the importance of meeting the needs of all four quadrants in the built environment. Integral sustainable design values the measurable and performative requirements of sustainable design and gives equal value to the unseen and intangible aspects.

In the upcoming chapters we will look at biomimicry, an emerging design process to create sustainable solutions to complex human needs while reducing resources and environmental damage. Biophilia explores the deeper benefits of exposure to nature and natural elements that enhance health and well-being, connections to nature, and sense of place. The principles of resilience and adaption will then be explored using an integral approach to evaluate the performative, cultural, systems, and experiential aspects. An integral thinker is well equipped to lead and participate in a collaborative design process, and is well suited to pursue transdisciplinary design. Integral theory and integral sustainable design will be explored with case studies and strategies at the various scales from global ecosystems to the human scale of materials, finishes, and products in the end chapters.

Additional resources

The Four Worlds of Sustainability, Barrett Brown, http://nextstepintegral.org/wp-content/uploads/2011/04/Four-Worlds-of-Sustainability-Barrett-C-Brown.pdf

References

AIA, 2018-last update, AIA Knowledge Communities [Homepage of AIA], [Online]. Available: www.aia.org/pages/4856-aia-knowledge-communities [July 29, 2018].

BFI, 2018-last update, Buckminster Fuller Institute [Homepage of BFI], [Online]. Available: www.bfi.org/about-fuller/biography?gclid=CjwKCAjwy_XaBRAWEiwApfjKHoQAVUX7PybvG-HancxfKQhmObaMCtcFnDTO8wYC6KqFMBZ6H-aJWhoC_8sQAvD_BwE [July 29, 2018].

Buchanan, P., 2012. The Big Rethink: Integral Theory. *The Architectural Review*, pp. 1/25–25/25.

Eberhard, J.P., 2009. *Brain Landscape: The Coexistence of Neuroscience and Architecture.* Oxford and New York: Oxford University Press.

Goldhagen, S.W., 2017. *Welcome to Your World.* 1st edn. New York: Harper.

Hosey, L., 2012. *The Shape of Green.* 2nd edn. Washington, DC: Island Press.

Mallgrave, H.F., 2006. *Architectural Theory.* Malden, MA: Blackwell Pub.

McDonough, W., 1993. *A Centennial Sermon Design, Ecology Ethics and the Making of Things.* New York: William McDonough Architects.

McLennan, J.F., 2004. *The Philosophy of Sustainable Design.* 1st edn. Kansas City: Ecotone Publishing Company.

Pallasmaa, J., 2005. *The Eyes of the Skin: Architecture and the Senses.* Chichester and Hoboken, NJ: Wiley-Academy and John Wiley & Sons.

Rosenberg, N. 1982. *Inside the Black Box: Technology and Economics.* Cambridge: Cambridge University Press, p. 72.

Rosenfeld, A.H., and Poskanzer, D., 2009. *A Graph Is Worth a Thousand Gigawatt-Hours.* Innovations. www.energy.ca.gov/commissioners/rosenfeld_docs/INNOVATIONS_Fall_2009_Rosenfeld-Poskanzer.pdf edn.

Stipe, R.E., 2003. *A Richer Heritage: Historic Preservation in the Twenty First Century.* Chapel Hill and London: The University of North Carolina Press.

Trigg, D., 2012. *The Memory of Place: A Phenomenology of the Uncanny.* Athens, OH: Ohio University Press.

Wilber, K., 1977. *The Spectrum of Consciousness.* 2nd edn. Wheaton, IL: Quest Books.

Wilber, K., 2000. *A Brief History of Everything.* Boston; [New York]: Shambhala; Distributed in the United States by Random House.

6 Bio-inspired design

6.0 Introduction to bio-inspired design

Bio-inspired design is a term used to describe design processes that use biological forms or systems as a model or inspiration. There are two main bio-inspired design processes profiled in this chapter. Biophilia (bio = life + philia = love) theorizes that humans have an inherent connection and need for nature and natural living things. Biophilia also acknowledges the measurable health benefits from exposure to and connection with nature. Biomimicry (bio = life + mimicry = emulate or copy) is the practice of analyzing natural organisms or systems to solve complex human problems. It evaluates the time-tested "technologies" of nature as valuable guide for human invention. Both of these approaches "draw upon nature in different ways … show[ing] the diversity of inspiration we can derive from nature" (Bernett 2017).

6.1 Historical context of bio-inspired design

Natural organisms and systems have successfully existed, thrived, and have found solutions for the same kinds of problems that humans face using renewable energy sources and local materials without destroying the surrounding ecosystem. Before the advent of mechanical heating, ventilating, and cooling systems, humans studied nature for solutions to increase comfort, safety, and solve complicated problems.

Age of the Hunter-Gatherer – Humans looked to nature and natural systems as a guide for solving problems. Without manufacturing and transportation systems, they were also forced to use readily available local materials and renewable energy.

Age of Agriculture – Vernacular architecture was rooted in its place and was composed of local materials, often using form to mitigate climate forces.

Age of Industry – The advent of factories, assembly lines, and the division of labor. Performance was based on output or volume instead of quality or beauty. Interestingly ornamentation of the built environment, fabrics, and furniture was inspired by natural forms. Art reflected a romantic view of agrarian lifestyle and nature.

Age of Information – Early Green Design espoused a return to nature and natural forms. E.O. Wilson, a Harvard Professor, first proposed the theory of biophilia, human's inherent longing and need for nature. Janine Benyus was starting research on biomimicry, emulating nature and natural systems for solving complex human problems.

Age of Integration – Humans have begin to emulate nature and its systems as a way of solving problems that have eluded them for decades. The theory of biophilia is gaining more prominence and credibility due to the changes in science, neuroscience, and medical testing equipment that can provide supported findings and quantitative information on the benefits of nature. The Biomimicry Institute was founded to provide research and educational resources to promote biomimicry in order to solve complex human and scientific problems.

6.2 Nature and bio-inspired design

Bioinspired design acknowledges the deep effect and inherent need that humans have for nature as well as modeling our designs and processes after those we find in the natural world. Nature uses only resources that are plentiful and close at hand. There are no shipping or mining resources in the natural world. The chemical reactions found in nature occur at ambient temperatures only using rare resources sparingly. Yet nature has created some of the strongest and most durable surfaces and innovative, efficient, and resilient systems. Studying nature to understand the basis of these design strategies only makes sense. Approaching design this way allows us to create superior products, reduce resource use, and restore and even regenerate ecosystems, all while creating less waste.

6.3 Motivations and bio-inspired design

As society becomes "partners" with the natural world, motivations are shifting towards long-term thinking, and we are beginning to discover and develop new and improved design methodologies to better realize a future shared with nature. Bio-inspired design improves design solutions and the ecosystems that we inhabit. If the natural world that we need to sustain our life is in better health, it only makes sense that its occupants will be healthier.

Biophilia goes far beyond the aesthetic preferences expressed by the majority of designers. It addresses the inherent need for nature at neurological, chemical, and physical levels. Evidence-based design and neuroscience are providing supported findings to confirm this.

6.4 Sustainability values and bio-inspired

The sustainability values of People, Profit, Planet, and Place are central to the success of the human race. Bio-inspired design is a key to expressing our fundamental core values and will lead to a new way of viewing products, processes, and systems. Bioinspired design supports:

People by lowering toxic chemicals in the built environment, products, and the ecosystem;
Profit by reducing resources, extraction and initial costs, creating valued product, pollution remediation and mitigation processes;
Planet by creating less waste, pollutants, toxins, and ecosystem disruption;
Place by the restoration of natural ecosystems, more green space, and enhancing people's connection to nature.

6.5 Integral sustainable design and bio-inspired design

Looking at biophilia and biomimicry from an integral perspective reveals the important opportunities afforded by these approaches. Biomimicry seeks to study nature's system-based approach as a co-creative design and manufacturing approach, while biophilia's goal is to reconnect occupants with nature through form, materiality, and configuration (Figure 6.5a).

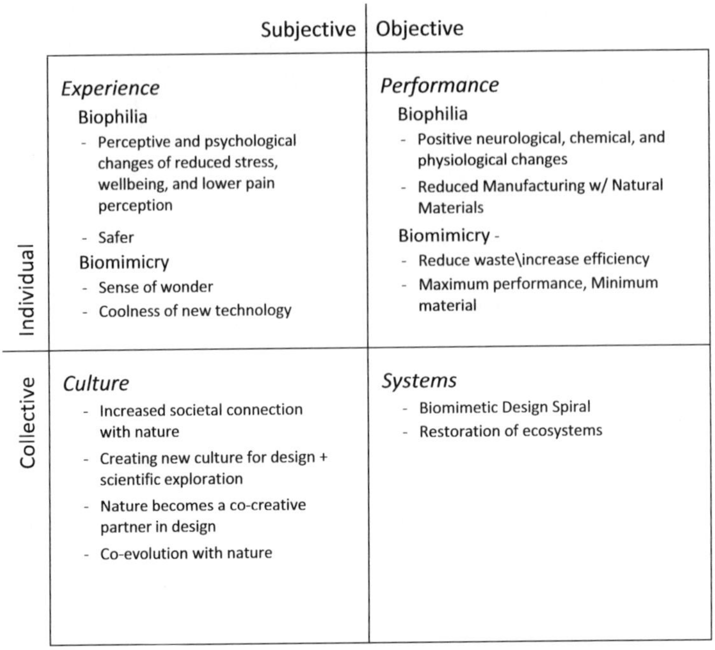

	Subjective	Objective
Individual	*Experience* Biophilia - Perceptive and psychological changes of reduced stress, wellbeing, and lower pain perception - Safer Biomimicry - Sense of wonder - Coolness of new technology	*Performance* Biophilia - Positive neurological, chemical, and physiological changes - Reduced Manufacturing w/ Natural Materials Biomimicry - - Reduce waste\increase efficiency - Maximum performance, Minimum material
Collective	*Culture* - Increased societal connection with nature - Creating new culture for design + scientific exploration - Nature becomes a co-creative partner in design - Co-evolution with nature	*Systems* - Biomimetic Design Spiral - Restoration of ecosystems

Figure 6.5a Bio-inspired design from an integral sustainable design perspective.
Source: Created and drawn by the authors.

6.6 Bio-inspired design

6.6.1 Biomimicry

Biomimicry is a term popularized by scientist and author Janine Benyus in her book *Biomimicry: Innovation Inspired by Nature*. Biomimicry is derived from Ancient Greek for bios or life and mimesis or imitation, and is sometimes referred to as biomimetics. The Biomimicry Institute defines biomimicry as "An approach to innovation that seeks sustainable solutions to human challenges by emulating nature's time-tested patterns and strategies" (Biomimicry Institute 2018). Put another way, it is a "Conscious emulation of life's genius" (Biomimicry Institute 2018). The process intentionally seeks solutions through the analysis of biological systems, plants or animals for well-adapted sustainable solutions to complex human problems. The goal is always long-term sustainability.

What biomimicry is not

The design approaches listed below are all valid, useful, and important to long-term sustainability, but they have a different focus than biomimicry.

Bio-morphic – having the form or aesthetics of biologic elements without consideration of process, function, or purpose. Example: Frank Lloyd Wright in Johnson + Johnson's open office design that replicated a Savannah setting with the use of "tree"-shaped columns and lighting or many wallcovering and fabric designs have organic or biomorphic patterns.

Bio-utilization – using biological elements in the manufacturing or as an element of a product. Example: Ecovative Design developed a compostable alternative to plastic foam by using agricultural waste like corn stalks or fungal mycelium from mushrooms.

Why is biomimicry important

Society is still influenced by the Industrial Revolution and the belief that the earth offers an endless supply of resources. By examining how natural organisms and systems address the same problems humans face, other ways to accomplish the same goals without mechanisms, harsh chemicals, or ecosystem damage can be developed.

Biomimicry requires humility and openness, and it may take a little more time at the beginning of the process but has the power to inspire our thinking and technology to go beyond brute-force techniques for solving problems. It can lead to the application of more sophisticated solutions using nontoxic and readily available resources and processes that we have been overlooking the whole time.

If manufacturing could be free of toxic chemicals and heat-intensive process or reliance on fossil fuels, think about how different our environmental conditions could be. No more toxic dumps, water and resource contamination, sick-building syndrome, or the need to destroy ecosystems when searching and mining for resources. Imagine the positive changes in switching from petrochemicals to renewable resources for workers, occupants, and the ecosystems we all rely on for basic life.

The process of biomimetic design

The spiral is a form that recurs in nature. It is found in sea shells, flower buds, the water draining from your sink, and the shape of entire galaxies. Because of this, the spiral was chosen to represent the analysis process used in biomimetic design. Developed by Carl Hastrich, an industrial designer, and his team in 2005, the Biomimicry Design Spiral is a step-by-step process that allows us to tap into nature's innovative sustainable strategies and our own creativity and imagination (Figure 6.6.1a).

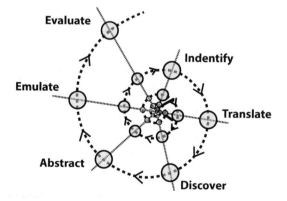

Figure 6.6.1a Biomimicry design spiral.
Adapted from (DeLuca 2016) via the Biomimicry Institute.

The Biomimicry Design Spiral is the foundation for biomimicry as we know it today. Hastrich took a standard design process, added the unique steps needed for biomimicry, and then, emulating one of nature's pervasive patterns, turned the process into a spiral.

These are the steps for applying the Biomimicry Design Spiral when you already know the problem you would like to solve (DeLuca 2016):

1   **IDENTIFY:** Create a list of functions or things you would like your design to do or ways for it to perform. Example: Reduce sonic boom from the Shinkansen Bullet Train when it exits a tunnel caused by the air pressure waves.

2   **TRANSLATE:** Translate those functions or terms into terms that make sense in the biological world. Example: Is there something in the animal kingdom that experiences drastic or sudden changes in air pressure?

3   **DISCOVER:** Research for species or systems that have successfully solved a similar problem, and then observe with an open mind and curiosity. Example: The kingfisher bird faces this problem all the time when it is going after fish in the water. Scientists observed and analyzed the beak and head shape.

4   **ABSTRACT:** Reverse engineer the principles or strategies discovered, and put them in terms that make sense to your profession or the project you are working on. Example: Notice the extreme angle and shape of the bill along with the half-round profile of the bird's head that allowed no water to be displaced when the bird enters the water after its prey. Theorizing that this could be applied to air pressure as well.

5   **EMULATE:** Use professional skills to create a design solution that uses one or more of the strategies that you have discovered. Example: Applied the angle of the kingfisher's bill and shape of its head to the front of the train. Tests were conducted with smaller objects first in similar conditions.

6   **EVALUATE:** Compare and evaluate your new design solution to your original design brief or solution. Evaluate your design solution against Nature's Unifying Patterns for sustainability. Reflect on ideas and lessons that have emerged as you went through the design spiral and plan how you will move forward with the next lap through the spiral. Example: Tests were conducted using bullets passing through tubes to replicate air pressure changes. Designers went on to study other bird species to solve other problems and make their design even more efficient.

The Shinkansen Bullet Train: The new 500 series bullet train reduces air pressure by 30%, stays below the 70-decibel noise restriction for residential areas, and reduced electricity use by 10% even though train speeds increased 10% over previous models. Additionally, customers favorably rate the train for comfort, which is due in part to decreased air pressure changes. (To learn more about the design of the Shinkansen Bullet Train and Biomimicry, you can watch this YouTube video "The world is poorly designed, but copying nature helps.")

Additional resources

The World is Poorly Designed. But Copying Nature Helps, YouTube, www.youtube.com/ watch?v=iMtXqTmfta0
Innovation Inspired by Nature, Janine Benyus

Nature's unifying patterns

10 LESSONS TO CONSIDER EVERY TIME YOU DESIGN SOMETHING

The last step in the design spiral references evaluating your solution against Nature's Unifying Patterns. These unifying or ruling principles are an expansion of what was originally published by Janine Benyus in her book in 1997 and are based on the research and discoveries made in recent years. The list and descriptions are adapted from an article developed and published by the Biomimicry Institute on their website in the Biomimicry Tool Box (Biomimicry Institute 2018).

1 Nature uses only the energy it needs and relies on freely available energy

   Natural organisms tailor their needs to the limited amount of energy available because running out means death. They use sunlight, rising air currents, dissolved minerals from deep sea vents, and nutrients from the plants and animals they feed on.

2 Nature recycles all materials

   Or as William McDonough expresses it in *Cradle to Cradle*, "Waste equals food." Everything is broken down so that it can be recombined and reassembled into something completely new, nothing is wasted or thrown away.

3 Nature is resilient to disturbances

   The natural world uses diversity, redundancy, decentralization, or self-renewal and repair to recover from disturbances or unpredictable changes.

4 Nature optimizes rather than maximizes

   Because energy is so important to a species survival, nature seeks a balance between the resources available and resources expended. There are checks and balances within its system, and it understands that growth for growth's sake can be harmful.

5 Nature rewards cooperation

   Even with competition, predators, and parasites, which may be harmful to the individual prey, the prevailing relationships in nature are cooperative. Many times, the benefits involve more than two organisms involved and provide checks and balances at a larger scale or systems level.

6 Nature runs on information

   Because there is a narrow range of optimal conditions, and to keep resources from being wasted, nature needs to monitor and respond when systems are near their limits. They use feedback loops to monitor conditions. Negative feedback loops slow processes down, and positive feedback loops speed up a process.

7 Nature uses chemistry and materials that are safe for living beings

   The chemical compositions used by natural organisms are all supportive of life's processes and do not deplete or contaminate surrounding areas. Life's chemistry is water based, is done with ambient temperatures, and uses a subset of chemical elements that allow biodegradation into useful parts when finished. Organisms live where they work, so not polluting is very important to their future quality of life.

8 Nature builds using abundant resources, incorporating rare resources only sparingly

   Because nature creates without transportation, it needs to find its resources locally that are abundant. The most common and abundant chemical compounds are from carbon, nitrogen, hydrogen, and oxygen. If rare minerals are used, they are what is found locally and are available without mining, processing, or shipping.

9 Nature is locally attuned and responsive

Survival increases when you can recognize local conditions, opportunities, and resources and can manage them well. Natural organisms and systems are present in a location because they have directly responded to the local environment. They use predictable cyclic patterns as an opportunity to adapt and flourish.

10   Nature uses shape to determine functionality

Nature uses shape and form, rather than added material and energy to meet functional requirements. Organisms can then thrive and accomplish their work with the minimum of resources.

Examples of biomimicry

SHARKLETT TECHNOLOGIES – ANTIMICROBIAL SURFACES

Repels bacteria without the use of chemicals. Developed after microscopically studying the diamond-patterned topography that prevents bacterial attachment to a shark's body. This is being applied to all types of surfaces in the healthcare industry.

VELCRO

Invented by George de Mestral, a Swiss inventor, in 1955 after a walk in the woods with his dog noticing how well the burrs stuck to the dog's fur and his own clothes. He analyzed the burrs under a microscope that revealed it was covered with hooks that grabbed any looped object and realized that this principle could be replicated in a cloth-like material. Velcro was born.

CALTECH: SPINNING VORTICES OF FISH TAILS INSPIRE WIND FARMS

An aeronautics and bioengineering Professor and fluid mechanics expert from Caltech noticed the spinning vortices that trail fish when they are swimming in schools. He worked with students to explore using the same pattern when placing Vertical Axis Wind Turbines (VAWTs) in groups. In simulations, energy production went up ten times, and initial tests show dramatic improvement to energy production.

Grouped wind farms using VAWTs decrease land usage, noise, and visual impact, and they can be placed in situations where 300-feet tall Horizontal Axis Wind Turbines cannot be placed (Fesenmaier 2013).

FARMING BASED ON PRAIRIES

The Land Institute is working to dramatically change agriculture by using natural prairies as a model. Deep-rooted plants, or perennials, that return year after year can produce the same amount of yield as the current shallow-rooted plants that are prone to weeds. This could not just maintain but also improve soil quality and water resources. With deep-rooted returning plants, other species can also be introduced that will reduce "weeds" or competing plants.

6.6.2 Biophilia

What is biophilia

Biophilia is a theory first put forth by Harvard Professor E.O. Wilson in 1984 in his book *Biophilia Hypothesis*. It asserts that humans have a biological, inherent longing and need for connection with the natural world and its processes. This need and fascination goes beyond aesthetic enjoyment of the natural world and influences emotional, cognitive, and even spiritual development. It states that humans have an integral and reciprocal relationship with nature and the natural world. Stephen Kellert and Elizabeth Calabrese put it this way: "Biophilia and biophilic design necessitates recognizing how much human physical and mental wellbeing continues to rely on the quality of our relationships to the world beyond ourselves of which we remain a part" (Kellert and Calabrese 2015, p. 21).

The main objective of biophilic design is to create healthy and supportive spaces to increase occupant health with ecologically sound environments. The goal is to provide opportunities at all scales for occupants to be exposed to natural living elements, processes, or those based on their patterns. The fields of neuroscience, evidence-based design, and environmental psychology are conducting research that confirms the theory of biophilia. These professions are showing that exposure to nature, natural elements, and animals does improve physical, mental, and emotional health and increase efficiency and productivity. Designing the built environment based on these strategies supports people while conserving or restoring the site and surrounding ecosystems.

Biophilia is NOT just putting potted plants in a space or landscaping around a building, having plenty of windows, or including a green wall. These are all great design strategies, but the principles of biophilia go much deeper and are more comprehensive. As the American writer and naturalist Henry Beston put it, "Nature is a part of our humanity and without some awareness and experience of the divine mystery, man [or woman] ceases to be man [human]" (Kellert and Calabrese 2015, p. 22).

Why is biophilia important

According to the Environmental Protection Agency (EPA) Report on the Environment (ROE) on July 26, 2017, the average person spends 90% of their time indoors (US EPA 2018). This is a big change from even 50 years ago. If we are deeply affected in a positive way from exposure to nature and natural elements, spending the majority of our time indoors is at the very least, not letting us reach out potential.

Place

The built environment has a deep effect on human consciousness, actions, and physical, emotional, and psychological health. John Paul Eberhard in his book *Brain Landscapes: The Coexistence of Neuroscience and Architecture* puts it this way, "... architectural design can change our brains and behavior. The structures in the environment ... affect our brains and our brains affect our behavior" (2009, p. XIV). Dylan Trigg, in his book, *The Memory of Place*, expresses it this way: "Although we fundamentally shape our surroundings, ultimately place exists independently of human life, in turn shaping us" (Trigg 2012, p. 2). He goes on to talk about how it also becomes part of our permanent memory and can even change our sense of self. These are all the basis for what is referred to as "Place," or a sense of belonging and attachment to our physical world, the construct where we live and experience existence. This is relevant because we need to acknowledge that all of our design decisions have a very deep and lasting effect on the users of our products or occupants of a space that they may not be able to articulate.

Health

Roger Ulrich, in his now famous research between 1972 and 1981, profiled sets of almost identical patients with the same surgical procedures in a hospital in Pennsylvania. The major difference between patients was the view outside of their hospital window:

> ... patients in rooms with outdoor views were discharged on averaged 0.74 days earlier, with less negative remarks from the caregivers, [easier to care for], and required less pain medicine the first 5 days of recovery than those in the other rooms.
>
> (Eberhard quoting Ulrich 2009, p. 168)

The financial implications alone are huge, not to mention the positive effect on employee retention and stress, along with the personal benefits of healing faster with less medicine. This should be impacting the way we design all buildings, but especially healthcare facilities. Ulrich continued his research and has greatly influenced the fields of evidence-based design, environmental psychology, and modern healthcare design.

Physiological

Scientists in the fields of neuroscience and evidence-based design pursued extensive research in the years since Ulrich's initial studies. They have supportive findings that quantify the positive benefits resulting from exposure to nature, natural elements, and natural materials within the built environment. In *Biophilic Design*, their research found, "People living in proximity to open space report fewer health and social problems, and this has been identified independent of rural and urban residence, level of education and income" (Kellert et al. 2008, p. 4) and "... even briefly viewing nature settings can produce substantial and rapid psychological restoration from stress" (Kellert et al. 2008, p. 91).

FP Innovations is a not-for-profit research and development institute that specializes in the creation of scientific solutions in support of the Canadian forest sector. Their study references over 110 separate publications profiling the health benefits of viewing nature, natural light, indoor plants, wood, nature sounds, and visual fractals, which are elements with the same form, but are of different sizes or scales – salt crystals are a great example. Their study was aimed primarily at healthcare facilities, but it included findings from office and educational spaces as well. The reduction of stress and pain, positive neurological, emotional and psychological effects, and hormonal and circadian rhythm balancing were just some of the most prevalent benefits. We will go into more detail of the specific performative aspects of their study in the Health + Well-Being chapter.

Kellert et al. also conducted experiments using animal contact and care as therapy with boys aged 9–15 years in a residential facility with Attention Deficit/Hyperactivity Disorder (ADHD), Conduct Disorder (CD), or Oppositional Defiant Disorder (ODD). The study showed long-term positive changes to behavior and impulse control. These positive changes extended into multiple experiences outside of the therapy sessions or when they were in direct contact with the animals the boys were caring for (Kellert and Wilson 1993, p. 185). How could this change treatment of these conditions? What would be the personal benefit in reducing medication, and improving conditions for caregivers?

Productivity

In studies conducted within office settings, it was found that "… natural lighting, natural ventilation, and other environmental features result in improved worker performance, lower stress, and greater motivation" (Kellert et al. 2008, p. 4). Also, "Contact with nature has been linked to cognitive functioning on tasks requiring concentration and memory" (Kellert et al. 2008, p. 4). This information tells us that not only are we affecting human health and making life more enjoyable but potentially affecting profits and financial health as well as supporting employees. How should we be changing the way that we design based on this information?

Social and cultural connections to nature

It has been shown that communities with more open spaces and natural features that are cared for have a "… more positive valuations of nature, superior quality of life, greater neighborliness, and a stronger sense of place than communities of lower environmental quality. These finding occur in poor urban as well as more affluent and suburban neighborhoods" (Kellert et al. 2008, p. 4). *Biophilic Design* also profiles research results indicating that "… trees were associated with higher levels of attention and self-discipline, less violence and other aggressive behavior, lower crime rates, and better interpersonal relations" (Kellert et al. 2008, p. 111). This knowledge is reshaping the architect's, designer's, planner's, urban designer's, and civil engineer's approach their projects.

Integrating biophilic elements

Stephen Kellert was a Professor Emeritus of social ecology and senior researcher at Yale University. He worked with many of the top researchers in related fields exploring the benefits of biophilic design and how they could be incorporated in architecture, landscape design, and urban planning. He has authored over 150 publications on the subject. Elizabeth Calabrese is a licensed, practicing architect specializing in biophilic design and has taught biophilic and ecological design at University of Vermont for over 20 years.

In 2015, they co-authored an article in *The Practice of Biophilic Design*, which outlines the basic attributes of biophilic design and its importance to occupant health and well-being, or physiological effects. Below is a brief overview of their research and is meant to give a deeper sense of biophilia's importance and how it can be easily included in your projects. The opportunities and strategies are endless and only bound by our own creativity. See the Additional Resources section below or Chapters 10–14 for more detail.

Attributes of biophilia

Direct experience of nature

This involves direct visual or physical contact, and/or brief momentary experiences. These need to be meaningful, diverse, and include movement.

LIGHT – ***Physiological effects:*** circadian rhythms, passage of time, wayfinding, aesthetics, reduces eye strain. ***Architectural strategies:*** clerestory windows, glass walls, windows, skylights, atriums, reflecting surfaces or colors.

AIR – *Physiological effects:* thermal comfort, productivity, sensing variations through-out seasons and the day. *Architectural strategies:* operable window; passive air flow; thermal dynamic air movement; heating, ventilation, and cooling (HVAC) system cycling; outdoor areas; pergolas; porches or covered areas attached to interior spaces.

WATER – *Physiological effects:* relieves stress, masks sound, enhances health + performance, multiple sense activator. *Architectural strategies:* interior or exterior fountains, views of large bodies of water, interior waterfalls, constructed wetlands, aquaria.

PLANTS – *Physiological effects:* purify air, increase concentration, creativity, perfor-mance and productivity, reduce stress, improve health and comfort. *Architectural strategies:* abundant local species vs. exotic, ecologically connected, large scale, atriums, constructed exterior landscapes, viewes through windows. Single or isolated plants rarely have the same benefits.

ANIMALS – *Physiological effects:* calming, connection to natural world + past history, positive behavior, increase verbal communication in therapy sessions. *Architectural strategies:* feeders in landscape, green roofs, tree + plant species to attract in exterior landscape, aquaria, aviaries, webcams or video, spotting scopes, visiting animals.

WEATHER – *Physiological effects:* experience nature + passage of time or seasons, mental stimulation, part of survival instinct. *Architectural strategies:* direct exposure to outside spaces, operable windows, porches, decks, balconies, pavilions or colonnades, gardens, large expanses of glass.

NATURAL LANDSCAPE + ECOSYSTEM – *Physiological effects:* comforting, user preference over artificial or human-dominated landscapes, evolutionary. *Architectural strategies:* functional + variable systems, forest glades, constructed wetlands, green roofs, aquatic environments, direct views accessed by observation platforms, windows, or direct interaction.

FIRE – *Physiological effects:* source of physical and emotional comfort, light, relaxa-tion, reduce stress, fractals. *Architectural strategies:* fireplaces, hearths, outdoor spaces with fire or flames, varying warm light, light with color + movement.

Indirect experience of nature

This involves representations of nature or natural forms, process, systems, or other aspects of the natural evolving natural world.

IMAGES OF NATURE – *Physiological effects:* intellectually + emotionally satisfying, preference with occupants or users. *Architectural strategies:* photographs, paintings, sculpture, murals, videos of plants, landscapes, water, animals; should be repeated, thematic + abundant; single or isolated images have less impact.

NATURAL MATERIALS – *Physiological effects:* stimulating, reduce stress, lower blood pressure, promote healing, improve creative performance. *Architectural strategies:* visual + tactile applications, finish materials, building elements, exposed structure, furniture, interior elements, exterior elements, wood, stone, leather, cotton.

NATURAL COLORS – *Physiological effects:* connection to nature, soothing. *Archi-tectural strategies:* muted tones of rocks, soil, plants, foliage, sky, location specific.

SIMULATING NATURAL LIGHT + AIR PATTERNS – *Physiological effects:* circadian rhythms, psychological + mental stimulation, reduce eye strain, increase comfort, concentration + well-being. *Architectural strategies:* varying light sources, full-spectrum light sources with roof monitors, skylights or clerestory windows, movement and variability in lighting, varying airflow, temperature, or humidity to simulate outside conditions.

NATURALISTIC SHAPES + FORMS – *Physiological effects:* occupant preference, lower blood pressure, positive impact on mood + attentiveness. *Architectural strategies:* leaflike forms, animal forms, or fractals in fabric, finishes, forms, structural or architectural forms.

EVOKING NATURE – *Physiological effects:* occupant preference, reduces stress, improves mental engagement. *Architectural strategies:* abstracted principles or forms from natural elements, roof lines, windows, finishes, fabric, textures, furniture, interior and exterior materials cohesion.

INFORMATIONAL RICHNESS – *Physiological effects:* reduces stress hormones, improves mental engagement. *Architectural strategies:* richness of details or materials, organized and evolving spaces or forms, material textures, structural composition, lighting, ornamentation.

AGE, CHANGE + PATINA OF TIME – *Physiological effects:* sense passage of time, connection to seasons and daily cycles, positive illustration of adaptation + aging. *Architectural strategies:* naturally aging materials, allow weathering, natural finishes, intertwining of building and site.

NATURAL GEOMETRICS – *Physiological effects:* observer preference, positively impact perception and physiological stress response. *Architectural strategies:* self-repeating but varying forms, sinuous form, fractals, hierarchy in scale, golden ratio, Fibonacci sequence.

BIOMIMETIC FORM – *Physiological effects:* observer preference. *Architectural strategies:* patterns + textures from nature, lighting form, architectural elements, finishes, patterns, materials, systems and structures modeled on nature or biological forms.

Experience of space and place

Elements or attributes of the built environment that inspire an emotional, intellectual sense of attachment or belonging to a physical location.

PROSPECT + REFUGE – *Physiological effects:* feelings of safety + protection, reduce stress, improve concentration, reduce boredom, irritation + fatigue. *Architectural strategies:* long views, areas of protection especially from behind or above, views between rooms or areas of activity, vistas to outside, sheltered areas, balconies, nooks, transparent materials, furniture, dropped ceilings, blinds, shades + screens, interior + exterior importance.

ORGANIZED COMPLEXITY – *Physiological effects:* observer preference, reduces physiological stress. *Architectural strategies:* exposed structure, floor plan, exposed mechanicals, interior + exterior form, materials + finishes, color + texture, circulation patterns, need to be variable + diverse, plants if varied.

INTEGRATION OF PARTS TO WHOLES (FRACTALS) – *Physiological effects:* reduces stress, creates intrigue. *Architectural strategies:* central focal point, sequential linking of spaces, spatial boundaries, vernacular forms columns, finishes + materials.

TRANSITIONAL SPACES – *Physiological effects:* provide sense of passage + transition, add mystery, light + air variability, stimulate pleasure response. *Architectural strategies:* vestibules, porches, pergolas, gateways, courtyards, colonnades, hallways, interior + exterior, landscaping.

MOBILITY + WAYFINDING – *Physiological effects:* reduce stress, fatigue + irritation, improve comfort + sense of security. *Architectural strategies:* clear lines of sight, clear pathways to entry + egress, transparent materials, open plans, elevated planes, views, landmarks, finish + material variability, lighting, architectural configuration, interior glass, location of stairwells at exterior building.

CULTURAL + ECOLOGICAL ATTACHMENT TO PLACE – *Physiological effects:* improve mental engagement, increase happiness, connection to community + place, encourage conservation, reduce stress. *Architectural strategies:* vernacular forms, indigenous materials, forms, and vegetation, courtyards, views, gardens, interior + exterior.

For more information, see the resource list at the end of the chapter. For applying biophilic principles at urban, building, and human scale, see Chapters 11–14.

6.7 Conclusion and a look ahead

The use of bio-inspired design strategies should be a basis for integral sustainable design. It greatly affects the way we approach design, leading us to innovative design solutions. In Chapter 7, we will look at the ways it influences resilience, and in Chapter 8, we go deeper into the positive effects we see when natural elements are introduced into the built environment at a deep level. At the global scale, bio-inspired design influences the way countries view their natural resources and ecological systems Chapters 11–14 will include bio-inspired design and products which change the experiential qualities of the built environment, productivity, and health and well-being.

Additional resources

Biomimicry

Biomimicry Institute, https://biomimicry.org/
Michael Pawlyn TED Talk, www.ted.com/talks/michael_pawlyn_using_nature_s_genius_in_architecture
Janine Benyus TED Talk, www.ted.com/talks/janine_benyus_biomimicry_in_action
Biomimicry: Innovation Inspired by Nature, Janine M Benyus et al.
Nature's Unifying Patterns website, Biomimicry Institute, https://toolbox.biomimicry.org/core-concepts/natures-unifying-patterns/
Learning from Prairies how to grow resilient, Biomimicry Institute https://asknature.org/strategy/plant-species-diversity-creates-long-term-stability/#.WwW670gvyUl

Biophilia

14 Patterns of Biophilic Design. Terrapin Bright Green, LLC.
The Biophilia Hypothesis, Stephen R. Kellert, Elizabeth F. Calabrese
Biophilic Design, Kellert et al.
The Practice of Biophilic Design, Stephen R. Kellert, Elizabeth F. Calabrese
Biophilia and Biomimicry: What's the Difference? Terrapin Bright Green www.terrapinbright green.com/blog/2017/02/biomimicry-versus-biophilia/

References

Benyus, J.M., 2002. *Biomimicry: Innovation Inspired by Nature*. New York, NY: Perennial.

Bernett, A., February 14, 2017. *Biomimicry versus Biophilia: What's the Difference?* Terrapin Bright Green.

DeLuca, D., 2016. *The Power of the Biomimicry Design Spiral*. The Biomimicry Institute. https://biomimicry.org/biomimicry-design-spiral/edn.

Eberhard, J.P., 2009. *Brain Landscape: The Coexistence of Neuroscience and Architecture*. Oxford and New York: Oxford University Press.

Fesenmaier, K., 2013. *Caltech's Unique Wind Projects Move Forward*. California Institute of Technology. www.caltech.edu/news/caltechs-unique-wind-projects-move-forward-39703 edn.

Goldhagen, S.W., 2017. *Welcome to Your World*. 1st edn. New York, NY: Harper.

Kellert, S.R. and Calabrese, E.F., 2015. *The Practice of Biophilic Design*. www.biophilic-design.com/.

Kellert, S.R., Heerwagen, J. and Mador, M. 2008. *Biophilic Design*. 1st edn. Hoboken, NY: John Wiley & Son, 2008. Print.

Kellert, S.R. and Wilson, E.O., 1993. *The Biophilia Hypothesis*. Washington, DC: Island.

The Biomimicry Institute, The Biomimicry Institute. Available: https://biomimicry.org/what-is-biomimicry/ [July 21, 2018].

Trigg, D., 2012. *The Memory of Place: A Phenomenology of the Uncanny*. Athens, OH: Ohio University Press.

United States Environmental Protection Agency, July 16, 2018, 2018-last update, Report on the Environment Indoor Air Quality [Homepage of United States Environmental Protection Agency], [Online]. Available: www.epa.gov/report-environment/indoor-air-quality [July 21, 2018].

7   Resilience + adaption

Designers of resilient projects plan for flexibility, self-reliance, and change by assisting buildings and cities to adapt to new uses, catastrophic weather events, climate change, rising sea levels, power outages, or just normal changes that occur over time. The underlying process for resilience and adaptation is to understand that change is inevitable and that our basic design approach needs to build in the systems and strategies required to respond to those changes. In addition, we need to think of our buildings lasting for 100–300 years versus the current norm of 25–50 years. This means that projects need to be able to adapt to a potentially broad range of changing conditions. This chapter will introduce the concept of resilience and adaption as a means to accomplish these goals.

7.1 Historic context of resilient design

Historically, creating buildings required heavy sacrifices of time, energy, natural resources, and money. Because of this, most buildings were constructed with the best materials that were locally available, using the best practices because they needed to last "forever" and be able to adapt to multiple needs. From the Hunter-Gatherers to the Industrial Age, populations were more attuned to their specific climate and built to mitigate larger climate forces out of their control. For example, in the tropics, shelters were made from light, naturally available, prevalent materials because they were destroyed often, and it was easier to clean up and rebuild. In Amsterdam, where most of the city is below sea level they built dams, levees, and containment areas to deal with water when it floods and build with materials that are resistant to moisture.

In the Age of Agriculture, the ancient stepwells of India offer some of the earliest examples of resilient design, where very large and very deep "open wells" that captured and stored water from the rainy seasons to be used later in the dry months. They are a clear example of infrastructure doubling as civic space, a concept that has come into vogue for today's sustainable design projects (Figure 7.1a [next page]).

Resilient buildings are designed with looming threats in mind. Frank Lloyd Wright's design of a hotel in Tokyo is a simple example of resilience where he knew that an earthquake would eventually hit the city and so he designed the foundations to "float" allowing the building to be one of the few major structures left intact after the earthquake of 1949 (Tenner 2011). In fact, his giant fountain in front of the hotel became the primary source of water used to fight the fires. Wright actually created a resilient structure, but it was eventually torn down to make way for another structure, showing how design and culture impact the life span of a building. In other words, it's beauty and cultural significance were not strong enough to transcend the forces of new world views.

Figure 7.1a Chand Baori (stepwell) at Abhaneri.
Source: Wikipedia. By Chetan [CC BY-SA 3.0 (https://creativecommons.org/licenses/by-sa/3.0)],
from Wikimedia Commons. https://commons.wikimedia.org/wiki/File:Chand_Baori_(Step-well)_at_
Abhaneri.JPG

7.2 Environment and resilience + adaption

For many years, much of society has seen nature as the enemy, building gigantic sea walls
and levies to prevent flooding. Today, sustainable design is starting to seek a partnership
with nature. This can translate to designs that work with nature to address flooding and
other disasters. The citizens of Rotterdam in the Netherlands have dealt with living below
sea level for centuries and have recently found ways to work with nature rather than re-
strain it by providing retention areas that function as social and activity spaces during non-
flooding times. Resilient sustainable design looks for ways to mitigate the effects of nature
and provide human safety, while conserving ecological and human resources.

To provide buildings that shelter and protect us we need to use environmental goods and
services. The earth is not an endless source for our use and exploitation, and as such, there
is great value and importance in what has already been manufactured and built. Human
capital and materials involved in building new structures are a significant investment.
Renovating and repurposing buildings through adaptive reuse capitalizes these previous
investments and saves materials and energy. This allows more resources for future generations and means that we have to design new buildings with the capability to be reused over
and over again.

7.3 Motivations and resilience + adaption

In Chapter 3, we discussed the differences between short-term and long-term thinking and
the difference between profit and prosperity (long-term profit). If we start to expand our
view of profit, to include waste, materials that are already manufactured, or the land we
extract resources from, we suddenly have a very different view of resilience, or adaptation

or historic preservation. Resilient design and adaptive reuse may at first glance seem to go against the idea of profit since renovations may have a higher initial cost or take a little longer. For many years historic or structurally sound buildings have been torn down, simply because it was easier and faster to replace them. But if we look deeper and understand the importance of historical preservation on our personal and social well-being, or personal safety and long-term viability, we can see there is more benefit than purely monetary profit. We really are doing what is good, or right, or best for society, the ecosystems, and long-term profitability.

7.4 Sustainability values and resilience + adaption

Resilience, adaptive ueuse, and historic preservation promote sustainability by taking what is already there and acknowledging its value and importance on the multiple levels of the Quadruple Bottom Line (QBL). The sustainability values of the QBL are listed below:

People: The built environment reflects our values and cultural heritage, and this is very important for social cohesion and a sense of belonging.
Planet: By evaluating embodied energy, reducing landfills, and reusing existing materials this benefits ecosystems, saves resources, and reduces energy costs.
Profit: The initial cost of design and evaluations may be slightly higher than typical construction, but the savings from not having extensive storm damage, or recurring replacement or renovation costs far outweigh the initial design cost. There may also be substantial cost savings during construction with adaptive reuse projects.
Place: Reusing existing structures and planning for structures to have a longer life span contributes to a sense of place by creating connection to history.

7.5 Integral theory and resilience + adaption

Achieving resilience requires a different approach than the traditional sustainable design processes. The goals are slightly different when survivability in a crisis or an extended life span becomes the focus. This chapter covers the major strategies in achieving resilient design. Figure 7.5a outlines some of the major approaches and characteristics needed to achieve a resilient design.

7.6 Bioinspired design and resilience + adaption

Nature always diversifies and has redundant systems making it robust. It does not put all its hope for survival into one system. A maple tree, or any plant, sends out hundreds of seedlings to continue to reproduce the species. Ecosystems like the prairie are made up of many different species of plants that serve the same function, like fertilizing the soil or controlling predators. Nature relies on diversity and redundancy to counteract the effects of fires, blights, or climate change. If one of those events occurs, other species are able to take over and the ecosystem remains intact and resilient. By looking to nature for guidance, we can plan our cities, neighborhoods, and buildings to respond to natural disasters and extreme weather events in a way that reduces catastrophic damage.

	Subjective	Objective
Individual	*Experience* - Beauty + Aesthetic connection to past = care + preservation - Phenomenological aspects of Historic buildings - Place making	*Performance* - Reduce landfill volume - Reduce CO_2 emissions - Reduce Embodied Energy - Maximize access to Fresh Air + Day Light - Architecture: disposable v permanent
Collective	*Culture* - Social Value - Socially Cohesive communities - Cultural Identities - Historic Preservation - Vernacular Forms	*Systems* - Work with, not against the climate - Create favorable microclimates - Use Life Cycle Analysis Processes - Employ Passive Systems - Redundant Mechanical Systems - Self-sufficient energy and water systems - Structural System - Long-life, Loose-fit Approach + Shearing Layers

Figure 7.5a Resilience design strategies within the integral framework.
Source: Created and drawn by authors.

7.7 Resilience + adaption

7.7.1 What is resilience + adaption?

The definition of the word "resilience" is the ability to spring back into shape or to have the ability to recover or withstand change or difficult conditions. The definition of the word "adaptable" is the ability to adjust or be modified for a new use. The words pliable, supple, or hardy are good and important adjectives that describe a resilient or an adaptable building, community or urban setting.

Resilient designs account for flexibility, self-reliance, and change by giving buildings and cities resources to adapt to new uses, catastrophic weather events, climate change, rising sea levels, power outages, or just normal changes that occur over time. Resilient sustainable design integrates a long-term view of the building or product and acknowledges that the original design or use will change. It considers the specific site, the urban and societal context, and historic or cultural forms along with the current use. It provides diversity and redundancy, includes passive systems where possible, and looks at potential disposal and component reuse.

Currently, in the U.S. the expected life span of a building is as low as 25–50 years. History tells us that it is possible to design buildings that will be structurally sound and useful for hundreds or thousands of years. We need to begin asking how our buildings, products, and infrastructure can facilitate alternate uses for 50, 100, and 250 years in the future in addition to its current use. This requires us to think about how things change over time and how we can design projects that account for those changes.

7.7.2 Why is adaption and resilience important?

Ability to change

Frank Lloyd Wright said, "The sins of the architect are permanent" (Brand 1994, p. 66). That means the design choices that we make today will affect society for many years to come. Change is a huge part of societies and businesses, and homes and our design decisions need to allow room for that inevitable change.

Kevin Lynch, a city theorist says, "Our most important responsibility to the future is not to coerce it, but to attend to it ..." (Brand 1994, p. 185). Understanding that any design project will change over time brings a new dimension of thought to the traditional design process. This is more complicated and deeper than providing room for future expansion, though that can be part of it. This is about making decisions that allow the most flexibility in the purpose or function of the building without major structural changes. Factory buildings of the early 1900s are a great example of this and have facilitated the current resurgence in adaptive reuse projects. These industrial factories are easily being made into the condos, apartments, or retail establishments. The high ceilings, large windows, and open floor plans allow the structure to remain viable for 100s of years and support many different purposes.

Resource consumption and waste

Buildings in the U.S. consume approximately 39% of the energy resources. They consume 72% of electricity and 13% of potable water, and produce 39% of CO_2 emissions. By making small changes to the amount of energy used by the buildings stock exponential energy savings are possible (DOE 2011). Reducing the energy demand of buildings makes them more resilient in times of economic distress or energy shortages.

The U.S. building stock is approximately 300 billion square feet (sq ft). It is estimated that every year 5 billion sq ft is built and another 5 billion sq ft is renovated. As we saw above the "viable" life of a building in the U.S. is very short. By 2035 it is projected that about 75% of the existing square footage will be renovated. Construction and demolition waste is taken to landfills, which are the second largest human-made source of methane. Keeping construction materials out of landfills also reduces their volume and size as well as the methane produced by them. Maintaining landfill capacity for the future is part of a resilient approach.

A full understanding of all the energy used and the environmental impacts of each building product are extremely important in our decision process. The calculations need to start from raw material extraction through delivery to the jobsite, installation, expected life span, disposal, and reuse. The term that is used to express all the energy involved in any material or product is embodied energy. The process used to determine the long-term impacts of any material is called life cycle analysis (LCA). See Chapter 14 for more detailed information.

Land usage

The ability to change and grow with different occupants or different uses not only saves energy with initial resources used to build the building but can also affect land usage. If a building's structure is easily adaptable the need for new construction is reduced and a new site is not required. This reduces urban sprawl and helps to maintain green space and undeveloped habitat. The preserved habitats allow for resilience because they provide important ecosystem goods and services and because they help deal with natural disasters.

Brownfields, which are previously developed sites with possible soil contamination, can be restored, preserving habitat by not building a new building on a green field. New buildings on brownfields improve the environment along with the social and cultural fabric of an area. For example, SteelStacks in Bethlehem, PA, is an abandoned steel factory that sat vacant for many years as a blight to the community. It is now a vibrant performing arts center that hosts concerts and festivals all year long. The main smelting furnaces have been preserved, lighted, and are a backdrop for the many outdoor performance areas. It reminds everyone of the great economic and cultural contribution the steel mills provided for many years to the town and the American steel industry. This is an example of adaptability at work. Vast amounts of material have been saved from going into landfill, and the cultural and historic character of the area is preserved. The portions of the steel mill that still exist have been used as sets in major films bringing more income to the community.

Climate change + extreme weather events

We are seeing an increase in extreme weather events, droughts, and flooding. Earth scientists tell us that these will be increasing in the coming years. If sea levels continue to rise, there will be dramatic changes in many of the most populated areas. By designing to address climate change, catastrophic weather events, rising sea levels, power outages, and flooding lives can be saved and the amount of repairs required after a disaster will be easier to make. This gets people back into their homes and businesses sooner, and saves time and money.

Some strategies to increase resilience to extreme weather events are

- Raising mechanical rooms and occupied areas well above flood levels;
- Following structural guidelines for building in seismic areas if required;
- Building with materials that can withstand water, wind, or other climate forces specific to the site;
- Installing diverse and redundant mechanical and electrical systems;
- Using passive daylighting, air flow, shading, and heat gain that is appropriate to the climate not only reduces energy use on a day-to-day basis, but can provide comfort when utilities go down in an event of an emergency;
- Incorporating water containment areas within typical site amenities or functions. These could include parking, landscape, or parklike areas that can be cultural additions, yet also contain water away from structures or utilities.

Social value

The built environment greatly affects us on personal, societal, and cultural levels, this is sometimes referred to as "Place." In his book *Memory of Place*, Dylan Trigg expresses it this way:

> As bodily subjects we have a relationship with the places that surround us ... over time these places define and structure our sense of self ... Place is at the heart not only of who we are, but also of the culture in which we find ourselves.
>
> (Trigg 2012, p. 1)

Environmental psychology and environmental design are two fields that study the effect of buildings or urban environments on their occupants. Their research supports Trigg's observation and should alert us to the important and interrelated aspects of our design decisions.

The buildings and structures we are designing today should be viewed as a gift to present and future generations, and the world at large. These structures should be considered an integral part of our infrastructure and cultural fabric, not disposable shells to be demolished every time contemporary style or our needs change. Buildings of differing age, architectural styles, and historic forms enrich the aesthetic value of an area and can clearly illustrate the cultural and historical progression of a society.

The built environment also contributes to social interactions and shared values. For example, in the city of Rotterdam, water from rising seas and increased flooding is a big issue because most of the city was built below sea level. They have addressed the issue head on, not by building higher walls to hold the sea and rivers, but have designed places that can act as retention areas. These areas designed as parks, water sport centers, or parking lots that enhance daily life for all the citizens every day. By adding these areas, the city has noticed a dramatic change in the social structures, especially in formerly blighted areas. Property values have increased, buildings have been restored, and there is more involvement and support for climate change efforts because of the city's positive, proactive approach and how residents have benefited from the efforts.

Economic value

Basic common sense tells us that if we use more materials, we are using more financial resources. The most expensive parts of a building are the site work, foundation, roof, load capacity of structure, mechanical systems, and plumbing. These are also the most durable parts of a building and the most difficult to change.

If the structure, envelope, foundation, and roof can remain structurally sound and are designed to allow for flexibility, changes within the building itself can take place for hundreds of years and a great percentage of construction cost can be saved. Many older buildings were also designed with excellent daylighting and natural ventilation systems. These can be utilized moving forward to reduce the required energy consumption, mechanical requirements, and energy costs.

The choice to build so that the structural integrity of the building will have longevity or to renovate an older existing building may have a higher initial cost. The original investment, or initial cost, should always be weighed against the expected life of a building, maintenance, and operating costs. For example, better performing windows cost slightly more than their lesser counterparts, but will last much longer; provide better insulation; save on energy costs; look better; and be easier to use. The initial cost averaged out is much less for a better quality product.

The decision between renovating and replacing is not always an easy or automatic one. We need to holistically analyze all aspects of the site, social setting, and surrounding ecosystem along with the structural integrity of the building, so we are making the best choice for the client, occupants, and community.

Human comfort

Resilient buildings should be designed to leave options for the occupants to make changes that better suit the current functional needs. This means allowing space in the floor plan for changes or additions, allowing user controls for daylighting, heating and cooling, and placing structure in such a way that interior partitions can be added or subtracted easily.

Summary

It is our responsibility to carefully and holistically analyze the multiple options for every aspect of the buildings that we design to create a regenerative, truly sustainable structure with the potential to last hundreds of years – beautifully enhancing the community it serves. We need to acknowledge future additions and changes as something expected. As careful and thoughtful as we are when designing the initial building, things happen that we can't foresee. This means that instead of thinking of a building as a finished, completed, or static, our aim should be towards flexibility or fluidity.

7.7.3 How is adaption and resilience achieved?

The remainder of the chapter is mostly geared toward architecture, but the general principles and approach can be applied to all areas of sustainability.

Resiliency starts with considering a broader base of information from multiple perspectives. It also requires thinking of buildings as a permanent part of the infrastructure and community. This is a "long-life, loose fit" approach. We need to change our thinking to understand that "Future preservation means that the building is not only built to last, but ... has the freedom to adjust and the ability to even change direction entirely is preserved" (Brand 1994, p. 185). This means that our aim should be towards flexibility or fluidity. The structural components of a building should be considered as a living, breathing, resilient platform to facilitate an ever-changing set of human activities and world climate conditions.

The trick is to frame the question regarding building a new structure in a larger context, so instead of leaping to the conclusion "We need a bigger [or another] building," explore the more general problem, "We need to handle our growth" (Brand 1994). If we decide that a new building is really the best answer, then there are general design strategies that will help our designs to be more resilient.

7.7.4 Strategies for achieving resilience

Understand shearing layers

"Shearing Layers" is a term used by Stewart Brand (1994) in his book *How Buildings Learn* to describe the various elements of a building, classifying their permanence and showing that a structure is in the state of constant change. Understanding these layers gives insight into creating a resilient and adaptable building. The six broad categories that he calls shearing layers are summarized below:

- **Site** – The geographic setting and the legally defined lot. This element is eternal.
- **Structure** – The foundation and load-bearing elements. These are the building and have a life between 30 and 300 years or more.
- **Skin** – Exterior surfaces that now change approximately every 20 years because of fashion and technology could last much longer depending on materiality choices.
- **Services** – The working guts of a building include communication; electrical and technical wiring; plumbing; heating, ventilation, and air conditioning (HVAC) systems; and elevators. These wear out or reach obsolescence every 7–15 years. Buildings have been demolished early simply because these systems are too deeply embedded to be replaced easily.

- **Space plan** – The interior layout encompasses non-load-bearing walls, floors, ceilings, and finishes. Commercial spaces can change every 3 years, whereas a residence may only change every 30 years.
- **Stuff** – Furniture, equipment, and personal items. This layer can change monthly or daily.

The most expensive parts of a building are the site work, foundation, roof, load capacity of structure, mechanical systems, and plumbing. These are also the most durable parts of a building and the most difficult to change.

We can also see from above that the shortest lived parts of a building are its use, interior partitions, technology, and finishes. These are the easiest to change as well, but often are what is given the most attention.

How do we keep these easily changeable parts of a building from informing our design too much?

1   Start with a long view of the building, business, or activities
2   Realize they will change
3   Leave some decisions at the end
4   Allow space and ability for users to change the building to meet their specific requirements.

Scenario planning

One method for increasing our chances of creating a building that will "… remain always capable of offering new options for its use" (Brand 1994, p. 185) is to use a process called scenario planning in the very initial stages of a design. Scenario planning is a process used in business administration to help visualize what future conditions or events are most likely to happen, what the outcomes or effects would be like, and then to examine various ways to respond to them.

Stewart Brand applies scenario planning to building design because "All buildings are predictions … [and] … All predictions are wrong" (Brand 1994, p. 183). One of the first questions to explore when asked to design a new building is as follows: Do we even need a building at all? And if so, how can it be converted into a flexible tool for the owner, not a potential prison.

This is a very different way to think, but it is a crucial first step to help clients and owners think through the issues they are having with their company and come up with an effective solution that addresses the larger general problem. For example, is the question about a building or about handling growth? The building is then treated as a strategy rather than an automatic solution. The stakeholders, everyone involved in and/or effected by the building, explore the driving forces that will be shaping their organization at the large scale, a more specific smaller scale, and at the societal level.

The scenarios are then analyzed according to their strengths (S), weaknesses (W), opportunities (O), and threats (T) or SWOT. SWOT analysis evaluates both the internal and external forces that could affect the success of any plan and tries to mitigate or capitalize on them from the beginning of the process.

Both positive and negative scenarios or options are explored. "The goal is to develop scenarios that are both plausible and surprising" (Brand 1994, p. 182). This approach ends up examining a building from so many perspectives that gaps which are normally overlooked become evident. This approach can open many solutions that would not come to the surface in the typical linear programing analysis. It allows issues to arise when it is still very easy to make changes, because the project is still in the initial stages and is represented on paper, not concrete and steel.

Does it always give us a perfect plan? No, but it helps "... expand our thinking, provides a shared language, invites divergent views, and can even bridge conflicting factions ... In the end this approach leads to more versatile buildings" (Brand 1994, p. 185). Greater input and examination at the beginning of the design process creates a more flexible building designed for adaptation later, rather than looking only at the immediate use which most likely will change in the next 10–30 years.

Listed below are some examples of the kinds of issues and opportunities that scenario planning accounts for:

- Extreme climatic events such as floods, droughts, and forest fires
- Extreme economic shifts such as a depression or an economic boom
- Extreme changes to the social makeup of a local context such as gentrification or abandonment
- Energy crisis
- Water shortages
- Power outages
- Social upheaval and unrest
- Food shortages
- Influx of refugees

Stakeholders – what are they and why are they important

Stakeholders are anyone, or any organization or community that is, or will be, affected by your design decisions. Stakeholders also include any professions that will be utilized in the design or construction process of the project including the owner and occupants. Typical stakeholders would be owner/client, building occupants, architect, engineers, designers, suppliers, construction professionals from all trades, general contractor or construction manager, and maintenance people along with neighbors and community members.

Including all the stakeholders along with the design professionals and contractors at the beginning of the project allows important information to be gathered and discussed. This reduces the time and cost associated with changes, because these can be addressed early in the design process rather than after construction has started. A sense of ownership is created, even from those who may oppose your project. Stakeholder involvement can be especially important when working on sensitive or potentially controversial projects.

The cost of changes in both time and dollars rises exponentially as the design process moves forward, so it makes sense to make changes while the design is still on paper rather than after production has started. Engaging stakeholders and the integrated design process is covered in more detail in Chapter 9.

7.7.5 Site evaluation

The first and most important thing to increase the resilience of any project is to evaluate the site. The specific site should influence all decisions because, as noted above in the shearing layers, it is the most permanent part of any building or community. Beyond the basics of size, topography, codes, zoning, and utility location there are other factors to consider.

Soil composition studies are done to assess the bearing capacity that determine the type of foundation required. Tests should also be completed to assess whether there are any contamination issues that need to be addressed or would designate your site as a brownfield. The site should also be examined for wetlands, streams, underground water, natural or invasive vegetation, and how water and air move across the site. This information aids in placing the building and hinting at possible configurations to best capture the renewable resources and inform the use of passive systems like natural ventilation, shading, and daylighting.

Understanding the general climate for your geographic area can inform not only passive systems, but also what type of active or mechanically operated systems you use. For example, Phoenix, Arizona has on average 299 sunny days compared to Seattle, Washington that has 152, making solar collection more viable in Arizona than Seattle. To fully understand the climate you need to know the following information for each season: sun angles, average temperature, wind direction, landscape features, and trees. This information along with any surrounding structures will help you determine the microclimate, or how the general climate forces are affected by the specific conditions on your site.

If there is a chance to assist the client in the selection of the site, be sure to evaluate how close public transportation and other amenities are and how accessible it is for pedestrians or vehicles. These factors can be very important if you are pursuing a rating system certifications or just want to make your project is more pleasant for the occupants. If the site is already selected, make sure the building along with the entrances is placed as conveniently to these things as possible. The location of the building on the site can preserve any greenfield space, or allow for restoration and provide open space. The existing historical and cultural context of the surrounding community can inform material and configuration selections so any new construction acknowledges and creates cohesion with the existing community. Studying the vernacular architecture is also very useful for this and can inform passive strategies that mitigate local climate forces as well. Knowing the goals for the client's cultural and social interactions along with all the information above is vital to shaping design directives for a resilient structure that will serve the community for 50, 100, or 200 years.

7.7.6 Mechanical systems and technology

As Stewart Brand states above, technology and mechanical systems are replaced approximately every 7–15 years. In residential buildings this could be much longer, but in either type of construction any technology or mechanical systems do not work at all if there is no electricity. This means in catastrophic weather events, power disruptions, or just in the normal course of life, there will be times when these systems will not function. The idea of resilience is to provide diversity and redundancy so that the occupants can have basic comfort and can carry out basic functions when there are power disruptions and that when technology docs change, it can be easily changed or adapted.

Passive strategies are one of the best ways to provide basic function and comfort. Passive systems are anything that happens without an outside power source like electricity, automation, combustion, or any other power source. Passive systems commonly incorporated into the built environment are daylighting with windows or roof monitors, natural ventilation, shading, and passive heat gain in colder climates. If the overhangs of a building are properly designed, they will prevent the hot sun from entering the building or window while still allow light to enter, regardless of if there is power or not. Resiliency thrives on low tech!

Active strategies are mechanical systems that do require some power source or mechanical means. This would be any type of HVAC system, solar panels, any automated shading, or ventilation system. These are all excellent ways to increase efficiency and reduce resource consumption, but will do no good if the power grid is down for whatever reason. To use solar panels during a power outage, batteries can be used to store energy directly on-site, or systems can be set up at the meter base to take them offline and divert the energy to batteries.

Ecodistricts, which are covered in Chapter 12, require the consideration of diversifying power within a city. They look to have local sources of power, or micro-grids that are broken down by districts within the city to prevent large metropolitan areas from losing power from an isolated breech. Ecodistricts can also include shared food production, shared stormwater collection, and sewage treatment as well.

An excellent example of a resilient building that uses passive systems is the National Building in Washington, D.C. It was designed by General Montgomery C Meigs and built in 1887 as the Pension Building for the U.S. government. This was close to 60 years before air conditioning was available, and 30 years before electricity was a common part of buildings. The configuration of the building allows daylight to penetrate deep into the center of the building, as well as provide adequate lighting to the offices at the perimeter. We can also see the synergetic relationship between the center atrium and clerestory windows that allow light and the natural ventilation system. Air is brought into the building along the exterior of the building at lower windows. Then as the air in the atrium warms it rises to the top to be exhausted by the clerestory windows. This in turn creates negative pressure, pulling in cool air from the exterior in a continuous cycle, the temperature and air velocity within the building are now being controlled by how many perimeter windows are opened or closed. Figure 7.7.6a is taken without any artificial lighting, notice how the entire space is daylight from clerestories placed high in the building.

Figure 7.7.6a The National Building Museum.

By RadioFan at English Wikipedia, CC BY-SA 3.0, https://commons.wikimedia.org/w/index.php?curid=35454825. https://en.wikipedia.org/wiki/National_Building_Museum#/media/File:National_building_museum_columns.png

Structural considerations

Structure is one of the most enduring layers of a building and its design will affect the ability for future changes, ultimately affecting its preservation and sustainability for the rest of its life. Some examples are described below:

- How could the structural components be designed to withstand or accommodate future dramatic changes in weather patterns, sea levels, or unforeseen world events?
- What are the traditional and/or vernacular building methods, materiality, and massing? How can these inform our choices?
- How can my sizing, materiality, and placement be the best for the current owner, yet allow flexibility for future changes in purpose, use, surrounding density, climatic changes, and occupancy loads?
- How could the structural elements be designed to accommodate future changes such as solar panels, a green roof, or increased occupant load that would reduce structural changes or the need for renovations?

Using a structural grid system is one way to increase the adaptability and resilience of a building. It does not have to be orthogonal grid as seen below in the organic form of the Real Goods Market Solar Living Center designed by Sim van der Ryn (Figure 7.7.6b [below]). Even with the organic form the structure has rhythm, purpose, and flexibility for future modifications (Figure 7.7.6c [next page]). The structure is an independent element allowing the occupants freedom of use.

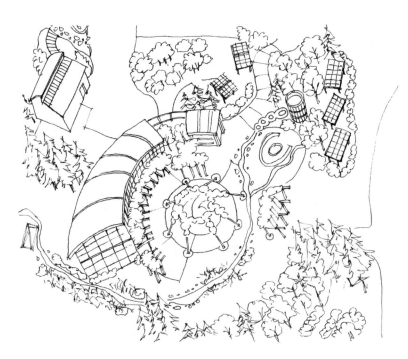

Figure 7.7.6b Site plan of Real Goods Solar Living Institute, designed by Sim van der Ryn.
Source: Structural Special Topics, courtesy Routledge.

Figure 7.7.6c Interior of Retail Store at Real Goods Solar Living Center, designed by Sim van der Ryn.

Source: Structural Special Topics, courtesy Routledge.

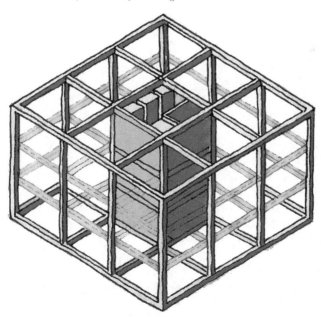

Figure 7.7.6d Left 3D illustration of structural grid with central mechanical core, SOM *Intelligent Densities | Vertical Communities.*

Source: Structural Special Topics, courtesy Routledge.

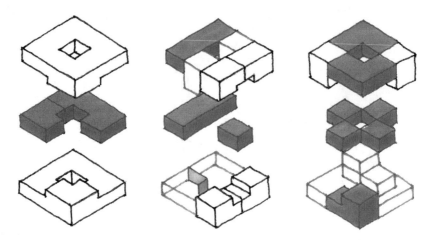

Figure 7.7.6e Center and right isometric of flexible structural grid, SOM *Intelligent Densities |
Vertical Communities.*
Source: Structural Special Topics, courtesy Routledge.

Another example of a flexible structural grid system was designed by SOM Architects. They conducted extensive studies of London to address the increasing need for sustainable, flexible housing addressing multiple housing types without using greenbelt land or creating more urban sprawl. To accommodate the varied housing needs and create future flexibility SOM came up with the concept of a module. The general overall grid (Figure 7.7.6d [above]) allows the building to be broken down into modules of varying sizes and volumes. These volumes could then accommodate a variety of amenities, functions, or housing types (Figure 7.7.6e [above]).

Many buildings are torn down simply because it is too costly or difficult to update their mechanical or technology systems. A centralized core was designed for all the "working parts" of the building. Because of the central location, large chase, and conduits, all the mechanicals, plumbing, electricity, and technology are easily accessible. SOM also found that using interconnected varying size modules with amenities placed throughout the building "… allow[s] the building to come alive and help its inhabitants to develop a strong sense of community" (SOM 2015, p. 32).

Considering how the structural system of a building can add beauty, meaning, and articulate the structure's purpose while it relates to the culture of its location are consistent with the basic tenets of architecture and add depth to human existence. This can be accomplished in many ways. Exposing the structural elements or using materials that allow for greater opportunities of form are just a few.

7.7.7 Embodied energy | life cycle analysis | carbon emissions

Embodied energy + life cycle analysis

To fully understand which type of building or finish material is the most sustainable in each specific situation it is vital to look at the whole process of creating each material. This process is called life cycle analysis (LCA) and measures embodied energy.

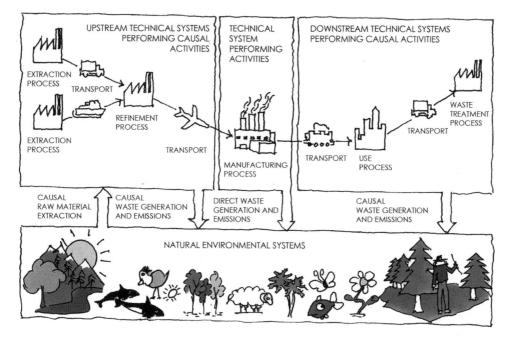

Figure 7.7.6f A graphic representation of LCA.
Source: Structural Special Topics, courtesy Routledge.

Calculating embodied energy examines each step of manufacturing from resource extraction through final product delivery including transportation, construction, and building maintenance. It also considers life expectancy, end of life disassembly, reuse, or disposal (Figure 7.7.6f [above]).

LCA is a process used to analyze the wide-ranging environmental footprint of a building, product, or material, including aspects such as energy use, global warming potential, habitat destruction, resource depletion, and toxic emissions.

All the energy, raw materials, and emissions to the atmosphere, water, and soil for each step of production are recorded and combined into a flow chart. This allows us to see the true cost of producing any material, product, or building. Just as important as recording and calculating all the figures is a consideration for any negative impact on the soil, air, and water quality at the site of extraction. We also need to examine how the community and surrounding ecosystem has or will be affected aesthetically, physically, and socially by all the steps above even though there may not be facts or numerical figures to put on a spreadsheet.

LCA is one tool to aid in understanding the complexity of sustainability and shape our design decisions. Other benefits of utilizing LCA, as cited by an American Institute of Architect's study, are as follows:

- Choosing between building design or configuration options
- Choosing between building structural systems, assemblies, and products
- Reducing environmental impacts throughout a building's life cycle
- Improving the energy performance of entire building
- Mitigating impacts targeted at a specific environmental issues

This integrative analysis and design process should take place very early with all professions involved. As an emerging designer you can introduce this new process.

7.7.8 Resource conservation

Conserving the amount of raw materials used and therefore embodied energy is a large part of resilience and adaptation achieved primarily through reusing existing buildings or building materials. This keeps millions of tons of viable materials out of landfills. Another strategy is doing more with less, known as dematerialization. Thomas Jefferson, in his design for the dormitories at the University of Virginia, used a very thin serpentine form to make a self-supporting garden wall. This created a striking visual structure that was strong and saved bricks (Figure 7.7.6g).

This was possible because he understood the material properties of brick and construction methods and used them to his benefit. Having this type of extensive knowledge is why

Figure 7.7.6g Serpentine garden wall at the University of Virginia designed by Thomas Jefferson.
Source: Structural Special Topics, courtesy Routledge.

collaboration is so extremely important. We will be learning more about the integrative design process in Chapter 9.

7.7.9 Climate change + catastrophic events + extreme weather

Catastrophic weather events are becoming more frequent, and it is projected that these events will continue to become more common. We are also seeing sea level rises that will have a great effect on cities, communities, and buildings as the coast line changes. All these factors need to be considered in the site and design evaluation, even if they aren't issues now. The way we approach these changes will make all the difference in the social fabric and built environment of any area. For example, the city of Rotterdam was originally constructed by reclaiming land from the rivers and the ocean and is a busy seaport that serves Northern Europe and England. Most of the city is below sea level and is also the watershed for the same area. Dealing with water and flooding has been an issue for them since the beginning of their history. Rotterdam's approach is unusual because instead of building higher walls, dredging the river, or building dams or levees to keep the water out, they are providing opportunities for the water to be safely collected in dispersed areas around the cities. The urban designers have devised lakes, parks, garages, and plazas that are designed to hold excess water when needed, but enhance life and provide areas for sports or social activities during the other times. As a result they have also seen urban blight reverse, real estate values rise, citizens becoming more active in outdoor and water sports. In addition the people of the city are positively engaged in resilience and climate change reduction efforts.

7.7.10 Aesthetics + beauty

Beauty goes beyond pure aesthetic preferences. Studies in medical facilities have shown there are also neurologic and psychological benefits that result from our material selections, the availability of natural views, and exposure to art or beauty. This shows that our choices have a deep impact on occupants.

Beauty and transcendence are foundational to architecture and joy in human existence, and should therefore always be considered. Stewart Brand, in *How Buildings Learn*, observed that buildings that are beautiful are loved by the community and, as a result, are renovated and cared for. Reducing the amount of waste going into landfills and increasing our connections to history, culture, and the local community. Lance Hosey in his book *The Shape of Green* states, "... if it's not beautiful it is not sustainable" (Hosey 2012, p. 7). Here is a simple equation:

Beauty = Love + Connection = Care + Historic Preservation

= Heritage = Resource Conservation

As we are selecting the shape, form, and material of our buildings or products we should be asking ourselves:

- Do our choices reinforce or deny the purpose, history, or present culture?
- Will they be viable in the future?
- Are they beautiful?
- Will they be easy to match or coordinate with if renovations or changes occur?

7.7.11 Adaptive reuse + historic preservation

The term "adaptive reuse" refers to changing a building to function in an entirely different way or for an entirely different purpose. For example, a 1900s factory changed into a mixed use retail and apartment complex, or like above, a steel mill into a performing art center.

Historic buildings in particular can remind or reinforce our cultural past and help us know who we are, individually and as a society. It provides a sense of place and even belonging, allowing us to feel part of a community which meets one of our most basic needs, as shown by Maslow's hierarchy of needs (4.4.4). By allowing buildings from different time periods, residents or visitors have a visual and tangible reminder of the passage of time and succeeding world views.

Hundreds or thousands of structurally sound buildings are being demolished, increasing the waste taken to landfills simply because it is more expedient to tear them down. This is wasting a huge amount of embodied energy. Structures built before the 1950s and the advent of air conditioning were all designed with passive ventilation, shading, and solar gain strategies specific to their geographic location. These passive strategies are still present, viable, and are excellent means of saving energy and bringing in natural light. The biggest challenge in adaptively renovating older building for use is in incorporating the current codes and contemporary technological requirements. These challenges can easily be overcome with planning and collaboration with the appropriate professionals.

7.7.12 Frameworks

Stewart Brand's rules for strategic building designers

This framework is the result of summarizing the principles of Stewart Brand's book *How Buildings Learn: What happens after they are built* (Brand 1994, pp. 186–187).

- Favor moves that increase options;
- Shy away from moves that end well but require cutting off choices;
- Work from strong positions that have many adjoining strong positions;
- Overbuild structure;
- Provide excess services capacity;
- Separate high- and low-volatility areas and design them differently;
- Work with shapes and materials that can grow easily, both interior and exterior;
- Use materials from near at hand – they'll be easier to match or replace;
- Medium-small rooms accommodate the widest range of uses (120–150 sq ft best);
- When in doubt, add storage;
- Shun designing tightly around anticipated technology – we overestimate technology in the short run and underestimate it in the long run;
- Design loose and generic around high technology;
- The most convenient form of expansion is cellular;
- If necessary, postpone design decisions and leave them to the eventual users;
- Loose fit provides room to adjust. Perfect fit is fleeting;
- Need to retain the freedom to shrink as well as to grow.

In summary, we need to analyze the specific site, its cultural connections, and the most permanent elements of a structure along with the most flexible or fluid elements of the design directives. Then create an expanded, holistic plan to best serve the present and future generations or the "long now." This is the cornerstone of resilient sustainable design. Structures designed this way offer greater intrinsic value, decrease the demand for and production of new building materials, and divert vast amount of debris from landfills. Using this information to inform our designs we can move us toward truly adaptive, resilient and sustainable structures, communities, and products.

Additional resources

A Richer Heritage: Historic Preservation in the 21st Century, Richard Stipe

Applying Resilience Thinking: Seven Principles for Building Resilience in Social-Ecological Systems, Stockholm Resilience Centre

Basic Information | LMOP | US EPA: Methane Emissions from Landfills U.S. EPA, www.epa.gov/lmop

Buildings Energy Data Book, Department of Energy, https://catalog.data.gov/dataset/buildings-energy-data-book-6d4d2

Copenhagen Solutions for Sustainable Cities, https://international.kk.dk/sites/international.kk.dk/files/Copenhagen%20Solutions%20for%20Sustainable%20cities.pdf

Estimating 2003 Building-Related Construction and Demolition Material, U.S. EPA *Amounts. www.epa.gov/smm/estimating-2003-building-related-construction-and-demolition-materials-amounts*

How Buildings Learn: What Happens after They Are Built, Stewart Brand

Stewart Brand, TED Talk, www.ted.com/speakers/stewart_brand

Stewart Brand, "The Long Now," TED Talk, www.ted.com/talks/stewart_brand_on_the_long_now

Stewart Brand "What Squatter Cities can Teach Us," TED Talk, www.ted.com/talks/stewart_brand_on_squatter_cities

Sun, Wind and Light, Mark DeKay

The Big U, www.dchamberlinarchitect.com/travel-north%20america-united%20states-utah-zion-VISITORS%20CENTER.htm

U.S. Energy Information Administration, Independent Statistics & Analysis, www.eia.gov/tools/faqs/faq.php?id=86&t=1

U.S. Energy Information Administration, www.eia.gov/

U.S. Environmental Protection Agency Landfill Methane Outreach Program, www.epa.gov/lmop/project-and-landfill-data-state

References

Alexander, C., Ishikawa, S., and Silverstein, M., 1977. *A pattern language: towns, buildings, construction.* New York: Oxford University Press.

Augustin, Sally & Fell, David, 2015. *Wood as a Restorative Material in Healthcare Environments.* Job#301009845. Canada: FP Innovations.

BAYER, C., 2010. *AIA Guide to Building Life Cycle Assessment in Practice.* 2010: The American Institute of Architects.

Brand, S., 1994. *How Buildings Learn: What Happens after They're Built.* New York, NY: Viking.

Department of Energy, 2011. *Buildings Energy Data Book.* Washington, DC: Office of Energy Efficiency & Renewable Energy.

Hosey, L., 2012. *The Shape of Green: Aesthetics, Ecology and Design.* 2nd edn. Washington, DC: Island Press.

Skidmore, Owings and Merrill, Inc. 2015. *Intelligent Densities | Vertical Communities*. NLA Breakfast Talk July 2015. London: NLA London's Center for the Built Environment.

Stipe, R.E., 2003. *A Richer Heritage: Historic Preservation in the Twenty First Century*. Chapel Hill and London: The University of North Carolina Press.

Tenner, E., 2011. *How Tokyo's Imperial Hotel Survived the 1923 Earthquake*. Boston: The Atlantic Monthly Group.

Trigg, D., 2012. *The Memory of Place: A Phenomenology of the Uncanny*. Athens, OH: Ohio University Press.

8 Health + well-being

8.0 Introduction

We now have quantifiable information from medicine, science, neuroscience, and business showing that the built environment greatly affects people. Health and well-being go deeper than physical health and include social interactions, culture, equity, and experience along with the ecosystems that support everything we do. The first step in creating environments that support their occupants and restore physical, psychological, and ecological health is to acknowledge and understand this.

Health and well-being at the most basic level are about not damaging people or the environment with shelters meant to protect us. All our decisions have long-term consequences to the health of the people who live and work in the spaces we are designing. In this chapter We will explore some of the recent research on air quality, materials, natural elements, and social aspects like food deserts. We will also be looking at the rating systems that are already in place providing strategies and goals to improve the built environment.

8.1 Historical context of health + well-being

Age of the Hunter-Gatherers – Medical treatment was administered by shamans using herbs and natural elements. Their approach mixed spirituality and the desire to connect with nature and the cosmos to realign the body with health. Early sacred architecture was made with locally available materials and connected people to their geographic location and the cosmos. Stonehenge may have been a "centering force" for people to orient themselves with the cosmos and with each other – a psycho-spiritual healing space.

Age of Agriculture – The origins of western medicine were formed by the Egyptians and then the Greeks as they began to analyze physical problems and create a systematic way to diagnose, formulate prognosis, and develop a code of ethics. The Hippocratic Oath was written in the 5th century BCE and directly inspired what physicians swear to today. Scientific examination and classification began around this time, and healing practices began to rely on treatments based on science and human intellect.

Age of Industry – The separation between humankind and nature widened with the development of modern medical techniques. Factories were being built and cities were being established which lead to denser population and high mortality. "Disease flourished as a result of many factors: overcrowding, poor sanitation, dilapidated, badly designed, or non-existent sewers; lack of potable water; and streets filled with rotting refuse" (Healing 2017, p. 254). Water sources were close to and sometimes intermingled with sewers or industrial pollutant sources. "In 1880 the urban areas in the United States had a 50% higher mortality rate due to infectious diseases" (Healing 2017, p. 253). This situation contributed to the realization that the built environment greatly affects physical health.

Germ theory was discovered, and sanitation measures were employed that greatly improved the quality of life for millions and lowered mortality rates significantly. While this approach saved millions of lives, it led to isolating the causes of disease and treating symptoms with synthetic chemicals instead of promoting preventive health of the whole person that values spiritual and emotional components.

Age of Information – By the 1960s, the rise of ecology began to affect both healthcare and architecture as the worldview began to shift towards the environmental impacts on ecosystems and human health. In 1984, Roger Ulrich conducted his landmark study providing quantifiable evidence that the built environment can greatly affect healing, stress reduction, and the occupants (Ulrich 1984). His continued research is the basis for evidence-based design which considers the whole body, mind, and emotional factors in treatment strategies.

Age of Integration – A rise in the understanding of our interconnected relationship with the earth and its ecosystems has promoted the use of alternative medicine and sustainable design practices. Acknowledgment and use of holistic treatments using food, vitamins, exercise, massage therapy, and yoga along with medications are becoming more common. The medical field is starting to acknowledge the use and value in the practices that previously were excluded from "western" medicine. Preventative medicine is receiving more focus as relationship between the built environment and human health is moving to the forefront, especially via the use of evidence-based design and environmental psychology. Architecture is starting to follow with buildings that look to employ holistic design methods and blend with the site. The understanding that the built environment can positively affect human health, efficiency, and productivity has become more widely accepted, and these principles are being applied over a greater scope of project types.

8.2 Health + well-being and environment literacy

The impacts on human health from the larger natural environment are more impactful than those of the built environment. Without clean air, water, or soil to grow food, human health and well-being will be greatly compromised.

The approach of many sustainability proponents is the view that whatever happens with global warming and sea level rising, the earth itself will be fine. It will be the human beings that suffer and may become extinct. Humans need a very narrow set of environmental conditions to be able to maintain life. Without these, or the ability to artificially manufacture them, we will not be able to survive. As we saw in Chapter 2, all the natural elements are interconnected and affect the viability of the others.

Health and well-being look at designing the built environment in such a way that we can be supportive and restorative to human occupants and the ecosystems. It also looks at the deeper issues of social equity, social interaction, and culturally rich environments to meet the deeper psychological and philosophical needs of the occupants.

8.3 Health + well-being and motivations for sustainable design

The general understanding of health-focused design as a preemptive healing opportunity is growing, and clients are increasingly demanding projects with a clear focus on human health. Whether this is tied to the desire for healthier employees as an empathetic goal or if it's to increase productivity or profit – a self-interest goal, more and more, design is focused on health.

Perkins + Will, an interdisciplinary research-based architectural and design firm, is exploring new directions in community preventive-based healthcare. Their research showed that clinical care only accounted for between 10% and 20% of health, while healthy behaviors, like eating properly, not smoking, and getting enough exercise, accounted for between 30% and 50% of health (Alkan 2014). They also discovered that a person's physical environment is responsible for between 10% and 20% of health (Alkan 2014). Because of these figures, they are proposing programs that partner with community organizations focused on keeping citizens healthy to avoid medical care by holistically addressing the environmental and socioeconomic causes along with healthy lifestyle behaviors.

Thinking across scales helps us see the negative health impacts of our choices at the global scale, coming from climate change, or at the microscopic scale, from coal-burning emissions that enter the bloodstream directly through skin.

Professional obligation

We have a personal and professional obligation as design professionals to serve individuals and society by creating safe, beautiful, supportive environments acknowledging and addressing all areas of human perception and understanding. These include efficient performance, beauty and aesthetics, cultural connections and equity, and ecosystem integration within all contexts.

The American Institute of Architects (AIA) and the Association of Collegiate Schools of Architecture (ACSA) have just established a Design + Health Research Consortium to advance university-led research related to the correlation between the design of the built environment and health. The consortium has determined "The healthiness of places where people live and work is influenced by the way they're designed and the people who design them. Therefore, architects have a huge impact on health and wellness" (Eytan 2016). The consortium's purpose is to examine ways that architectural education can be changed to better meet these objectives. Perkins + Will puts it this way, "Ultimately, the integral perspective calls for a whole system shift from a disease-centered to a healing-centered model for health care" (Alkan 2014).

8.4 Health + well-being and sustainability values

The best health care, whether through the design of its facility or the actual medical delivery, is strengthened by using a quadruple bottom-line approach where economic, environmental, equity, and experiential values underlie important decisions. For example, the trend towards making hospitals more aesthetically attractive, the introduction of outdoor garden spaces, improved interior design of spaces, and private rooms with more daylight, is made not just as marketing strategy to get more patients, but it also helps people heal faster and retains staff. These do provide an economic benefit to the hospital, but they also address, experience, and connect with nature; create equity; and promote the health and well-being of patients and staff.

The creation of these projects also needs to consider the environmental aspects of construction, and maintenance of the physical facility. Perkins + Will in their study, *A Vision and Planning Framework for Health Districts*, proposes smaller facilities integrated into neighborhoods increasing access and equity of healthcare. The quadruple bottom line of healthcare and the built environment organizes and highlights key core values that need to be addressed in a sustainable design project. The social bottom line is always at stake. Studies show that the Richest 1% of men live 14.6 years longer on average than the poorest 1% and for women it's 10.1 years on average (Dizikes 2016). Allowing greater access to all communities could be important in changing these figures.

8.5 Health + well-being and integral sustainable design

Health and well-being are something that should be considered in every project, not just healthcare projects. There are some very general integral theory considerations that apply to all projects, as well as very specific requirements that need to be considered per project type. Specific requirements can be found researching best practices and interviewing clients, employees, and occupants. Each type of project will have its own specific conditions or requirements that should be included. In the process section below, we will look at how to use integral theory to analyze stakeholder concerns for any project type.

The items in each quadrant of the chart in Figure 8.5a (below) should be directing our guiding principles, project goals and objectives, and design decisions. Having a holistic understanding of the client, occupant, project type, and stakeholders allows for solutions with deeper meaning. Deeper meaning supports a connection to place, greater efficiency, and the desire to care for and preserve the built environment as well as increase occupant health and well-being.

In the table shown in Figure 8.5a, general sustainable design considerations to promote health and well-being for all occupants are organized within the four perspectives of integral sustainable design.

	Subjective	Objective
Individual	*Experience* - Physical, acoustic and emotional comfort - Sense of Privacy - Sense of control - Sense of Place - Psychological + subconscious needs - Multi-sensory experience - Beauty/Aesthetics	*Performance* - Reduced toxins at all scales - Efficiency for occupants - Energy and resource reduction - Reduce stress levels - Specific performance based goals as per program/discipline - Increase windows + access to Nature - Increased air flow
Collective	*Culture* - Create Cultural Vibrancy + Inclusion - Social Equity - Historical Connection - Social and Cultural Connection - Connection to Nature and Natural Cycles - Transparency and accountability	*Systems* - Interconnectedness with site, district, urban and global scale - Passive Systems for daylighting and natural ventilation - Active Systems to support health - Living Systems to support health - Human systems: Policies and procedures to support health

Figure 8.5a Health and well-being from an integral perspective.
Source: Created and drawn by the authors.

8.6 Health + well-being and bio-inspired design

Biophilic design

Biophilic design principles can be introduced at multiple levels from pure aesthetic choices or views to help patients and family be distracted from stressful situations to the configuration of architectural form to continuously improve physiological, psychological, and emotional health. Research at Crown Sky Garden at Lurie Children's Hospital in Chicago, Illinois has found that "When inpatients spend time in a "healing garden" their heart rates slow and their cortisol and stress levels fall … physiological effects are measurable within just *twenty seconds* of exposure" (Goldhagen 2017, p. 142, italics in original). The ability to integrate biophilic principles and strategies to improve health and well-being is only limited by your own imagination. The benefit of such strategies extends to cognitive function, increased creativity, reduced stress, higher pain tolerance, greater productivity, and increased rates of healing.

Biomimicry

Biomimicry affects or is related to health and well-being because when systems, products, or buildings are designed to restore the ecosystem that supports all life, humans benefit. There are many products that directly affect human health within the healthcare industry based on biomimetic design. Patient bed stands, door hardware, sink faucets, and counters are made of a copper alloy that resists 99.9% of bacteria 24 hours a day without the use of chemicals or constant disinfectant.

8.7 Health + well-being and adaptation + resilience

Structures and communities that respond better to extreme weather events or climate change keep people safer and reduce deaths. Natural materials and passive systems allow fresh air and light into buildings even without power, but are also providing better conditions every day of the year. We will see that more natural light increases the health of all occupants and promotes healing and learning.

Urban agriculture cultivates, processes, and distributes food locally allowing a higher quality of life all the time, but easier access to food supplies during disaster when other means may be cut off. The same holds true for utilities, water, and waste treatment that can function at a district level.

8.8 Health + well-being

According to a recent study on buildings and health, "The indoor built environment plays a critical role in our overall well-being because of both the amount of time we spend indoors (~90%) and the ability of buildings to positively or negatively influence our health" (Allen 2015). As more research is conducted on the impacts of indoor environments and health, sustainable designers are spending more time working with medical professionals to design spaces and places to improve the health.

When looking at the overall operating costs for running a business, 90% of the costs are related to employee's salary, benefits, and healthcare. The American Institute of Architects (2018) in their Design + Health Consortium talk about "The cost [of poorly designed buildings] to health and productivity is too high to continue to ignore. Just 10 percent of a building's operating costs are attributed to energy, maintenance, and mortgage." Designing the built environment to be supportive and restorative increases productivity, reduces injuries and illness, lowers healthcare costs, reduces sick leave, improves morale and job satisfaction, and elevates the health and quality of life for everyone.

The medical profession is discovering just how damaging long-term stress can be to the human body (Kopek 2009, p. 40). As sustainable designers, we have the opportunity to consciously designing the built environment on every level to reduce stress, improve health, and be beautiful. This shows empathy and respect for all populations, socioeconomic levels, and racial groups, since it's a form of equity to allow everyone access to the benefits of good design. Just as a chain is only as strong as its weakest link, so to, a society only prospers when all its members prosper. We also know that satisfaction with life transfers to how we treat others which has huge social implications.

Google and other large corporations are using well-designed buildings to attract and keep the best talent. Hospitals are doing the same thing because employees are constantly in high-stress situations. Spaces that function well are aesthetically pleasing, and include views to help reduce employee turnover rates and lower training costs.

Every detail and aspect of the built environment affects us physically, psychology, and subconsciously. It also affects energy performance, the quality of our social interactions, the surrounding community, and ecosystem. The following is a brief overview of the major health and well-being considerations in the design process. These apply to all building types including healthcare facilities, offices, schools, and residences.

Indoor air quality

Air is the one element that is always present and something we can't live without. The choices of interior finishes, construction products, furniture, or types of ventilation systems all affect the quality of the air we are constantly breathing.

Construction materials and finishes emit fumes, chemicals, or toxins, long after they are installed. This is called off gassing. Some examples of materials that off gas are particle board, plastic laminate, carpet, and any vinyl products. Office supplies like whiteboard markers, whiteouts, or design markers also add toxins to the air. If there isn't adequate exchange of fresh air into the space, these toxins will remain in the indoor atmosphere for varying lengths of time and be inhaled. The surrounding site affects indoor air quality (IAQ) as well. In dense urban areas, emissions from vehicles or surrounding manufacturing units may reduce the air quality before it even reaches the interior. In this case, fresh air needs to be filtered by as it is introduced into the mechanical systems.

Mechanical systems and their maintenance can also affect air quality. Legionnaires' disease is a type of pneumonia caused by bacteria. It is spread through aspirated water, most often through improperly cared for air-conditioning systems. The disease can be fatal. To reduce CO_2 and other toxins from any enclosed areas that humans occupy, air needs to be exchanged. Codes mandate minimum air exchange rates based on the number of people occupying the space or the type of space and its exposure to moisture or contaminants.

Joseph G Allen, Assistant Professor at Harvard University, specializes in the dynamic relationship between the built environment and health. He leads the Healthy Buildings program at Harvard's Center for Health and the Global Environment. Allen along with his colleagues conducted research to determine the effect of ventilation and fresh air on nine different types of higher-order cognitive function (Allen et al. 2016).

The research simulated three different IAQ levels based on code compliance and minimum LEED IAQ Assessment Credit and commonly found air quality conditions in conventional buildings. The three different IAQs are as follows:

- **Conventional Building** = 20 cfm exchange | 1,400 ppm CO_2 | VOCs meeting LEED IAQ Assessment Credit | 50% outdoor air | code compliant
- **Green Building** = 20 cfm exchange | 945 ppm CO_2 (LEED min) | no VOC | 50% outdoor air | LEED + code compliant
- **Green + Building** = 40 cfm exchange | 550 ppm CO_2 | no VOC | 100% outdoor air

To insure the test, subjects were unable to detect the differences in air exchange rates, the flow of air from the HVAC system remained at the same rate (40 cfm) and the exchange rates, and CO_2 levels were maintained with the percentage of outdoor air introduced into the spaces. On "Conventional Building" days, common materials, finishes, and office products were added to the air ducts to reach VOC levels commonly found in conventional buildings. Twenty-four participants worked in the environmentally controlled spaces for six full days. Each day different levels of CO_2, VOCs, or outdoor air were simulated in the closely monitored spaces. Everything else in the spaces remained the same.

The simulations revealed: "On average, cognitive scores were 61% higher on the 'Green' building day and 101% higher on the two 'Green+' building days than on the 'Conventional' building day" (Allen et al. 2016). The information from this study was then used to analyze conditions in seven different cities, in different climate zones across the U.S.

> This study looked at four different HVAC system strategies and showed that doubling the ventilation for improved productivity costs between $14 to $40 per person per year in all the investigated climate zones. When energy-efficient technologies are deployed, the cost is between $1 to $18 per person per year in all investigated climate zones.
>
> (AIA 2018, p. 13)

Natural light

Natural light is light that comes from the sun and is also referred to as daylight. When daylight hits the skin, it triggers a chemical reaction that produces vitamin D in the human body. This vitamin is vitally important in bone health, immune systems, hormone levels, and sleep cycles. Natural light, even though our eyes can't detect it, contains all colors of the color wheel and is referred to as full-spectrum light. There are light bulbs that now can replicate full-spectrum lighting, and they are used in light therapy, retail applications, or anytime that true color rendition is desired.

Circadian rhythms is the term for the natural cycles that the human body experiences in a 24-hour period, and it comes from Latin meaning "about a day." Circadian rhythms are responsible for hormone levels, sleep cycles, and stress levels. Hormones and nerve chemicals are released from various parts of the brain and are stimulated and controlled by the level, timing, and quality of natural light. The most influential and important type of light is the bright sunlight when we wake.

Our bodies physically need light. Lack of sunlight or prolonged exposure to artificial light, especially fluorescent light, can dampen moods in most people or lead to a form of depression called *Seasonal Affective Disorder* (SAD). Daylight regulates the production of melatonin, the chemical that regulates sleep and energy levels throughout the day, and helps regulate stress hormones. When sunlight enters the eyes, it can trigger hormonal and neurological changes that are important for general health. It has the ability and affect the moods of shoppers, sales rates, lengths of hospital stays, and healing rates. Daylight is still the preferred treatment for some types of skin diseases or neurological conditions.

Color of light is important for health and neurological reasons, but it has also been observed to influence human behavior. Specialized interiors such as retail spaces, restaurants, spas, schools, and medical facilities require more intensive studies into the best colors and qualities of light to support the desired activity level or response.

In a study conducted by Architectural Digest in 2006, design-show participants and their interactions were monitored with three identical rooms. Each of the rooms had their door painted either red, blue, or yellow and was lit inside with the same color. Participants were observed and wore heart monitor bracelets as well. The study showed that the majority of people perceived the blue-lighted room as calming, and red- and yellow-lighted rooms as stimulating. Food and Beverage consumption was twice as high in the yellow room, even though people reported feeling hungry and thirsty in the red room (Goldhagen 2017, p. 146).

It has also been found that different color lights stimulate different levels of dopamine as well as causing people to interact with each other and their surroundings differently. "Retail environments with natural daylight attract more customers and keep them longer … one supermarket moved to a location with multiple skylights…. [and] had a 40% increase in sales" (Goldhagen 2017, p. 145). "Children in properly daylit classrooms focus better, retain information better, behave better and score better on tests" (Goldhagen 2017, p. 146).

Natural light in the healthcare industry has been shown to make dramatic comprehensive improvements: Recovery time is reduced, less-pain medication needs to be administered, symptoms of depression are relieved, and workers are more satisfied, reducing turnover rates for medical workers, a serious problem in the medical field. "Cardiac intensive care unit patients on the sunny side of buildings have different outcomes than people on the non-sunny side … [they] spent less time in the unit and mortality rates, in general are lower…" (Augustin and Fell 2015, p. 8). There is also research being done by the University of Colorado and the University of Edinburgh indicating that exposure to sunlight, indoors or outdoors, could reduce the risk of heart attacks or lessen the damage afterwards (Augustin and Fell 2015, p. 10).

Exposure to daylight has been found to significantly shorten the length of hospital stays for psychiatric patients as well. A study in Alberta, Canada compared patients hospitalized for all types of depression and found that patients in brighter rooms were discharged two and a half days sooner than patients with lower light levels (Sternberg 2010, p. 50). A hospital in Milan, Italy compared patients hospitalized for bipolar disorder and found that patients in east-facing rooms were discharged three and a half days earlier than patients on the west side of the building (Sternberg 2010, p. 50).

Physical health in general is also affected. Exposure to daylight changes our heart rhythm, the interval between beats, and the ability to recover faster, all leading to less stress and more focused cognitive functioning. Vitamin D, produced by sunlight hitting the skin, is essential for calcium absorption improving bone formation, and it also strengthens the immune system and macrophage cells that rid the body of inflammation and infection.

Individual control of varied levels and types of lighting is a very simple thing to incorporate into a design. Within any business, work, or home setting, multiple tasks require varied light levels for productivity and comfort. The multiple benefits include the following:

- Reduce eye strain due to glare,
- Increase visibility for computer screens,

- Increase productivity,
- Accommodate varying levels of complexity,
- Support varying age levels or sight impairments.

As we age, our eyes require higher levels of lighting to do the same things. Visual impairment, special eyesight requirements, and even technology or computer screens are also reasons that occupants can benefit from individual lighting controls. Certain illness, injury, or diseases can affect how we see light. It is important for these people that there are different types, levels, and sources of lighting which can be individually controlled and have dimming capability.

Summation: Clearly, the design of the built environment plays a significant role in making sure that building occupants have access to the best health possible. Collaborating with medical professionals or using well-established metrics for daylighting as well as daylighting simulations early in the design process is a key to meeting the health requirements as described in this section.

Nourishment

Food is the body's energy source. A proper diet maintains the balance between the amount of energy or nutrition consumed (calories) and the amount of energy used for mental, physical, and emotional activities. The other important aspect of nutrition is making sure that a balanced combination of vitamins, minerals, and proteins be consumed through eating a variety of fresh fruit, vegetables, meat, and grains that are free from additives or chemicals.

Importance of good nutrition for physical, mental, and emotional health and development is more evident than ever. Research is starting to indicate that diet can greatly reduce a person's chance of getting certain types of cancers. "Foods that appear to decrease the risk of cancer include fresh fruits, vegetables, and grains" (Kopek 2009, p. 51). Low-fat sources of protein should also be included with a reduction of processed or chemically altered food sources. Deficiency in nutrition can be linked to many psychological conditions, such as depression, and other medical conditions. Without adequate nutrients and calories, the body doesn't have the energy to perform physical or mental tasks.

Food deserts

Studies related to obesity, diet, and health are starting to reveal that socioeconomic and racial factors play a part in access to nutritious, healthy fresh food. A term that has been used in the evaluation of nutrition and access to healthy food is "food deserts." Food deserts are defined as residential areas where there is limited access to affordable nutritious food. This usually means a full grocery store within a mile, in urban settings, to a ten mile radius in rural settings. Living in a food desert increases the risk of obesity and poor diet, while people living closer to a full grocery store tend to eat more fresh fruits and vegetables. Initial attempts to correct this have shown that the "...perceived access to healthy food improves, [but] diet quality and body mass index (BMI) do not" (Cooksey-Stowers et al. 2017, p. 1).

Food swamps

A Food swamp is another recent term that has emerged from research in the same geographic areas. Food swamps are "...areas with a high-density of establishments selling high-calorie fast food and junk food, relative to healthier food options" (Cooksey-Stowers et al. 2017). There is a connection to people living in food swamps and obesity, especially if they are restricted in their access to public or private transportation. Research is still investigating the raised rate of obesity within lower-income and other racial populations. The research is aimed at investigating how education, cultural norms, and fast food content may influence diet and exercise habits.

Urban agriculture

Urban agriculture is sometimes referred to as urban farming or urban gardening, and is the practice of cultivating, processing, and distributing food in an urban setting. It can be done by individuals, communities, businesses, for-profit or for-community service. Many feel that it could be helpful in eliminating food deserts. The benefits of urban agriculture are that it creates fresh food locally available, reduces transportation cost and CO_2 emissions, and dependence on shipping or other geographic locations. Communities or individuals producing their own food increase the resilience of an area in the case of disruption to food distribution systems.

Farming many times uses locations such as abandon lots, rooftops, or vertical walls, or abandoned warehouses are repurposed for hydroponic growing. Hydroponic growing uses a water-based solution rich in nutrients instead of soil. In addition to producing food and introducing urban dwellers to natural cycles, urban farming also connects people to their communities. It educates people about natural growing systems, emphasizes the importance of a healthy ecosystem, and increases the appreciation of fresh food. In some instances, it may be the first time for someone who has grown up in an urban environment to be exposed to how food is grown or produced. Gardening can also be an exercise or activity encouraging people to get outside and benefit from natural daylight and the benefits of being in contact with nature.

Integration of food production with design

The Living Building Challenge (13.9.3) requires a certain percentage of food production on-site for all projects. The size of the project determines how much food is to be grown. This is an exciting development in sustainable design. Including food production into the mental map of sustainable design sounds like a more challenging task, but the rewards for building occupants and the larger environment are well worth the effort.

8.8.4 Water

An adult's body is made up of approximately 60% water; every cell uses water; and without it, we would quickly die. Our body loses water through sweating, breathing, and elimination. Because of this, it is imperative that we replace what was lost. Water is vitally important to all living things. Age, body size, physical condition, activity level, and climate or geographic area all affect the amount of water our body needs to stay healthy.

We need clean water to survive, but it must be free of pollution and toxic substances. The term for water that is safe to drink is "potable." Potable water is in limited supply and is considered a closed system, meaning that the supply never increases as it cycles through the ecosystem. The natural means of water purification happens when water is evaporated, falls as rain, and is then absorbed into the earth. It then filters through the deep layers of the earth over a period of time to be gathered into veins under the earth's surface. Traditionally, wells were dug and later drilled to access the sources of purified water. Depending on what is contained in the earth that filters the water, or the amount of surface water that penetrates these veins, the water source can contain chemicals or bacteria that make it unsafe for human consumption. To mitigate this and provide for the growing demand of water, purification systems and central water supplies have been created.

Purification systems for central water systems filter out sediments and pollutants, and then disinfect the water, chlorine is the most used disinfectant to maintain purity of the water for human consumption. Many cities or districts also add fluoride to their water. The value of adding fluoride to regular drinking water is debated by some people as unnecessary and even harmful. Many private water purification systems rely on ultraviolet light to kill harmful bacteria. Research is being done by a company in Copenhagen, Denmark to create a filtering membrane which can remove pollutants, salt, and bacteria without using chemicals. The membrane is based on a naturally occurring protein in cell walls that allows water to pass through it with the exclusion of everything else.

Water can be in short supply due to drought, climate change, or geographic location. Collection and conservation strategies can be incorporated into the built environment or site. Water-efficient fixtures, stormwater management, gray water usage, and efficient manufacturing procedures should always be incorporated. Living systems, which treat wastewater with plants and living organisms similar to a wetland, should be considered whenever possible. Keeping stormwater on-site can help with irrigation and to mitigate drought conditions. Unfortunately, our need and desire for water have created a huge problem for waterways and oceans.

Water also has restorative and medical benefits that have been used for centuries. In medicine, hydrotherapy is used for relaxation and to treat burn victims, musculoskeletal disorders, spinal cord injuries, and physical therapy for sports injuries. It stimulates the flow of blood to speed up healing, assists in reducing inflammation, and relaxes muscles to relieve tension and stress. The sound of water in fountains, streams, or even soundtracks reduces stress. The sound of water is also an effective way to mask unwanted noises and has a calming effect when heard. Research shows people preferred pictures of urban environments with water over those without.

For the sustainable designer and engineer, clients are increasing requesting that the design team address the issue of water quality. The Well Building Standard requires regular testing of water to ensure that it meets a high standard of purity.

Physical fitness

Obesity is a rising problem in the U.S. According to the Center for Disease Control (CDC), obesity levels have risen from 10% in 1986 to almost half the U.S. population being in the 30%–34% range in 2018, with some of the southern states with an obesity rate over 35%. Obesity and lack of exercise have been shown to be important indicators in overall health and disease prevention. Obesity has been shown to contribute to stroke, cancer, asthma, heart disease, and diabetes. It also affects blood pressure, cholesterol levels, some psychological disorders and it can cause breathing problems, sleep apnea, or problems during pregnancy.

Incorporating opportunities for exercise or physical movement within normal daily life can make a big difference in combating obesity. The New York Health Department published a report evaluating obesity rates. They found that Manhattan's obesity rate was only

10%–14% in contrast to the rest of the state where rates ranges from 15% to 24%. It is proposed by most researchers that the urban environment fosters physical activity. "If one were to overlay a map of urban sprawl onto a map of obesity, it would almost be a perfect match" (Sternberg 2010, p. 264).

Because the places we inhabit shape our behavior, the design of urban environments can encourage or discourage physical activity with their aesthetic and organizational choices. Walking is one of the most basic and easiest forms of exercise. "Urban design studies have shown that something as minor as the variety of details and finishes … can encourage walking…" (Sternberg 2010, p. 265). Sidewalks and the proximity of the walkway to parks or natural areas can also encourage walking. It seems that the more interesting the surroundings, or larger variety of things to do, the greater the chance that people will walk. Safety concerns can be met with street lights, sidewalks, surrounding shops, proper maintenance, and being where other people are using the street.

Movement should also be considered part of the daily routine when designing interior environments. Activity can be increased by using adjustable furniture to allow occupants to switch between sitting and standing. Locating amenities or shared resources in a central location increases the need for movement throughout the office and encourages collaborative interaction. Interior stairs can be easily accessible and attractive to make them more convenient. Exterior walkways can be landscaped to be more attractive, and parking areas can be moved farther from entrances creating a more parklike setting, of course creating options for those with physical limitation.

The health benefits of exercise range from physical to emotional and short term to long term. Exercise relieves symptoms of anxiety and depression, and promotes psychological well-being. It has been shown to reduce the risks of heart attack, colon cancer, and stroke. It increases blood flow to speed up healing and the elimination of toxins, insulin, and estrogens associated with breast cancer along with elevating energy levels. Exercise is also crucial for healthy bones, muscles, and joints, and it relieves the pain of arthritis and reduces falls among older adults. Exercise increases muscle mass to prevent bone or disc damage and aid in maintaining proper body position. Research shows getting up and walking around for two minutes out of every hour can increase your lifespan by 33%, compared to those who do not. Dr. James Levine (2014), co-director of the Mayo Clinic and the Arizona State University Obesity Initiative, and author of the book *Get Up! Why Your Chair Is Killing You and What You Can Do About It*, actually recommends that you be up and moving for at least ten minutes out of every hour.

Ergonomics

Ergonomics is an applied science that studies the human body and how it interacts with products and spaces to promote efficiency, safety, and health. It studies the human body and movement to determine the best way to design a product or space, and is extremely important in every area of design to prevent long-term physical damage. Ergonomics when practiced at the building scale is mostly the job of architects, interior designers, and engineers. At the product scale, ergonomics includes industrial design as well as other professions.

In the workplace, we perform the same tasks for extended periods of time. This is referred to as repetitive movement. Our bodies were meant to handle a variety of actions and be moving frequently, not continuously repeating the same motion. Sitting at desk all day and using a keyboard can cause long-term or permanent damage if our bodies are not in a proper position. Many professions or jobs have typical job-related injuries caused by doing in the same motion or task over and over.

Some general rules for designing to prevent repetitive movement injuries are as follows:

- Joints should be in a neutral position. While typing, the elbows should be at 45° angle and the wrists should be aligned with the forearms, not bent. The neck should be in neutral position, not looking down or up continuously.
- Work surfaces and chairs should be easily adjustable to meet the specific physical requirements of a wide range of users.
- Counters used for standing work should be high enough to eliminate bending for long periods of time.
- Lighting should be consistent, focused on the task, and not cause glare.

Workstation adjustability, chair design, variety, and flexibility in motion to provide muscle relief are important in any work situation. Any time a person has the ability to slightly alter the way their body moves there is a chance to increase muscle strength and decrease damage done by constant repetitive movement.

Ergonomic design increases productivity, reduces medical and medical leave costs, and prevents long-term physical damage. Ergonomic design principles should be applied to residential, commercial, and industrial projects. For more info, see *Fitting the Task to the Human: A Textbook of Occupational Ergonomics* at the Resources section at the end of the book.

Sound | Acoustics

Sound is an important part of our environment, and it is how we communicate, or listen to music, or know there is a breeze outside by hearing the leaves rustle. Sound waves keep us safe by alerting us that there is a car coming down the street before we can see it. But when the volume is too loud, or there are too many sources, it can become hazardous or distracting, and we refer to it as noise.

Sound waves do not need a clear path of travel and can travel through objects or be reflected off surfaces. The ability of sound to travel through an object is determined by its density. For example, sound will travel through air at approximately 1,115 feet per second, but it travels through steel, which is much denser than air, at 16,404 feet per second (Kopec 2009, p. 202). This is important to remember because in the built environment the materials used in construction can make a big difference in both positive and unwanted sound transmission.

Sound is measured in decibels (dB) referring to its loudness. The higher the decibel level, the louder the sound. A quiet conversation could be between 60 and 65 dB, whereas loud shouting would be approximately between 80 and 85 dB, or the noise of a jet engine would be between 120 and 150 dB. Sound is also measured in Hertz (Hz) which refers to its pitch, or fluctuations and vibrations made by the sound wave. The closer the sound waves, the higher the frequency. High-frequency sounds are perceived as louder and more annoying, like nails on a chalkboard. Most human speech is between 2,000 and 5,000 Hz (Kopek 2009, p. 203).

Permanent hearing loss or damage is possible from exposure to noises over 80 dB from either sudden loud noise or exposure over time. OSHA (Occupational Safety and Health Administration) has enacted codes and restrictions that limit decibel level, frequency, and duration of exposure to high-decibel noise. Even if there is no permanent hearing loss, excessive or continuous noise can be disruptive and cause other psychological reactions. It stimulates the autonomic nervous system, or the "fight or flight" response. This includes raised blood pressure, accelerated heart rate, contracted blood vessels, slow digestion, increased muscular tension, and hormone changes.

Office workers report that "...conversations of other people are what they find most disruptive" (Kopek 2009, p. 206). In this case, loudness of the conversation was not the main cause, but it was the content because, even without the intention to listen, words act as intellectual stimulation. This type of distraction interferes with complex mental activities like interpreting, analyzing, or synthesizing information, or even with physical dexterity skills. "Young children are particularly affected when noise inhibits their ability to hear [clearly] ... because they do not have the vocabulary and experience to fill in meanings when words are missed" (Kopex 2009, p. 206). When noise is a constant problem, language skills or vocabulary may suffer, and some studies believe that attention deficit disorders may be the result of continual noise exposure. Typical disrupting noise sources include, but are not limited to communication, transportation, productive activities, HVAC equipment, and entertainment.

Acoustic specialist can be called in for specialized situations, but there are many strategies that can significantly improve the acoustic quality of a space and do not require specialized engineering skills. The four basic categories to mitigate acoustic disturbances are as follows:

- **Space Planning** – Arranging spaces that produce noise away from areas of learning or concentration, or placing spaces away from external sources of noise. For example, placing classrooms on the side of the building away from the road.
- **Blocking** – Obstructing the path of sound wave travel. This could be accomplished with landscape features to redirect road noise, gaps in continuous surfaces such as walls to stop sound wave transmission, or large barriers that are commonly found along highways.
- **Absorption** – Creating areas that will absorb sound waves instead of reflecting them. This could be done using acoustic ceiling panels, carpet on the floor, cork on the underside of tables, and soundproofing batts within wall structures.
- **Sound Masking** – Providing consistent, nonintrusive, or sound-canceling waves to counteract or disguise distracting noise. Water fountains, or a sound-masking system that produces a slight sound of static or a hum obscures distractions so that other unnecessary noises become difficult to be heard and are easily ignored.

In contrast to the sounds reviewed above, natural- or biological-based sounds, such as birds singing, crickets, running water, or ocean waves, have been shown to be restorative, comforting, and healthy. These types of sounds can help to mask distracting noise and improve health. For more information, see Chapter 6.

Thermal comfort

Thermal comfort is one of the basic human needs. It's the main reason that we began to build structures. Temperatures need to be within certain limits for human beings to physically survive. Thermal comfort is based on the combination of five environmental factors:

- Physical activity
- Amount of clothing

- Temperature
- Humidity
- Air current (or wind).

We can tolerate much higher temperatures with a breeze or when the humidity is lower. Conversely, if there is high humidity, even relatively low temperatures in the summer can become unbearable. If it is winter and you are hiking or cross-country skiing which are heavy aerobic activities, your body is producing a great amount of heat and you can be quite comfortable at −10° Fahrenheit with very little clothing on, but if you are standing still for any length of time in these conditions, you will become quite cold very quickly.

Thermal comfort is extremely important to occupant productivity and sense of well-being, job satisfaction, learning in a school environment, and healing in a healthcare environment. It is crucial to understand the activity level of occupants, the type of clothing they wear, typical humidity level, surrounding air temperature, and projected air movement. An interior environment where the occupants are engaging in a physical activity, like a gym or manufacturing plant, will require much different heating and cooling needs than a healthcare unit where the patients are mostly still with very little clothing on. These factors also apply to exterior spaces such as bus stops, concert venues, and public spaces.

Common contaminants

Contaminants are substances that when introduced into the human body cause unfavorable reactions, disease, harm, or death. These can be found in the landscape or within the built environment because of building products. Common forms of contaminants are as follows:

- Mist or fine droplets
- Vapors – gaseous form of substances
- Gas
- Smoke – small particles that don't burn completely
- Dust – fine, dry, solid particles
- Fume – Gas, smoke, or vapor
- Aerosol – droplets that remain dispersed in the air
- Live organisms – animals, insects, or microorganisms such as mold and bacteria

Asbestos – An extremely heat-resistant material that is resilient to wear with excellent noise-reducing properties. In the 20th century up until 1980, it was mostly used for its fire retardant properties or as binders in many common building products. It was commonly used for ceiling tiles, floor tiles, millboard, shingles, acoustical products, and pipe insulation. It was thought to be a "perfect product" until it was discovered that its microscopic fibers became permanently embedded in the lungs and caused lung cancer, mesothelioma, or asbestosis a chronic breathing disorder caused by the scarring of the lungs.

The EPA (Environmental Protection Agency) has tried to ban asbestos usage completely but has met with resistance. There are still a few products that contain asbestos, but in the majority of applications, it has been replaced. The biggest danger comes from older or existing facilities where asbestos may be exposed or crumbling resulting in the fibers being airborne. Any suspicious materials should be inspected and removed by certified professionals. In some states, asbestos is considered a hazardous material and can only be legally removed by a licensed professional.

Lead – A naturally occurring element on the earth's surface. In the 20th century, it was added to paint to extend its protective properties and to gasoline to increase fuel efficiency. It was also used in water pipes. It is estimated that "about two-thirds of the houses built in the United States built before 1940, half the houses built between 1940 and 1960, and a lesser number between 1960 and 1978 have lead paint in them" (Kopek 2009, p. 86).

Lead is highly toxic to humans when ingested and is particularly dangerous to young children with developing brains and bodies. The most common routes of exposure are eating paint and breathing in lead dust. Lead poisoning can be hard to detect because of its gradual accumulation within a body. Lead poisoning affects the whole body and can result in lower intelligence, kidney damage, and speech-, language-, and behavior-related problems. It is recommended that any building constructed before 1978 be tested by a professional for lead vapors.

Mercury – Some of the paints developed to replace lead-based paint contain a chemical called phenylmercuric acetate or PMA. PMA was found to emit mercury vapors. In 1991, the EPA prohibited the use of PMA in latex paints. Fluorescent light bulbs and batteries also contain mercury which is the reason they should always be disposed of properly. Appliances and some switching devices also contain mercury and should be handled by an appropriate professional.

Biological Contaminants – Living things can be harmful to humans. They can range from single-cell organisms to large rodents like rats. Bacteria, viruses, fungi, and protozoa can all be present in the built environment. Biological contaminants can be present in any building type, so cleanliness and disinfection are especially important in all public and healthcare facilities or any food preparation areas.

Chemical Pollutants – These are commonly the result of trying to solve a problem or safety issue which creates another problem or issue, as in the case of fire-retardant materials used in children's sleepwear, toys, and household furniture. The surfaces are sprayed with or the products contain polybrominated diphenyl ethers (PBDEs). It has since been shown that this chemical accumulates in the body and harms the developing brains of infants. The European Union banned the use of PBDEs, but the U.S. EPA is still studying its long-term effects. ———

Volatile organic compounds (VOCs) are another example of chemical pollutants. These chemicals are an inherent part of a product or building material that continues to release the chemical compounds into the air after installation in what is known as off-gassing. Off-gassing can occur over time and is compounded by other building materials that absorb VOCs and multiply their effect. A common VOC is formaldehyde which is used as a binder in particle board, new carpet, vinyl flooring, and upholstery. VOCs are also present in adhesives, paints, solvents, synthetic fabrics, foam padding, cleaning products, and office or art supplies to name a few.

Some of the specific conditions that are attributed to prolonged exposure to VOCs are asthma, cancer, birth defects, neurological problems, and allergies. Some nonspecific symptoms that have been reported are drowsiness, cough, fatigue, headaches, blurred vision, and joint and muscle pain. It is important to research and understand everything that is being specified or used in a project. If a rating system is being used in addition to local and national building codes, there may be restrictions on VOC levels and various methods of reporting that are required.

Carbon Monoxide – A colorless, odorless gas that is a byproduct of burning fuels that contain carbon. It can be given off by water heaters, space heaters, furnaces or boilers, and wood- or propane-burning units. It can also result from leaking chimneys, or car exhaust from attached garages. Carbon monoxide can be fatal as it removes oxygen from the body. Some of the lesser symptoms are similar to the flu and make it hard to diagnose. The best practice is to equip all spaces with carbon monoxide detectors that are regularly checked and serviced.

Radon – A naturally occurring radioactive gas that is found in soil in many parts of the U.S. It is odorless, tasteless, and colorless. The only way to detect its presence is through testing. It can enter buildings through cracks in the foundation, gaps in piping, or construction joints. Radon damages the lungs and is thought to be the second leading cause of lung cancer. Testing is the best way to address radon contamination, followed by complying with any remediation procedures recommended by a certified radon professional if it is found in the building.

Place

Our body has the ability to sense and gather information about our surroundings that go beyond the traditional five senses, or even cognitive consciousness. This means that we as humans are always gathering information from our surroundings, and our brains are always processing and being influenced by things that we can't always articulate verbally, or point to in a quantitative way. Researching further illuminates that these impressions can become a permanent part of our view of self and influence our behavior. In looking at bio-inspired design, we were presented with many benefits that happen automatically without the occupants' conscious decisions or actions. This should tell us that there is more to the built environment than visual, tactile, auditory, or olfactory sensations. This concept may be a little far out from where you are at this point, but it is really the underlying essence of design and the architectural professions. The areas of study and parts of the professions that cover this are environmental psychology or phenomenology. The Academy of Neuroscience for Architecture is an organization that also explores the built environment and the neurological changes that occur in human brains as a result of experiencing architecture.

What it boils down to is that whether or not we believe, or realize, or try, we are creating place. The only question is what type of place and what type of messages are we sending. This ties into equity, providing equal value and access to all populations, which we have talked about in other parts of the book. This is also related to the cultural and ethnic values of a society or business or the beauty and aesthetics of a space. And it is very much related to sustainability because if people value a place or building they will care for it, and preserve it and continue to invest in and use it for hundreds of years. Speaking about importance of place in architecture, Dr. Ashraf Salama (2007), the Founding Head of the Department of Architecture and Urban Planning at Qatar University, writes, "…the built environment [is] a two-way mirror … it conveys and transmits non-verbal messages that reflect inner life, activities, and social conceptions of those who live and use the environment." If we really understood that "…the environments being designed today are shaping how people live and function in the future" (Roberts 2013, p. 2), how would it change our design decisions? What message do we want to convey and how would we like to shape the future?

Place can affect more than our deep subconscious. Medical and scientific advances of the late 20th century have established a connection between the brain and the immune system. Along with this connection is the assumption that "…physical places that set the mind at ease can contribute to well-being and those that trouble the emotions might foster illness" (Sternberg 2010, p. 10). This includes psychological safety known as prospect and refuge, which is the ability to be able to survey the physical environment for planning or security from a place that is protected. In bio-inspired design, we discuss many of the elements and strategies that can add to healing environments.

Below are just two examples of how understanding occupants to create place can benefit special populations or those with certain medical conditions.

Alzheimer's Case Study

Elizabeth C. Brawley, ILLAD, AAHID, CID is an interior designer who has done extensive research, regarding the effects of the built environment in caring for Alzheimer's Patients. Her work has revolutionized the healthcare industry. She observed through her studies that by reducing the volumetric massing of the interior spaces, using indirect full spectrum lighting, and providing first floor exterior views with foliage patients exhibited many positive changes. Brawley also found that by grouping the patient rooms in smaller numbers around a small living area and placing a nurse at a residential desk in this room, it reduced the anxiety, aggression and disorientation of the patients dramatically.

Assisted Living Case Study

Neurological research is changing the way we are thinking about assisted living facilities. Research and testing has found that nerve cells can be regenerated. When testing mice with neurological conditions mimicking Alzheimer's or dementia the mice were able to access old memories and sprouted new nerve cells because of exposure to enriched environments and physical activity. Mice with the same conditions in empty, barren cages did not. It is not clear yet if the benefits were from the exercise or the mental stimulation, but "…There is no question that enriched environments coupled with moderate exercise can help to preserve memory and improve mood [in assisted living facilities]."

(Sternberg 2010, p. 168)

Recently designed assisted care facilities are starting to employ these principles, along with creating a sense of place familiar to patients. They have found that creating a "Main Street" or "Town Center" has been very successful in encouraging patients to spend most of their time outside of their rooms, reducing isolation. The public areas replicate store fronts, brick walls, faux balconies, with street lights and gathering areas. The exterior walls are lined with windows that allow sunlight and views for orientation to the seasons and time of day. Features like clock towers, water fountains, and indoor gardens act as landmarks so that residents can wander safely within the constructed "outdoor town center" and still know where they are. Some examples would be Maggie's Centres designed by Frank Gehry or The Village in Waveny Care Center, New Canaan, CT by Reese, Lower, Patrick & Scott.

8.9 Health + well-being and process

Evidence-based design

Evidence-based design is an approach that bases its decisions on findings from the scientific study of human interaction with spaces, environments, materials, acoustics, and visual stimuli. This research can be done by observing occupants' reactions in real time, doing post-occupancy evaluations (POEs), or by using MRI and CAT scans to understand human reaction on a neurological level. Researchers look for deep-level connections between the physical world and the inner psychological and neurological workings of human beings. The goal is to establish supporting data on measurable changes in health, behavior, and learning capability directly related to the built environment.

Some of the initial concepts and studies started with professionals observing the difference in the social interaction patterns of patients in mental hospitals related to furniture positioning. They conducted experiments with different furniture arrangements and noticed that when the furniture was placed in small groups facing each other patients would be more likely to interact with each other than when it was placed along the walls facing the center of the room or in rows all facing the same direction. One of the now landmark studies was done by Roger Ulrich regarding healing rates of surgery patients with as identical attributes as possible:

> ...in a Pennsylvania hospital between 1972 and 1981...patients in rooms with outdoor views were discharged on averaged 0.74 days earlier, with less negative remarks from the caregivers, and required less pain medicine the first 5 days of recovery than those in the other rooms.
>
> (Eberhard 2009, p. 168)

Since 1972, many more studies have confirmed these original findings. This was one of the first studies that indicated healing can be affected by the "intangibles" such as daylight and the patient's connection to the natural world. Better patient satisfaction reports affect the overall culture of the hospital and the personal experience of healing, increasing the nurse's ability to care for the patients. A more pleasant experience for the nurses affects employee retention rates, which is a large concern in the healthcare industry. Using less medicine is better for a patient's body, but it also saves money for the hospital.

Evidence-based design includes research in materiality and the resulting neurological, physiological, and psychological reactions. The presence of exposed wood in the built environment has been shown to lower blood pressure, lower perception of pain, and increase concentration. These are just a few documented benefits of exposed wood in interior spaces.

Design process

Research and precedent studies

It is incredibly important to understand the purpose, client and occupant needs, and the full scope of activities that will take place in the buildings and spaces that we design. By researching and studying previously built or designed projects of similar type, we gain valuable insight into what works or what doesn't. POEs, as mentioned above, are also an extremely valuable tool for understanding how occupants really use and perceive the environments. Looking at between three and five projects that are similar in function or purpose will give you a good idea of established standards. It will also highlight positive and negative strategies that can help you create the best solution for your specific project. This should not be thought of as copying or cheating, but learning from those that went before us with the goal of creating the best and most supportive space for the occupants and society.

Examining similar projects in a structured way allows us to learn from past projects and inform our designs with current best practices, which is paramount to Integral Sustainable Design. Below are some basic questions to answer and areas to analyze as you start your design process. They cover the basics and allow room to include many other specifics.

Basic Information:
Project Location
Project Size
Architect
Landscape Architect
Designers
Number of Years in Operation
Staffing

Physical Characteristics:
Profile Furniture
Profile Lighting
Programmatic Interrelations, are they mandatory or preferable?
Security Features, both physical and emotional?

Additional Questions:
Are there any historical factors that drive this project?
Are there any gardens, landscaping or access to nature, views or daylight that are important?
Are there any sustainable features?
What was the process design based on? (Traditional client/architect process, stakeholders, collaboration?)
Any special considerations?

Please think deeply and answer the following questions based on your research:
Who uses the building?
What goes on in there?
What do we want them to be doing?
What are the typical design directives for the project type?
What human factors do we need to consider?
What environmental psychology principles or issues need to be addressed?
What evidence-based design information is available or needs to be researched?
How should your project interface or relate to the surrounding community and its site?
What type of experience is needed or desired?
How can it function best from a productivity, energy, or space planning perspective?

Synthesis + Analysis:
How does it feel to be there? (i.e. what is it like to be there from what you can see from the images?)
Patient | client | occupant reactions?
What do see as advantages to this project? (Things that would be great to replicate)
What are some areas of concern with this project? (Things we should look to avoid or address)
What best practices can we see or establish from this analysis?
Anything else you think would be interesting or helpful in understanding what makes these project successful?

By consistently answering the above questions for each of your case studies, you will gain in-depth information for your project type. You will start to see patterns and established approaches for your specific project type. These approaches or design solutions may be positive strategies that we can use to inform your design. Illuminating the negative or unsuccessful aspects of past projects can also be very informative as well, because it can keep us from going down the same path.

Integral theory project analysis

To create a holistic deeper understanding of project type and occupant needs, and to be truly sustainable, it is important to examine your project, client, and occupants from an integral perspective. The following are a brief summary of questions that can be used to examine the experience, performance, cultural, and system needs of your project. Please feel free to add any other that come to mind or you would like to explore. The deeper the research and analysis, the better the solution will be.

Experience Quadrant

What are the occupants feeling before they enter project?
What do we want them to feel when they are here?
Are there special psychological needs to be met?
Aesthetic or visual goals for the project?

Performance Quadrant

What are the factual statics of my project?
At what level of efficiency should the project/building function?
Sustainable goals for energy performance?

Systems Quadrant

What kind of systems are important to our project?
How should the building work as a larger system that houses all the project functions?
How is or should our project work as a smaller part of the community?
How should our building work to help productivity?

Cultural Quadrant

What culture are the people coming from?
What culture do we want to create in our project?
Are there historic elements that we need to consider?
How are we going to create a connection to nature or natural forces?

These should go on to inform the guiding principles for the project, establish goals and strategies, and design directives. The analysis and research phase of a project provides opportunities for connection and collaboration with all the stakeholders and should be used to foster alignment as much as possible.

The nine foundations for a healthy building

Joseph G. Allen and his colleagues (2017) of Harvard's T.H. Chan School of Public Health provide nine foundations for a healthy building. The report was motivated by collaborative interaction with academic, business, real estate, and design professionals. It summarizes numerous scientific and academic research in a way that is easily transferable to industry professionals and gives clear actionable strategies. For more information on the complete report, see the Resources section.

The nine foundations of a healthy building

1 **Ventilation:** Meet or exceed local code or rating system ventilation rates.
2 **Air Quality:** Choose supplies, office supplies, furnishings, and building materials with low chemical emissions to limit sources of volatile and semi-volatile organic compounds.
3 **Water Quality:** Meet the U.S. National Drinking Water Standards and test water quality regularly.
4 **Thermal Quality:** Meet minimum thermal comfort standards for temperature and humidity, and keep thermal conditions consistent throughout the day.
5 **Dust + Pets:** Use high-efficiency filter vacuums and clean surfaces regularly to limit dust and dirt accumulation.
6 **Lighting + Views:** During the day, provide as much daylighting and/or high-intensity blue-enriched lighting (480 nm) as possible while maintaining visual comfort and avoiding glare.
7 **Noise:** Protect against outdoor noises such as traffic, aircraft, and construction. Control indoor sources of noise such as mechanical equipment, office equipment, and machinery.
8 **Moisture:** Conduct regular inspections of roofing, plumbing, ceilings, and HVAC equipment to identify sources of moisture and potential condensation spots.
9 **Safety + Security:** Meet fire safety and carbon monoxide monitoring standards. Provide adequate lighting in common areas, stairwells, emergency egress points, parking lots, and building entryways.

Additional Strategies/Considerations: No Smoking Policy within 20 feet of the building. Incorporate design elements that promote activity.

8.9.5 Applicable rating systems

Governing bodies and codes regulate the built environment, ASHRAE and OSHA are federal national codes that all building must comply with. Other rating systems like LEED, Well Building, and Living Building Challenge (LBC) are voluntary.

ASHRAE (American Society of Heating, Refrigeration, and Air Conditioning Engineers) Standard 62.1: www.epa.gov/sites/production/files/2014-08/documents/ciaq-webinar-muller.pdf

It specifies minimum ventilation rates and other measures for new and existing buildings to provide IAQ to minimize adverse health effects. VOC levels and tobacco and e-cigarette smoke are also taken into consideration. Codes are minimum requirements per building type based on research and past experience. Requirements are mandatory and regulated by the federal government. Inspections are required to verify compliance before occupancy permits are granted.

OSHA (Occupational Safety and Health Administration): www.osha.gov/about.html

It is started by President Richard Nixon in 1970, as part of the U.S. Department of Labor. Its mission is to assure safe and healthful working conditions for working men and women by setting and enforcing standards and by providing training, outreach, education, and assistance. Laws and regulations for every aspect of major industries have been detailed, and employers are held responsible for providing safe working conditions. The regulations cover construction, maritime, and agriculture industries and other general business types along with state-specific mandates for worker safety.

Codes are minimum requirements per building type based on research and past experience. Requirements are mandatory and regulated by the federal government. Inspections are required to verify compliance before occupancy permits are granted.

International Well Building Institute: https://v2.wellcertified.com/v2.1/en/overview

Well Building Standard deals directly with the health and well-being of the occupants. Areas covered by the standard are air, water, nourishment, light, fitness, comfort, mind, and innovation. There are 102 performance metrics that are performance tested by a third party. The level of achievement is determined by the amount of strategies employed in each areas.

It seeks to address the physical systems within the human body in its directives and acknowledges the importance of the interaction between occupants and the built environment and health and well-being. The prescribed and optimized strategies are extensive. One of the criticisms of the Well Building Institute is that only 1% of population can afford to do it.

Living Building Challenge (LBC) (see 13.9.3) https://living-future.org/lbc/

LBC is a voluntary rating system. LBC advertises itself as the built environment's most rigorous performance standard. Buildings trying to achieve full Living Building Certification must go beyond Net Zero Energy and meet Net Zero Water over a minimum of 12 months occupancy. To achieve the Materials Petal, the chemical composition of every component of every product or mechanical system within the building needs to be examined to verify they are free from toxins given in LBC's Red List. The Red List comprises a list of 14 chemicals, including but not limited to, lead, cadmium, asbestos, PVC, and formaldehyde.

LEED (Leadership in Energy Efficient Design): https://new.usgbc.org/leed

The LEED system was established by the U.S. Green Building Council (USGBC). It is a certification process based on third-party verification of a comprehensive list of green building strategies. It is a point-based system that allows buildings to achieve a different level of rating depending on the amount of green building strategies used.

There are five different rating systems that address the unique needs of each building or project type. Each of the rating systems has a credit category that covers materials and resources, water efficiency, indoor environmental quality, and sustainable sites. Within each of the categories, there are specific goals. Each goal reviews the intent along with any requirements. Project members can choose the goals that are most applicable for each project.

8.10 – 8.14 Health and wellness at scales

Health and wellness is fundamental to human survival and foundational to comfort and sustainability. Understanding the basic principles of personal health and well-being should inform all our choices at every scales. Understanding that all our design choices have long-term effects on the occupants, society, and our environment is crucial for integral sustainable design.

8.10 Global scale health

Health and wellness at the global scale needs to acknowledge how climate change, changing weather patterns, and the ease of international travel lead to infectious disease outbreaks and infections of resistant viral strains. Graphic Information System (GIS) is helping in tracking and understanding these trends. Sanitation and healthcare for all social and economic sectors, as well as availability of healthy fresh food, will become increasingly important as we move forward in the 21st century (Sternberg 2010, pp. 270–271). For more detailed information, see Chapter 10.

8.11 Urban scale health

Cities with their dense development and social structure and easier access to healthcare and social services can have a positive effect on the people who live there. Walking is thought to be the basis of this. Studies have shown that people are more willing to walk if their cities have a greater variety of finishes, materials, landscape types, and routes to the same destination. Creating an environment that is inherently healthy physically, mentally, and emotionally requires considering how to make urban areas more walkable and aesthetically diverse. We also need to consider food sources, distribution, and education on nutrition that will promote health for all populations. For more information, see Chapter 11.

8.12 District and site scale health

Creating neighborhoods and districts that allow residents to live close to where they work and shop with easy access to restaurants and entertainment allows people to walk rather than drive. Having amenities conveniently located is important and saves on emissions. It also promotes physical, mental, and emotional health. A great example is Atlantic Station in Atlanta, created on a 138 Brownfield site in the heart of the city (Sternberg 2010, p. 278). For more detailed information, see Chapter 12.

8.13 Building scale health

Strategies employed at the building level have great potential to increase health and well-being. Natural light, materials, access to fresh water, and nontoxic building materials are just a few areas the designer needs to consider For more information, see Chapter 13.

8.14 Human scale health

Elementally the greatest interaction we have with the built environment is through our experience of its materials. The materiality of a space should engage all our senses and create a connection. This scale is also where any toxins would be directly transmitted. Material selections also impact the sustainability of a project and any long-term effects to IAQ, toxins, and aesthetics. Understanding the total embodied energy for each selection enables greater savings and longevity. For more information, see Chapter 14.

Additional resources

Academy of Neuroscience for Architecture: http://anfarch.org/

A Vision and Planning Framework for Health Districts of the Future, Perkins + Will, https://perkinswill.com/sites/default/files/ID%206_PWRJ_Vol0602_05_A%20Vision%20and%20Planning%20Framework%20for%20Health%20Districts.pdf

Building Evidence for Health: The Nine Foundations of a Healthy Building, Joseph G Allen: https://buildingevidence.forhealth.org/

Fitting the Task to the Human, Kroemer and Grandjean, 1997

Health, Sustainability, and the Built Environment, Dak Kopec, Fairchild Books, New York, 2009

Health, Wellbeing and Productivity in Offices, World Green Building Council, www.worldgbc.org/sites/default/files/compressed_WorldGBC_Health_Wellbeing__Productivity_Full_Report_Dbl_Med_Res_Feb_2015.pdf

Indoor Air Quality: A Guide to Understanding ASHRAE Standard 62–2001, Trane, American Standard, www.trane.com/commercial/Uploads/PDF/520/ISS-APG001-EN.pdf

View through a Window May Influence Recovery from Surgery, Roger Ulrich, www.natureandforesttherapy.org/uploads/8/1/4/4/8144400/view_through_a_window_may_influence_recovery_from_surgery.ulrich.pdf

Well Building Standard, www.wellcertified.com/

Wood as a Restorative Material in Health Care, FP Innovations, www.woodworks.org/wp-content/uploads/Wood-Restorative-Material-Healthcare-Environments.pdf

References

AIA, 2018-last update, Design & Health Research Consortium [Homepage of American Institute of Architects], [Online]. Available: www.aia.org/resources/78646-design--health-research-consortium [July 26, 2018].

Alkan, B., 2014. *A Vision and Planning Framework for Health Districts of the Future*. Perkins + Will. https://perkinswill.com/sites/default/files/ID%206_PWRJ_Vol0602_05_A%20Vision%20and%20Planning%20Framework%20for%20Health%20Districts.pdf edn.

Allen, J.G., Bernstein, A., Cao, X., Eitland, E.S., Flanigan S., Gokhale, M., Goodman, J.M., Klager, S., Klingensmith, L., Laurent, J.G.C., Lockley, S.W., Macnaughton, P., Pakpour, S., Spengler, J.D., Vallarino, J., Williams, A., Young, A. and Yin, J., 2017. *Building Evidence for Health: The 9 Foundations of a Healthy Building*. Harvard T.H. Chan School of Public Health.

Allen, J.G., Macnaughton, P., Satish, U., Santanam, S., Vallarino, J. and Spengler, J.D., 2016. *Associations of Cognitive Function Scores with Carbon Dioxide, Ventilation, and Volatile Organic Compound Exposures in Office Workers: A Controlled Exposure Study of Green and Conventional Office Environments*. Environmental Health Perspectives. www.ncbi.nlm.nih.gov/pmc/articles/PMC4892924/edn.

Augustin, S. and Fell, D., 2015. *Wood as a Restorative Material in Healthcare Environments*. 301009845. Vancouver, BC: FP Innovations.

Cooksey-Stowers, K., Schwartz, M.B. and Brownell, K.D., 2017. *Food Swamps Predict Obesity Rates Better Than Food Deserts in the United States*. National Center for Biotechnology Information. www.ncbi.nlm.nih.gov/pubmed/29135909/edn.

Dizikes, P., 2016. *New Study Shows Rich, Poor have Huge Mortality Gap in U.S.* Cambridge, MA: MIT News. http://news.mit.edu/2016/study-rich-poor-huge-mortality-gap-us-0411 edn.

Eberhard, J.P., 2009. *Brain Landscape : The Coexistence of Neuroscience and Architecture*. Oxford; New York: Oxford University Press.

Eytan, T., 2016. How to Transform Design and Health Research into Real-world Strategies [Homepage of American Institute of Architects], [Online]. Available: www.refworks.com/refworks2/default.aspx?r=references|MainLayout::init [July 24, 2018].

Goldhagen, S.W., 2017. *Welcome to Your World*. 1st edn. New York, NY: Harper.

Kopek, D., 2009. *Health, Sustainability, and the Built Environment*. New York: Fairchild Books.

Levine, J.A., 2014. *Get Up! Why Your Chair is Killing You and What You Can Do about It*. New York: St. Martin's Griffin (July 29, 2014).

Roberts, S.H., 2013. *The Implications of Integral Theory on Sustainable Design*. Philadelphia, PA: Philadelphia University.

Salama, A.M., 2007. Mediterranean Visual Messages: The Conundrum of Identity, Isms and Meaning in Contemporary Egyptian Architecture. *Archnet-IJAR International Journal of Architectural Research*, 1(1), pp. 86–104. https://archnet.org/system/publications/contents/4946/original/DPC1665.pdf?1384787259

Sternberg, E.M., 2010. *Healing Spaces: The Science of Place and Well-Being*. 1st edn. Cambridge, MA: Belknap.

Ulrich, R.S., 1984. *View through a Window May Influence Recovery from Surgery*. Gale Group. www.natureandforesttherapy.org/uploads/8/1/4/4/8144400/view_through_a_window_may_influence_recovery_from_surgery.ulrich.pdf edn.

9   Integrative design process

In previous chapters, we saw how Integral Sustainable Design is an expression of an emerging worldview based on integration with nature. Innovation in renewable energy sources such as solar and wind, and new communication methods like social media are drivers towards this new view of holistic, inclusive, and evidence-based design practices. The traditional design process is a linear, hierarchical approach with most of the design decisions already made before any of the other professions have input. The image is of the lone "genius" designer working by themselves to create a "masterpiece" with everything else fit into the design afterwards. This adds more cost and time because of changes made later in the design process. It also reflects "architecture" taking precedence over all other professions, the ecosystem, and occupant health and well-being.

Looking at Integral Theory and Integral Sustainable Design, we see that all things, seen and unseen, measurable and unmeasurable, are given equal value. In this chapter, we will be looking at integrative design practices, a design approach most often seen in sustainable design circles, but is becoming more widely accepted throughout all professions. This approach is based on the collaboration between all of the design professions, contractors, client, and stakeholders from the start of the project. It is also referred to as integrated design process, or concurrent design, or cocreative design. This design process is applicable at all scales of design from product to urban or global design.

This approach to design has been shown to create projects that are more holistic, have an interconnected vision of the social, cultural, and ecological aspects of a project, develops deeper levels of integration, and is the gateway to a better quality of life and sustainable future.

9.1 A brief history of integrative design process

Age of Hunter-Gatherer – The design process was the reflection of a direct relationship between the geographic location, its climate, and locally available materials. The main purposes were to provide shelter and to meet basic physical needs. The designs that we see reflect more organic forms. Form, configuration, and materiality were all based on what had been found to work best over time and had been passed down by oral tradition or by an apprenticeship model.

Age of Agriculture – The rise of the guilds, apprenticeships, and eventually the profession of architecture meant that design was becoming increasingly complex. The design process became more abstract, based on established proportions or mathematical formulas with established forms of ornamentation. Drawings and models were used to describe the projects to workmen prior to construction. This created the structures of marvel and beauty that still stand centuries later, but also brought about a distance and removal from the natural world. Attention was fixated on the object or building itself, not always its integration into the surroundings or geographic setting.

Industrial Revolution – The ideology of the division of labor precipitated the creation of specialties, and the design process became more fragmented, linear, and hierarchical. The separation between the architect and builder or the product designer and the manufacturer added complexity, time, and the need for explicit communication methods. Integration with nature was rarely considered because it was seen as only a resource to be used. The linear process of design and construction did have the advantage of being faster and less expensive than holistic artisan-based construction. The end user was often not included within the process.

Exceptions would be Bauhaus and the Arts + Crafts Movement. They employed an interdisciplinary and holistic model of design and production. Artists and craftspeople were recognized as stakeholders in the process, and the investment of their labor, care, and the resulting beauty was respected and valued. Frank Lloyd Wright's processes mimicked the master builder model of integrated design where he constantly considered the site and nature as an opportunity for integration in process and product.

Ray and Charles Eames were one of the first modern designers to reference all scales as part of design conceptualization. This is expressed in their movie the Power of Ten (Eames 1977). As the short video describes, we need to think beyond what we can immediately understand with our senses and also consider the macroscale of the ecological systems, as well as the microscale that includes occupant health and well-being as a result of our design choices. The Eames and Frank Lloyd Wright were starting to explore this in their designs.

Age of Information – Significant changes in the design process started to occur. Ian McHarg's *Design with Nature* laid the groundwork for an ecological understanding of design. It stressed a process focused on reintegration with natural world (McHarg 1971). McHarg used an overlay of maps to study the geological, social, and natural aspects of a region. This information was used to influence the development of the site and raise the awareness of environmental patterns at the macroscale. His approach changed the design process to include the evaluation of the local and regional ecology before the building process began. The goal was to integrate any new planning and design projects into the existing ecological and social systems (The Cultural Landscape Foundation 2018).

The Age of Information featured a greater sense of empathy for the end user and an understanding of the need for holistic analysis. In architecture and the emerging profession of interior design, the programming process became more important. Information regarding the specific use, spaces required, and their interaction was researched and analyzed prior to formal design process.

Age of Integration – Design processes evolved dramatically with the level and amount of research into the functional aspects of a building, the occupants, the performative effects of energy sources, and the materials used. Emotional, psychological, and experiential aspects are looked at in a deeper way with growing acceptance that these unseen aspects of design can greatly affect the people using the built environment. The profession of Environmental Design has advanced as a discipline which combines science, statistical data, and design to analyze the best practices for influencing positive outcomes in occupant well-being. The Academy of Neuroscience for Architecture was established to explore the connection between the built environment and perception and response.

All of design, but especially sustainable design efforts are focusing on systems, feedback loops, and integration. The shift in design focus shows an understanding of the interconnectedness of the world and a desire to holistically shape design for the betterment of human well-being. Design professionals are becoming better collaborators seeking holistic, systematic functioning by identifying ecological patterns and human experience.

Additional resources

Thinking in the Powers of Ten: Charles and Ray Eames, www.eamesoffice.com/the-work/powers-of-ten/

9.2 Integrative design process and nature

Integration with nature in design requires a different process. Our design decisions are what determines whether the built environment is harming the natural world and our social structures or whether we are regenerating or restoring them. When we look at the design from an integrative view, we realize that nature is as much a stakeholder as the end user or the community in a design project. We need insight and creativity to allow the built environment and human intervention to be a restoration force.

9.3 Integrative design process and motivations

The tension between the motivations of self-interest and empathy plays out in the design process. The traditional model of design seeks to "shed risk," a clearly self-interested approach. Integrative design is based on shared or reduced risk because multiple professions are evaluating the project from the beginning.

Human motivations can be conflicting and in some cases result in the destruction of ecosystems. Traditional design processes have also had a tendency to be self-interested or have a narrow focus. Often other discipline's needs or requirements were not given the same level of consideration.

Design process can also be a part of the solution, if it is more integrative and collaborative and its goal is to find holistic, regenerative, and inclusive solutions. Altruism and empathy are human motivations that can motivate the collaborative process. Empathy allows us to put ourselves in the position of the people we are designing for. Altruism is what motivates us to find the "best" possible solution(s) that holistically meets the needs of all. When this is our basis for design, collaboration and integration are the natural choices. The built environment is a complicated, interrelated system, and multiple disciplines are required to achieve holistically supportive solutions that address all its facets.

Building a team of like-minded individuals and firms that have already adopted the collaborative model of working can be the easiest way to achieve an integrated design model. The important thing with any team is the ability to be open minded and work towards shared goals.

9.4 Integrative design process and sustainability values

Integrative design is essential for realizing the four Ps of sustainability: People, Profit, Planet, and Place. All four goals are important individually, but need integrative thinking and design processes to achieve them as a whole. With integrative design, stakeholders and professionals from all four values work towards a common solution that addresses all values from the beginning of the design process. This process may seem cumbersome at the beginning stages, but in the long run time is saved by not having to redesign each time new information or professionals evaluate from their perspective.

The Brundtland commission used a collaborative, multidisciplinary process to write *Our Common Future*, and these same goals were used to establish the United Nations Sustainable Development Goals. Having goals that address social equity and empathy included in sustainable design directives may be responsible for the interest and desire for "public" architecture and socially responsible design.

The ability to appreciate and even participate in the integrative process requires being in the upper levels of Maslow's hierarchy of needs because it is based on collaboration, shared values, diversity, and partnerships.

9.5 Integrative design process and integral sustainable design

In Chapter 5, the meta-framework of Integral Sustainable Design was used to describe a dynamic, holistic, and multidimensional framework. The four perspectives used to describe an *intentional* design process include a focus on performance, systems, experience, and culture – the four quadrants. The integrative design process (IDP) is perfectly aligned with the meta-framework of Integral Theory because its underlying goal is to equally value input from all professions and stakeholders. Because each of the project participants comes to the design process from their favorite perspective (quadrant), it is vital that a facilitator is in charge, who is well versed in group dynamics and does not have an emotional tie to any particular solution. They can then be the objective organizer, observer, and recorder of the process that allows all views to have equal input and can provide insight to underrepresented aspects of the project. Integral Theory can guide the team to set ambitious goals in all four quadrants, allowing a more comprehensive holistic evaluation process. The process creates solutions that are more inclusive, holistically connecting not only environmental but also social, experiential, and financial goals.

9.6 Integrative design process and bioinspired design

The core principles of bioinspired design are based on the integration of multiple disciplines, professions, and sciences, establishing a new paradigm for design. Biomimicry, the study of nature and natural forms to find solutions to complex human problems, relies heavily on the collaboration between the design and science professions. Without an organized process and open-minded collaboration, finding synergies and applications will be difficult. Biophilia, human's inherent longing and need for nature, has advanced as a viable and important design strategy, and it requires the collaboration between scientists, environmentalists, and neuroscience.

9.7 Integrated design process and resilient design

The design process for resilience requires a long view of the possibilities for the business, building, district, urban, and global scales. In order to have such a broad view, it is essential to have a large base of professions, business, planners, and the client along with the employees and community involved in the analysis process. IDP is a model to spur effective collaboration. Scenario planning, a process used to evaluate possible future directions for the business and building, is a part of the integrated design process because it involves holistically evaluating the project and its growth, business, and design from multiple perspectives. Scenario planning, resilience, and adaption are discussed in greater detail in Chapter 7.

9.8 Integrated design process and health + well-being

Designing for inherent health and well-being in all building types, functions, and scales requires input from multiple and varying professions. Traditional linear design models are beginning to change in favor of a more collaborative process. This move towards greater levels of collaboration allows a wider set of professionals and stakeholders to be involved in the initial design process. The fields of evidence-based design and environmental psychology have made great contributions by providing research and evaluating strategies that increase the health, productivity, happiness, and well-being of occupants. Rating systems like Living Building Challenge (LBC) and the Well Building Standard also focus on the experiential and health aspects of the occupants.

9.9 Integrative design process

IDP is a collaborative design process involving all stakeholders at the very beginning of the design process. Design directives, goals, and areas of concern are expressed by all parties, enabling strategies to be explored and vetted through a multi-lens perspective and steered towards long-term, sustainable design. IDP is based on the premise that building projects are more successful and more holistic, and innovative solutions are reached when we work collaboratively.

When used at the beginning of the process, IDP allows for dynamic value engineering and sustainable design strategies long before final decisions are made, thereby leading to greater efficiencies and less frustration in the process. The integrated design process can take many different forms that work at different scales. Some of the common styles are Integrated Project Delivery (IPD), the American Institute of Architecture's (AIA) process, co-creative design, charrette, or IDP; Leadership in Energy and Environmental Design's (LEED) model are several of the most popular and visible types.

Figure 9.9a (next page) shows an Integral Sustainable Design analysis of the various aspects of IDP. It underscores the complexity and challenge of the integrated design process.

IDP is known as an Asset Competency Model, which means that it looks to set up win-win relationships that involve and focus on holistic sustainable design solutions. The process looks to set up an outcome-based design process with transparency. Some of the benefits of employing integrative design practices are increased community buy-in, visionary and innovative solutions, reduced costs of long-term projects, reduced change orders during construction, focus of professionals on a solution instead of their own requirements or personal goals.

The core principles of the process are best summed by the phrase coined by Bill Reid with the addition of the term "equity" added to the end:

Everyone engaging everything early with equity

Contained in that simple quote are some key ideals:

Everyone refers to an inclusive design team and a comprehensive group of stakeholders. The stakeholders include neighbors, code officials, end user, client, all construction and engineering professions, maintenance staff, and anyone else touched by the project.
Engage means reflecting a motivation to reach out and connect with all the disciplines and stakeholders to discover new and innovative solutions that meet the needs of the project. So often design professions work in isolation within their discipline.

Subjective | Objective

	Subjective	Objective
Individual	*Experience* - Individuals change their mindset toward openness and acceptance - Experience buy in - agency - **Feeling of being heard and having influence** - View Nature as a stakeholder	*Performance* - **Maximize benefit - minimum time investment** - High upfront financial investment for improved project flow later in the process
Collective	*Culture* - Create a culture of collaboration - Shared vision through Guiding Principles - Equity in design process - Respect for nature - Respect for all disciplines - **Inclusive engaged stakeholder engaged process** - Transdisciplinary Process	*Systems* - Streamlined non-linear processes - Cohesive teamwork processes - **Living systems thinking** - Cyclical design - Thinking holistically

Figure 9.9a The IDP within the Integral Sustainable Design framework.
Source: Created and drawn by the authors.

Everything means holistically analyzing and including all aspects of the project at all scales. System thinking is a very useful way to approach this area.

Early means considering all aspects of the project from the beginning so that all perspectives can be considered. Understanding the maintenance issues can be really helpful when initial decisions on finishes, configurations, or space allocations are being discussed.

Equity means establishing a climate of respect, value, and giving a voice to everyone from the beginning. Input needs to be considered valid regardless of the person's profession, social standing, gender, or race. All views, ideas, and solutions are to be considered.

The reason it works so well to include all stakeholders, even those outside the associated professions, is that if we live within the built environment or use products, we can be considered designers. Whatever our chosen profession, or not, we all possess valid information on how products, services, and the built environment affect us and the environment. By allowing nonprofessionals in the process, valuable information is gained, precisely because these people aren't biased or trained to look at a special aspect of a project. Herbert Simon, the author of *The Sciences of the Artificial*, expressed it this way:

> Everyone designs who devise courses of action aimed at changing existing situations into preferred ones.
>
> (Simon 1996)

This is not always an easy process and can cause conflict in the beginning or with people not used to working in this way. The facilitator of the process helps the biased to see broader, the repressed to feel empowered, and to help everyone transcend traditional linear ways of working. This is best done with clearly established roles, responsibilities, outcomes, procedures, and "rules of engagement" set up at the beginning. To summarize, participants need to be willing to let go of control, be supportive of others and their views, support a culture of trust and mutual respect, and trust the process.

Predesign process

The predesign process includes many of the traditional aspects of research and analysis. Because ecology is a stakeholder in the process, in-depth research is completed in the support of discovering the best possible ecological design. Because project stakeholders are engaged early in the process, it is essential to show that their needs have been researched prior to the start as well. The predesign process includes the following:

- Guiding principles,
- Case studies and benchmarks,
- Environmental goal setting and rating system selection,
- Programming,
- Site inventory and analysis,
- Budget setting and analysis.

Guiding principles or touch stones

Guiding principles are overarching aspirational goals that are designed to push the team towards higher ideals. Guiding principles may be difficult to fully accomplish, but they are there to remind the team why the project is happening. The "why" of a project is critical because without a strong aspirational push for higher levels of sustainability, the team can quickly resort to taking the path of least resistance because of financial or time constraints. Guiding principles can be already established by the client, or they may need to be discovered and established. They are critical because they describe the shared values and vision for the project in simple, easy-to-use terms.

Here are some examples of guiding principles from a recent Living Building Challenge Project designed by Re:Vision Architecture for the Lancaster Conservancy in Pennsylvania:

Examples of Guiding Principles
- Create the healthiest environment imaginable at the urban forest center
- Instill an environmental ethic through safe and fun interactive experiences with nature
- Be the best example of integrated holistic sustainability through land conservation, restoration, preservation, and green building (Living Building Challenge Project)
- Educate and inspire about nature, stewardship, land protection, and sustainable design through the building, site, program, and process
- Connect the community and the Conservancy
- Infuse a sense of beauty into every decision

These should become a part of the design brief, allowing all participants to understand the underlying and aspirational goals of the project. The design brief is an important cultural document that sets the stage for an effective project because it answers most of the factual, aesthetic, experiential, cultural, and aspiration goals of the client.

Benchmarks and case studies

Without specific project goals, it is difficult to achieve ambitious sustainable goals, and the tendency of human nature is to take the path of least resistance. Goals with accountability push the team to work harder, be creative, and innovate. Many times, rating systems are used as a preset collection of specific project goals that have been researched and tested, and for the most part are holistic and ambitious. Rating systems have built-in benchmarks with their scores or levels or number of petals. Scores, though seemingly superficial, do serve as a shorthand as metrics for measuring and inspiring success.

The client is very familiar with how they need to have their own business or project operate and will provide specific information. To help get a better understanding of the project type and any special features or goals that are required, additional research will be required. This additional research allows for valuable insight, an intimate knowledge of the project type, and an understanding of the best practices currently in use as well as things to avoid. In a way this is collaborating with the sustainable design industry in general. By analyzing the projects of similar type, or in similar geographic locations, or that have used similar approaches, we can learn so much that will improve our designs. Case studies and the analysis process are covered in more detail in Chapter 8.

Environmental goal setting – ratings and metrics

Sustainable design projects are very complex, and it can be difficult to determine and understand all of the aspects needed to reach a holistic solution. It is also important to have a direct measure of success. Rating systems are an important tool in achieving this. They have been researched, vetted, and contain strategies and methods that are the result of collaboration between many different professions involved in sustainable design. Rating systems can also be a helpful tool in educating and creating buy-in with clients and stakeholders. They are concise and direct, and include metrics that act as a scorecard. Because many clients and stakeholders are more familiar with metrics or numbers, it becomes an easy way to translate the levels of sustainability.

There are criticisms of rating systems as being too restrictive and as just adding to the cost of a project, or being an overly complex administrative process. Specific goals or strategies within rating systems are often criticized. Projects can achieve true and holistic sustainability without rating systems, but rating systems help to codify our deep sense of empathy for the planet and its future. They are tools to help us organize our thoughts and make decisions that benefit not only the current project but also the future generations. Prior to the release of the LEED Green Building Rating System in 1999, there was little agreement on what a green building or sustainability really was (Kriss 2014).

Choosing the best rating system for your project is also something that should be decided collaboratively. The decision is based on many factors, some of which are project type, geographic location, sustainability goals, time frame, financial considerations, and professionals that should all be in alignment with whichever rating system you select. All rating systems have attributes that need to be fully understood and fit your projects goals. A rating system can be the following:

First Party Certified or **Third Party Certified**: If the organization providing the product or service offers the assurance, it is *first party certified;* if it is an expert unbiased organization that independently evaluates, it is considered *third-party certification.*

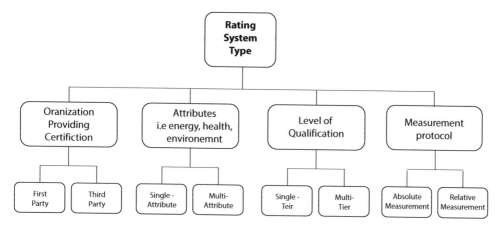

Figure 9.9b Organizations of rating systems.
Source: Created and drawn by the authors.

Single Attribute or **Multi-Attribute**: certification for only meeting one attribute, such as indoor air quality, or requirements to meet multiple attributes like net-zero energy, water, and waste as a minimum basis.

Single Tiered or **Multitiered**: when all requirements of a rating system need to be met to obtain certifications, it's a *single-tiered system*; when multiple levels of certification are obtained to meet the qualifications pertaining only to a specific level, it is a *multitiered system*.

Absolute or **Relative**: they can either require definite standards of compliance, like net-zero water, or have a relative measure, such as use 20% less water.

Figure 9.9b (above) shows all the different organizations and attributes of rating systems.

If everyone built to the high standards of rating systems, the net positive impact on the environment would be transformational. In addition, rating systems serve as an educational tool to move general knowledge closer to sustainable design and can influence zoning and code officials to change their standards towards higher performance. Table 9.9a (next page) shows a partial list of rating systems, standards, and reporting structures by scale.

Programming

The program or project requirements are achieved through collaboration with the end user and the client. Case study research can also help add insight to best practices, and adjacencies to increase efficiencies and user experience. Programming typically includes the type and number of spaces, size or square footage requirements for each space; furniture, equipment, and technology requirements; adjacencies, or what each space should be near; mechanical, air exchange, or environmental requirements; daylight, view requirements; and relation to public access. These are just a few of the considerations to review with the client in the programming stage. It is imperative these requirements are defined and articulated prior to the collaborative design or any design drawings or ideas for the project.

Table 9.9a Rating systems, standards, and reporting structure by scale (partial list)

Global scale	*Regional + urban scale*	*District + site scale*	*Building + interiors scale*	*Human scale products + materials*
Sustainable development goals by the United Nations	LEED for cities CBECS (Commercial Buildings Energy Consumption Survey)	Ecodistricts LEED ND 2030 Districts Envision Sustainable sites initiative	Multiple LEED rating systems Building Research Establishment Environmental Assessment Method Passive House Living Building Challenge Architecture 2030 Energy star Green Rating for Integrated Habitat Assessment Well Building Standard Fit well	Cradle to cradle Bifma Greenguard Declare

Source: Created and drawn by the authors.

Research, inventory, and analysis

In sustainable design, site inventory and analysis have become an important part of the initial design process because it builds a strong foundation for integration of natural systems into a project.

Site inventory is the process of collecting information about a site without making any conclusions or forming opinions. The information collected typically includes the climate data, sun paths, geological and landscape features, topography, plant and animal inventory, surrounding structures, transportation services, historic or cultural information, views, proximity to amenities, ecosystem information, and anything else that affects the site.

Site analysis is the process of interpreting and analyzing the information to make important decisions about the design project. It includes looking at the site from multiple scales, from the integral perspectives, and over time. Opportunities and challenges are identified in this process, and sometimes different building location and site design options are explored.

Inclusive design teams

There are many terms used for collaborative design team structures, and all of them actually have slightly different meanings. Clearly establishing the flavor of the team dynamics at the beginning is critical in order to choose the right style and manage expectations of the team. Table 9.9b (next page) describes each type of design team structure with their strengths and weaknesses.

The design charrettes

The most likely process embedded within IDP is the use of a design charrette. The term "Charrette" originated in the 1800s when design students at the École nationale supérieure

Table 9.9b Overview of the different collaborative design team structure

Type of team	Advantages	Disadvantages
Traditional	Expected and well-understood process	Silo-based thinking, competitive, risk averse, stressful
Interdisciplinary	Expected and well-understood relationships between disciplines – respect – are earned	Often territorial or one discipline dominates the others
Multidisciplinary	Spirit of cooperation and willingness to listen to perspective of others	All team members must buy into this approach or the process degenerates into a competitive model
Transdisciplinary	Holistic approach where disciplines are transcended to allow for creativity across scale and perspective. Empathy is both needed and augments by having different team members work within other team member's areas of expertise. New creative ideas can emerge from unlikely partnerships.	Difficult model to achieve. Requires a leap in mind-set to the Age of Integration. If one team member does not buy in the whole process easily falls apart.

Source: Created and drawn by the authors.

des Beaux-Arts in Paris worked in off-campus studio apartments. It was the process that a horse-drawn cart (charrette in French), would be sent to collect student projects and drop them off at the school to be graded behind closed doors. As was sometimes the case, the students who had not finished their work would literally jump on the cart and work feverishly to finish the project by the time the cart reached the school. Students found they did some of their best work "En Charrette" or "on the cart." The takeaway from this is the benefit of short working sessions. Today, the idea that design can happen in quick ideation sessions forms one of the cornerstones of the design charrette.

Design is traditionally seen as a slow, contemplative, linear process by a single person that works through many alternatives and iterations and leads to the final result. The modern day charrette seeks to temporarily remove professions from their silos, engage all stakeholders, present all project requirements at the beginning, and capitalize off the intuitive design process that is inherent with speed. The characteristics of a charrette are described as follows:

Speed – Generate as many ideas as quickly as possible, without judgment. Ideas will be vetted later by the group and then the entire charrette team.

Inclusivity – Everyone is included and valued equally: professionals, users, community member and any other stakeholders. Teams should be multidisciplinary. Allow contradictory views, plans, and designs. It's only paper, and these can many times spur innovative solutions and viewpoints.

Alignment – Seek to focus discussions and designs around the problem to be solved or unity of purpose.

Early Engagement – Generates feedback and inclusion early in the process when changes are still very easy and less costly. Creates ownership and equity by allowing all voices to be heard.

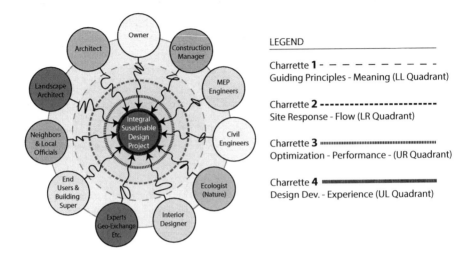

Figure 9.9c The Charrette model, adapted.
Source: R. Fleming (2013).

The result is usually a spirit of cooperation, alignment, and excitement as all stakeholders become aware of shared project goals despite differing approaches. New ideas and approaches are generated that can be explored further either in additional charrettes, or by the design team and client separately as the project moves forward.

Multidisciplinary teams

The second aspect of the design charrette is a multidisciplinary team made up of professionals involved in the project, along with stakeholders. Figure 9.9c (above) gives examples of the people that should be involved and the areas they represent. The main objective is to capitalize on the varied viewpoints and priorities that are represented. Traditionally, this caused conflict, but with the integrated design process and a new worldview of integration, it is celebrated because of the improvements that result to the project and personally.

Feedback loops

How we talk about a project says a lot about our values and our goals. With Integral Sustainable Design, the goal is to have a project that holistically integrates the performance, cultural, and ecosystem needs alongside the experiential and aesthetic needs. These goals not only necessitate multidisciplinary design teams but also accept their feedback as a way to improve the end result.

To achieve the best results and to help focus the thought process, the evaluation portion of the charrette exercise is divided into several steps which are directed by the facilitator.

Answer Clarifying Questions – Only questions that answer missed features or expand on previous explanations are taken. Others are either tabled for later in the process when the fit, or questions that debate decisions are required to be reframed and voiced at the appropriate stage.

Optimisms – Elements of a solution that have merit or should be pursued further are referred to as optimisms… Because we want to encourage as many varied ideas in the beginning as possible, this helps to create an atmosphere of acceptance that makes it possible. By not using the phrases "good" and "bad" we are again shifting focus from our personal bias to the problem.

Cautions – Again we aren't using the words, "wrong" and "bad," because of the perception, but more importantly because we don't want to limit our solutions. Cautions are things that need to be explored further or refined. There could still be real value in a strategy that presently conflicts with other goals. For example, a roof shape that optimizes solar collection, but will cause ice to form at the front entrance. Viewing issues as "cautions" allow us to expand our thought process and creativity.

Nest Steps – These are critical for moving forward in an organized and holistic way towards a solution.

All the Optimisms, Cautions, and Next Steps are written on large tablets in the front of the room so all can see. This reinforces their points, provides a record for future reference, and makes people realize their input really counts and is valued. This may seem like an overly regulated process that would stifle creativity. In reality, it keeps the discussion from degenerating into a discourse on personal design choices. Traditionally, feedback between disciplines typically focused on critically knocking down or eliminating solutions based on personal or professional bias.

Integrated Project Delivery

> Integrated Project Delivery (IPD) is a project delivery approach that integrates people, systems, business structures and practices into a process that collaboratively harnesses the talents and insights of all participants to optimize project results, increase value to the owner, reduce waste, and maximize efficiency through all phases of design, fabrication, and construction.
>
> (AIA 2007)

This system was developed by the American Institute of Architects and was intended to encourage the transition of traditional practice to a more integrated model – especially suited to achieving ambitious sustainable design projects.

In the MacLeamy curve shown on the next page, the two curves represent traditional design processes in comparison with the IPD process. The large lines show the most impact to project costs are changes made early in the design process and that costs related to change increase the closer to construction or completion they are. Because of the research and collaboration at the front end of the IDP, changes are made earlier, usually during schematic or design development phases. Changes at these stages are mostly made on paper and are far less costly. In the traditional process, changes are often made later during the construction document phase or after the project is under construction. This makes it much more difficult and expensive to make changes, because so much work has already been completed and other interrelated systems have already been designed (Figure 9.9d [next page]).

The AIA's version of IPD is more formalized than the charrette model we examined above, but still promotes all of the attributes of the integrated design process we have already covered. They have published a guide on Integrative Project Delivery and encourage all members to follow the collaborative and administrative procedures outlined. To date,

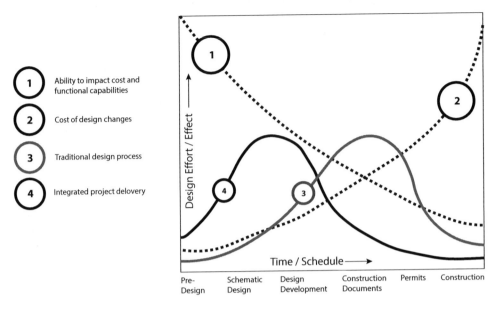

Figure 9.9d The MacLeamy curve.

Source: First appeared in *Paulson, Boyd C. 1976. "Designing to Reduce Construction Costs."* Journal of the Construction Division *102 (4): 587–592.*

the movement to adopt IPD approach in mainstream practice has been slow. The level of trust, collaboration, and shared risk are difficult for many to accept who are still operating out of an older, industrial worldview.

IDP and the LEED rating systems

IDP is part of the United States Green Building Council LEED rating system and is worth one point towards certification. It is a more formal and rigorous process that follows all principles mentioned above.

Participation is encouraged from the earliest stages of developing goals and performance standards. LEED breaks the process down into three stages:

Discovery – predesign – establishing goals, priorities, expands traditional design directives.

Design + Construction – Schematic design includes extensive exploration of building and site systems.

Occupancy, Operations + Performance Feedback – Post-occupancy evaluation (POE) measures performance in all areas, sets up feedback loops, and takes corrective actions to adjust actual performance to design goal projections.

The main object is to identify synergies and gain a better understanding of all the systems in the building, how they interact with the site and the synergies between credits, and how to maximize points. There is an established structure to deal with the flow of information and collaboration between the client, designer, engineer, and construction professionals.

9.10 Integrative design process and global scale sustainable design

The global scale is so large, and it is difficult to imagine design interventions at that scale. Design at this scale takes the form of policy making. The United Nations (UN) is creating global scale solutions by bringing large international groups together to develop comprehensive solutions to global problems. Policy is typically used to guide nations and international companies in the direction of sustainability.

The United Nations Sustainable Design Goals (SDG), which will be covered in more detail in Chapter 10, are an example of a collective or integrative process. The goals represent a holistic, comprehensive, and ambitious framework to organize and guide sustainable design efforts around the world. They were developed over a three-year period with literally hundreds of collaborators and hundreds more who had input into the process. The deliverable was not a building or product, but a document that will guide the efforts of hundreds of countries, which will then filter down to the other scales.

The SDG follow a similar pattern to most sustainable design methodologies covered in this book. However, the terminology is slightly different. In this book, we have used the term guiding principles to denote larger overarching goals that are not specific. In the case of the SDG's, the term "goal" is used in that sense as a larger overall objective. Underneath are more specific targets that direct action towards a something specific. Targets are also goals but more directed at a subset of the larger goal. Finally, indicators are used to measure how well the targets are achieved. This is a critical aspect of sustainable design – which relies on evidence as a feedback loop to determine the effectiveness of an approach over time.

Additional resources

Sustainable Development Goals, https://sustainabledevelopment.un.org/sdgs

9.11 Integrative design process and urban scale sustainable design

The development and design process at the urban, district, and site scales are highly regulated by policies established at a local-, state-, and countrywide basis. Many cities have sustainability departments that also help to influence the policy decisions. There are typically clear goals, requirements, and established processes for new or renovation projects at these scales. IDP is still a valid and viable approach. We can see from the iterative, transparent, and integrative process used by the United Nations to establish the SDG's that this process can work at any scale. The larger the scale of design intervention, the more important it is to have multiple disciplines and stakeholder views involved.

Geodesign is a new framework for sustainable urban planning which uses geographic information system (GIS) computer models to study the existing context for a project and to simulate the impacts of different design solutions. The principles of geodesign have their origins in the work of Ian McHarg's ecological mapping in Design with Nature and now are combined with current technology to analyze the resulting data digitally. GIS integrates many types of data and organizes it into maps and 3D imaging. GIS is a useful tool for understanding needs, measuring impacts, creating maps, analyzing performance, and engaging stakeholders (ESRI 2018). GIS can reveal deeper patterns, relationships, and outcomes that could impact the surrounding environment – both built and natural. Some cities use GIS as part of a smart cities approach to control energy usage, traffic, and emergency services.

In real time, the geodesigner can work with architects, landscape designers, and engineers to integrate their proposals into existing sites and model the results before they become reality to find any changes while still in the planning stage. This is IDP in practice because the existing city and the proposed growth can be quickly modeled in 3D to produce smart simulations and visualize and measure sustainable concepts. With new technologies like GIS and Building Information Modeling (BIM), it is easier to visualize information at multiple scales and quickly see the benefits or disadvantages of using one type of strategy over another. Not only does this increase the likelihood of meeting performance metrics, but also can help avoid painful value engineering cuts through the design process. Doing all of this analysis during the first phases of design means they can be included in the guiding principles and are more likely to be achieved within budget.

Additional resources

Geodesign Overview, www.esri.com/products/arcgis-capabilities/geodesign/overview#intro_
   panel
How Geodesign is Being Taught: www.philau.edu/msgeodesign/
Urban sustainability through strategic planning: A case of metropolitan planning in Khulna city, Bangladesh,
   Ashiq Ur Rahman, www.sciencedirect.com/science/article/pii/S2226585616300036
What is GIS, ESRI www.esri.com/en-us/what-is-gis/overview

9.12 Integrative design process and district, site, and building scale

As we mentioned above the development and design process at the district and site scales are regulated. At this scale, it's mostly by local zoning and building codes established by local and state governments. Many cities and counties also have sustainability departments who help to influence requirements. While there is still an established administration and inspection process, there is more freedom usually in the design process as long as the other requirements are met. IDP may have fewer stakeholders or professionals involved at this level, but because work is being done on a smaller more personal level, it is critical to get community input and buy-in.

Integrative design at the district, site, and building scale is much more "personal," meaning the relationship between the built environment, the ecosystem, and its effect on the behavior, culture, and perceptions of the residents is very tangible. Clients can be municipalities, community development corporations, city planners, and/or private developers. All would follow a similar process and have a great potential to make a difference to the communities with sustainable development.

The integrative process as the other scales follows the same path of early research, site evaluation, and stakeholder involvement. This usually occurs in a public design process after the research and evaluation is done with the client. Depending on the project and situation, the community can also participate in establishing the guiding principles. Once the community is involved in the initial planning and design, the rest of the design is collaboratively done with the professionals, end users, and client.

The current technological climate is a real advantage with IDP. The use of BIM is becoming more and more prevalent. BIM allows access to changes and documents in real time. It allows for each discipline to have immediate access to all the drawings in a networked fashion. BIM models include an extensive amount of information, metrics, and tools, so all the information required for the project is available in the model. When the lines representing say, kitchen equipment are highlighted, you are immediately able to see all the technical specifications for ordering, electric, plumbing, and any mechanical requirements. This is possible for any feature in the drawing. In addition to construction and specification information, there are tools and metrics that enable analysis with very few, if any extra steps for energy modeling, sun path and shading studies, interior daylighting levels, water use, embodied energy of materials, or R-values of building assemblies.

Life safety and legal liabilities with a constantly changing electronic "document" are still being worked out. But as a tool to facilitate the IDP, BIM is very valuable. Ultimately, as the new worldview continues to take hold and design teams become more collaborative and more integrative, the issues with risk management and control of information will be solved.

9.13 Integrative design process and human scale sustainable design

When the contractor, client, and end user are all involved in IDP even finish materials or exposed elements are addressed at the very beginning of a project. Sustainable options can be considered early. Items with longer lead times can be ordered, so the project can remain on schedule.

The contractor is also able to recommend any structural or substrate concerns or requirements that would affect the rest of the project. For example, should the building be made of timber, steel, or concrete? Early discussions about which systems can be procured locally and any significant opportunities to acquire salvaged materials or reclaimed lumber are held at this point.

Because design charrettes are dynamic nonlinear processes, it is quite common to see a breakout group work on materials while others are still working on the design itself. Because the performance goals are well understood by all team members, including the builder, discussions about efficient wall assemblies and the impact of high-quality construction can take place in a friendly cooperative environment.

9.14 Conclusion

At the core of IDP is a fundamental shift in mind-set to let go of control, to give agency to stakeholders, and to serve the needs of nature at all times. In Chapter 4, Maslow's hierarchy of needs was explored. IDP relies upon individuals moving towards self-actualization and transcendence in order to have the most success. It is the facilitators' responsibility to engage all the participants at their level of development without judgment. This requires a great level of self-awareness on the part of the facilitator.

Finally, it is critical to employ all the strategies and processes of integrated design, just not the terminology. It is up to the sustainable designer to hold the team accountable for its process and to use transparency at all times. Ultimately, the IDP is necessary because the traditional process is part of the problem. If we keep using the same strategies, we can't expect to have different results. Unsustainable design will continue to occur until we radically transform out processes and our mind-set. Until then, we will remain stuck in a repeating loop of buildings that generate too much pollution, use too much energy, and miss the opportunity to create projects that are healthy for the building user and for the society at large.

References

AIA National and AIA California Council, 2007. *Integrated Project Delivery: A Guide.* The American Institute of Architects. https://info.aia.org/SiteObjects/files/IPD_Guide_2007.pdf edn.

Eames, C. and Eames, R., 1977. *Thinking in the Power of Tens.* 1st edn. Chicago: Office of Charles and Ray Eames.

ESRI, 2018-last update, What is GIS [Homepage of ESRI], [Online]. Available: www.esri.com/en-us/what-is-gis/overview [July 23, 2018].

Kriss, J., August 6, 2014-last update, What is Green Building? [Homepage of USGBC], [Online]. Available: www.usgbc.org/articles/what-green-building [July 25, 2018].

McHarg, I., 1971. *Designing with Nature.* 1st edn. Garden City, NY: Natural History Press.

Simon, H.A., 1996. *The Sciences of the Artificial.* 3rd edn. Cambridge, MA: MIT Press.

The Cultural Landscape Foundation, 2018-last update, The Cultural Landscape Foundation [Homepage of The Cultural Landscape Foundation], [Online]. Available: https://tclf.org/pioneer/ian-mcharg [July 25, 2018].

United Nations, United Nations Sustainable Development Goals [Homepage of United Nations], 2016. [Online]. Available: https://sustainabledevelopment.un.org/sdgs [July 21, 2018].

10 Global scale sustainable design

10.0 Introduction

As a person working in the field of design, construction, manufacturing, etc., reading about sustainable design at the global scale may seem irrelevant or not applicable. The opposite is true. If we live on the planet earth, what happens at the global scale directly affects urban development, design, construction, manufacturing, planning, etc. Also, all the choices that we make are interconnected to the global scale. In Chapter 5, we learned about systems thinking and the reciprocal interdependence of all our actions at every scale. This chapter provides a foundation for large-scale sustainable design.

10.1 Global sustainability and space + time

The sustainable mindset has always been present in the human species. It took many forms throughout history, threads of empathetic actions woven into the fabric of humanity's relatively short existence on the planet. History books are replete with examples of hero after hero seeking to fight against injustice, or to "protect" nature from those who would do harm out of greedy intent. The long battle between self-interest and empathy has come to a new chapter, an inflection point in the story when the heroes and villain's discover a new pathway – one that offers reconciliation towards a common purpose – survival of the species.

Brundtland commission

The origins of global sustainability can be traced back to the birth of the United Nations (UN) in 1945. The UN was originally founded to deal with war and conflict and to promote peace in reaction to the atrocities of the Second World War. Over time, they began to take on other global issues. In 1984, The UN launched the World Commission on Environment and Development (WCED). Also called the Brundtland Commission. It was led by Gro Harlem Brundtland, the former Prime Minister of Norway because of her strong background in the sciences and public health. The goal of the organization was to create a shared vision, generate more awareness of sustainability, and develop implementation plans. *Our Common Future* was published in 1987. Chapters 3 and 4 cover sustainability motivations and values in detail.

This work led to the Earth Summit in Rio de Janeiro, Brazil in 1992. The conference led to the establishment of the Kyoto Protocol, the first international treaty regarding CO_2 emissions. The treaty states that there is consensus that greenhouse gases do in fact create global warming and that each country agreed to limit emissions of those gases. This protocol was signed by every country except the U.S. The signing of this protocol indicates an ability to use empathy across time and space to address a significant common problem.

Intergovernmental panel on climate change

In 1988, the Intergovernmental Panel on Climate Change (IPCC) was formed. Its main goals are to provide the world with an objective, scientific view of climate change, including its impacts. The IPCC issues regular reports that document the risk of human-induced climate change and its impact. Also, the committee reports on options for adaptation and mitigation strategies as a response to climate change.

The Earth Summit in 2002 in Johannesburg (Rio+10) further integrated sustainability into the culture of the UN. Sustainable development became an overarching goal for institutions at all levels and a further goal to integrate the goals into UN agencies and programs. Major outcomes of that conference include the Johannesburg Declaration and almost 300 international partnership initiatives meant to help achieve the Millennium Development Goals.

In 2012, the Rio+20 conference was held to further reconcile environmental and economic goals for the global community. These conferences reflected a deepening commitment to sustainable development as a guiding principle for the organization. With that comes, an increase in global awareness of sustainability and a growing adoption of sustainability principles by local, state, and federal governments all over the world.

10.2 Global scale sustainable design and environment literacy

The Paris climate agreement

The Paris Climate Agreement, signed in 2015, by almost every country in the world, is designed to keep the increase in global average temperatures to less than 2°C (3.6F), above pre-industrial levels, with a more ambitious goal of limiting temperature the rise to 1.5°C. On the surface, that sounds like a meaningless distinction. After all, how could 2°C have that much affect? Consider the impact of a two-degree rise in a healthy human being 37°C (96.8F) to 39°C (102.4F). For a human, the slight rise in temperature is enough to think about going to the doctor. An addition of 1°C rise now requires hospitalization. The diagram below illustrates the impacts of small changes in temperature, eventually leading to death. The planet is not much different. Small increases in temperature mean catastrophic climate impacts that will make life on earth very challenging for humanity in the near future (Table 10.2a).

The point here is to help visualize that slight changes in temperature are actually very meaningful and dangerous. Even though the earth is a very different system than the human body, it too, is sensitive to changes in temperature. Today, the current level of global warming is almost 1.5 degrees above preindustrial levels, i.e., "normal levels." The effects of a 1.5 degree rise in temperature are already evident and well documented in Chapter 2. The resulting climate change effects in the form of drought, forest fires, and sea level rise are already leading to loss of life and a reduced quality of life for many people, especially in coastal areas. The Paris Agreement is aimed at managing the rise in temperature to protect the earth from reaching even higher temperatures, which would change the planet in ways that would make human existence very difficult. The earth itself will continue to exist without us, so the fight against global warming is not about "saving the planet," but more about saving our own species.

Table 10.2a The impact of global temperatures as a metaphor from human temperatures

	Normal	*Concern*	*Danger*	*Life threatening situation*	*Death*
Humanity	37°C (98.6F)	39°C (102.2F)	40°C (104F)	40.5°C (105F)	41.5°C (107F)
Human actions	Healthy	Doctor's visit	Hospital stay	Emergency room visit	Death
Global impacts	Planetary climate conducive to the thriving of humanity	Climate change related climate events. Loss of life for some and lower quality of life for other	Significant climate change. Loss of life for some. Loss of property and lower quality of life for millions	Climatic tipping point is reached. Climate begins to spin out of control	The Earth's climate is no longer conducive to human life. Widespread suffering
Global average temperatures	Preindustrial temperatures (Holocene)	1.5°C Above preindustrial levels Paris Agreement Target	2°C Above preindustrial levels Paris Agreement Maximum	2.5°C	3°C

Source: www.scientificamerican.com/article/earth-flirts-with-a-1-5-degree-celsius-global-warming-threshold1/.

Despite the clear and present danger of global warming, and despite the signing of the Paris Agreement, the lack of progress towards addressing global warming is slow. At the time this book was written, The U.S. pulled out of this agreement citing a variety of reasons. Major corporations are making changes to their supply chains to reduce the emission of CO_2, but each year, there is a rise in overall CO_2 emissions and record temperatures are being reported each year.

In 2018, CO_2 levels reached another milestone at 410 ppm or parts per million (Kahn 2017). Like temperature, the discussion regarding CO_2 emissions faces the same difficulty. The increase from 250 ppm during the industrial revolution to 315 ppm in 1958 took approximately 100 years. Only 55 years later, we are at 410 ppm. Small changes in parts per million of CO_2 can have enormous consequences for the climate. 300 ppm is considered "normal" or beneficial to humans. 400 ppm is "bad" in that it creates global warming levels that are not conducive to optimal human life. 500 ppm is catastrophic, leading to extreme global warming and devastating climate change.

10.3 Global sustainable design and motivations

The indifference and apathy towards the world's environmental problems are an expression of basic human self-interest. Comfort, materialism, and socioeconomic mobility continue to be the primary driver of many human decisions. Oxfam is a global organization working to end the injustice of poverty. Their 2012 report on poverty offers the following:

Adding to the pressure created by the world's wealthiest consumers is a growing global 'middle class,' aspiring to emulate today's high-income lifestyles. By 2030, global demand for water is expected to rise by 30%, and demand for food and energy both by 50%.

(Raworth 2012)

A big part of the quest for comfort comes in the form of the desire for high-protein foods. The Brazilian, National Institute for Space Research has estimated that 65% of the cleared Amazonian land, or 45 million hectares has been utilized for cattle pastures (Yale University 2018). Clearing forests for cattle ranching triples the effect on global warming: Loss of carbon sequestration from deforestation; rise in temperature from loss of shade from millions of acres removed forests; and the new methane emissions from the cows, not to mention the loss in habitat and biodiversity. In Chapter 3, the tragedy of the commons was explored as a way to think about how humans often make decisions in their own best interest, but don't consider what happens when many people compete for the same resource and make the same decisions, in their best interest. The net effect is negative for everyone involved. In other words, as the earth's ecosystem becomes more and more compromised, the ability to continue to profit from the earth will diminish.

Despite the slow progress towards addressing climate change and the actions of self-interested countries, there is a vast movement of change in the world – a global movement towards sustainable development. The competing motivation of empathy and altruism is becoming stronger and driving many governments, corporations, and individuals to find and deliver real solutions. The Paris Agreement was one such example.

The precautionary principle approach

The precautionary principle was established in 1998 to create a new approach to environmental protection. The principle states, "when an activity raises threats of harm to the environment or human health, precautionary measures should be taken even if some cause and effect relationships are not fully established scientifically" (Silent Spring 2018).

The principle implies that there is a social responsibility to protect the public from exposure to harm, when scientific investigation is not conclusive or exhaustive in its assessment of risk.

> The principle critically shifts the burden of proof from the general public to the initiator of that public health or environmental risk. Instead of the public having to show they have been harmed, the initiator has to show that the activity, process, or chemical exposure is likely harmless.
>
> (Silent Spring Institute)

This use of social responsibility works in two ways:

1   We have a responsibility to **refrain from taking action** when scientific uncertainty exists.

   For example, the use of genetically modified foods; because we don't know for sure what the long-term effects to humans or plant species, we should wait to use this technology.

2   We have a responsibility to **take action** when faced with grave threats, even when scientific uncertainty exists.

   For example, climate change is an obvious threat that we face. Though, we may not have all the answers, we should take action, because the negative impacts of doing nothing are catastrophic.

The precautionary approach is used by many governments and by the UN. It was used in the Rio+20 Conference as a key rationale for action against climate change. Principle 15 of the Rio Declaration notes:

> In order to protect the environment, the precautionary approach shall be widely applied by States according to their capabilities. Where there are threats of serious or irreversible damage, lack of full scientific certainty shall not be used as a reason for postponing cost-effective measures to prevent environmental degradation.
>
> (United Nations 1999)

The approach has been adopted by the UN General Assembly and used in the development of the Montreal Protocol which addressed the growing ozone hole in the earth's atmosphere as a the result of spray aerosols. The world stopped using aerosols and the ozone hole did shrink, showing that global action to fight an environmental problem is feasible.

However, there are times when companies and governments may be threatened or they are resistant to change that benefits the environment. There are sectors that deny climate change, just like there were industries that denied cigarettes were responsible for health damage. In cases like this, "science" or lack of "scientific evidence" can be used to deflect blame and delay action.

This is why the precautionary principle is so valuable because it overrides confusion to justify action against a grave threat. The use of the approach in the establishment of sustainable development is underscored by a quote from Gro Harlem Brundtland who stated:

> In the face of an absolutely unprecedented emergency, society has no choice but to take dramatic action to avert a collapse of civilization. Either we will change our ways and build an entirely new kind of global society, or they will be changed for us.
>
> (Brundtland Commission 1987)

Sustainability, transcends space and time. Each individual is a microcosm of the entire world, and with that comes the opportunity and the responsibility to use the precautionary approach in your own life. Changing personal behavior to be more sustainable is one obvious application, but challenging the status quo is the second level. Choosing to incorporate sustainable design practices into every project in school or at work constitutes a "tiny revolution" in your life. When millions of other people also begin to change, a global revolution of thought and action comes into reality. And that is exactly what is happening in the world. As you will see in this chapter and in all the chapters to follow, change is happening everywhere, opening the door to a new worldview based on an agreed upon set of sustainability values. You just have to know how to see it. Chapter 15 will offer a roadmap to achieving this kind of change.

10.4 Global sustainable values

The work of the United Nations Brundtland Commission defined sustainable development in its publication *Our Common Future*. It was a seminal event in history and was addressed in detail in Chapter 4. Sustainable values can be tied to the different ideas about progress. "Progress" is typically used as a term to relate to "things getting better." "Better" is a relative term that is understood differently from different points of view.

Economic progress is a value that relates to the performance lens of sustainable design and it is usually seen as economic growth. New technologies, new housing developments, and new commercial developments – all geared towards increasing human comfort and profit reflect "progress." In this book, we've expanded and improved the concept of profit to the value of "prosperity," which has a longer-term and more holistic sensibility. The quest for economic progress, when pursued in concern with the values below can lead to a scenario where comfort, profit, and affluence can all continue as primary motivators of human behavior without destroying the environment.

Social progress is a value related to the culture lens where gaining equity and rights for underrepresented people is the goal. History is replete with examples of individuals and groups fighting for the rights of others. The term "progressive" is used to describe this form of progress and it includes gender equality, living wages, and inclusive and equitable organizations. Transparency and accountability are underlying traits of social progress. More on social equity is covered in Chapter 4.

Environmental progress is a value related to the systems lens of sustainable design. Without environmental progress, the values of prosperity and social equity can't exist. Remembering Maslow's Hierarchy of needs, our most basic forms of survival are completely reliant on a healthy planet that provides ecosystem services such as clean air and water, food, pollination, healthy soil, and temperatures within a habitable range for humans.

Experiential progress is a value related to the experience lens, not usually discussed in sustainability. The impact of design as an integral part of a sustainable future is often ignored because it's subjective. The role of beauty in sustainable design was covered in more detail in Chapter 5. Bio-inspired design addresses beauty indirectly as part of interacting with nature, in Chapter 6.

These forms of progress are often driven by the development of policies instead of physical design. This is a big shift in thinking for designers who tend to focus on the physical aspects of design. At first glance, policies and goals might seem like a very weak force, since they do not take physical form, and yet, they have the power to transform our world. Consider the impact of building codes (which are policies that become law). In the U.S., non-residential buildings require two means of egress, and multi-story buildings require an elevator and fire rated exits stairs. The policies themselves do not build structures, but they instill fundamental common values in built form. Policies themselves are like any technology. They can be used for evil or good, but in either case, they are a critical part of the sustainable design equation and therefore are part of an expanded view of "sustainable design processes."

10.5 Integral perspectives and global sustainable design

Integral sustainable design weaves all forms of progress into one comprehensive framework that has the potential to transform our current way of life. The values that underscore the different forms of progress lay the critical foundation to the formation of the emerging worldview of integration.

Global sustainable design is not drawn on paper, or visualized through 3D models on a computer screen. Instead, global sustainable design occurs mainly through ideas and words – in the creation of policy. Policy is not typically considered in design textbooks, and yet, it shapes design in very direct and important ways. The design of cars, for example, is heavily influenced by global and national policies that impact safety and fuel efficiency. Government mandated energy performance levels and environmental protection standards are another form of policy mandates with global consequences. Policy and codes shape the form and contents of products and the built environment.

In California, lawmakers are moving towards legislation that would require the use of photovoltaic panels on all new residential construction. Many countries recently announced the phasing out of combustion engines in new cars; in the Netherlands, by 2020, in Norway by 2025, and in India by 2030. All of these policies and many more are shaped by changes in global awareness regarding the state of the environment. The shift in public perception signals a worldview shift that is made more viable because of lower costs of emerging green technologies and the rise of renewable energy. The trend is only going to get stronger as governments, companies, and especially the UN support sustainable design strategies.

Global policy: the sustainable development goals

In 2015, The UN formally adopted the Sustainable Development Goals (SDGs) as a comprehensive way to attack the world's most pressing problems and emergent opportunities. The goals are part of a large UN initiative: Transforming Our World: the 2030 Agenda for Sustainable Development.

The goals are an evolution of the Millennium Development Goals, which sought to organize efforts by all states to pursue the sustainable development. The SDGs are organized into goals, targets, and indicators, and follow a similar pattern to most sustainable design methodologies covered in this book. However, the terminology is slightly different. The SDGs use the term "goal" as a larger overall objective. They use the term "shared values" which gets to the essence of the goal. This is the largest scale sustainable design initiative ever. The goals are explained in the *Getting Started Guide* (2015) as follows:

> Provide a shared narrative of sustainable development and help guide the public's understanding of complex challenges. The SDGs will raise awareness and educate governments, businesses, civil society leaders, academics, and ordinary citizens about the complex issues that must be addressed. Children everywhere should learn the SDGs as shorthand for sustainable development.

The goals, as stated above, provide a "shorthand" or easy way to identify the different aspects of sustainable development. This short hand is crucial to influence thoughts of people not deeply involved in sustainable design, and as an easy reference for directing priorities, intellectual discussions, and actions. The UN SDGs are gaining momentum and are already widely adopted by national, state, and local governments around the world. As stated throughout this book, common sets of principles, goals, metrics, and even procedures helps to move adoption of sustainability faster and deeper (Table 10.5a).

Table 10.5a Organization of the sustainable development goals

Topic	Goal	Targets	Indicators
Major topic of concern and high level aspiration	Measurable goals	Specific measurable sub-goals	Data collected over time to measure progress towards reaching goals

Source: Created and drawn by the authors.

Organization of the SDGs: values

In the case of the SDGs, the principles follow the typical sustainability values of People, Prosperity, and Planet. Notice the use of the term "prosperity" by the UN as a replacement for the word profit in the three 'P' of sustainable design. This is not arbitrary and reflects a broader sense of profit as defined by long-term profit, the associated affluence, and financial security. The fourth value, as defined in this book, **Place**, is not a major driver of the goals but does appear in a limited way, which will be outlined later.

SDG topics

The SDGs are divided into 17 overarching topics, like "Sustainable Cities and Communities" which is immediately followed by an overarching aspirational goal such as "Make cities and human settlements inclusive, safe, resilient and sustainable." These 17 topics/goals form the basis of sustainable development as defined by the UN (Figure 10.5a).

SDG overarching goals

The overarching goal is aspirational, meaning that it may not be reached in the short term but serves as a "push" or "pull" to organizations seeking to improve their efforts on behalf of a particular problem or opportunity. It is the reflection of one or more of the sustainability values of the organization. In buildings, the overarching goal being "Design the most efficient project possible" is open ended but encourages the team to think more ambitiously. In the case of the SDGs (2018), an example of an overarching Global goal can be found

Figure 10.5a The sustainable development goals.
Source: The United Nations, www.un.org/sustainabledevelopment/wp-content/uploads/2017/12/UN-Guidelines-for-Use-of-SDG-logo-and-17-icons-December-2017.pdf.

in Goal #11 Sustainable Cities and Communities: "Make cities and human settlements inclusive, safe, resilient and sustainable." This is an aspirational goal that helps the design team to remember what is important in a project, or the core values or guiding principles.

SDG targets

The "Targets" reflect sub-goals or tangible measurements. In the example, one of the targets for sustainable cities and communities is: "By 2030, ensure access for all to adequate, safe, and affordable housing and basic services and upgrade slums." Here, the goal is more concrete. It's easy to imagine how a designer might contribute to this target and by default to the large overarching goal.

SDG indicators

Indicators are used to measure how well the targets are being achieved. This is a critical aspect of sustainable design – which relies on evidence as a feedback loop to determine the effectiveness of an approach over time. It is important to note that all goals are tied to deadlines such as 2030. Setting deadlines is also a critical aspect of the overall planning process because it helps to force action. If goals and targets are set with no end date, the chances for action are decreased. Indicators are similar to the term "metrics" used in this book. An example indicator is shown below as it relates to the example of the target discussed of insuring adequate, safe, and affordable housing in the previous paragraph. The indicator or measure for this target is expressed as "Proportion of urban population living in slums, informal settlements, or inadequate housing." Here, we see a tangible way to measure change or progress towards meeting the larger target.

Additional resources

Sustainable Development Goals, https://sustainabledevelopment.un.org/sdgs

10.6 Bio-inspired global sustainable design

While there is definite global movement emerging for both the biophilic and the biomimetic design, the broader issues and opportunities of geo-engineering bear some attention. Geo-engineering or climate engineering are generally focused on attacking climate change. The approaches range from simply planting of millions of trees using saplings (afforestation), the carbon dioxide removal at the point source, or for a specific purpose like manufacturing, or carbon capture/storage projects (sequestration).

Afforestation

Wangari Maathai, born in Kenya, is considered a global hero in the fight against climate change and desertification. Ms. Maathai, who died in 2011, was a champion in the use of afforestation. She imagined a "greenbelt" spanning across the entire continent of Africa. In the many years of the Greenbelt Project, hundreds of nurseries were founded and over 200 green belts were established (Green Belt Movement 2018). It should be noted that gender livelihood was at the center of the movement focused on creating more opportunities for women to earn a better living. Her project exemplifies the Quadruple bottom line of values in action – a true sustainable designer.

Carbon sequestration

Carbon sequestration is naturally achieved by trees and plants. When they absorb CO_2 from the atmosphere, it is stored in their structure and oxygen is emitted. Afforestation, actively restoring the forests by systematically planting trees, can be used to attack the problem of excess CO_2 emissions. This approach also reduces global temperatures passively through shading. Sequestration of CO_2 through the use of active systems is also in development, it old coal mines to capture and store CO_2. It could be a possible "active" strategy to delay the effects of climate change until solutions to the problem of excess CO_2 emissions can be solved.

Clean coal

Clean coal is a generic term used for the goal of reducing or removing harmful pollution from burning coal at the point of its creation and preventing it from reaching the atmosphere. The goal is to capture and sequester CO_2 emissions as it is generated in the electricity producing plants. Research is still under way to prove this technology.

Geo-engineering

Geo-engineering refers to large-scale technological interventions in the earth's oceans, soils, and atmosphere with the aim of reducing the effects of climate change. Some of these strategies are potentially dangerous like the use of iron fertilization of oceans to generate the growth of phytoplankton, which can capture and store CO_2. In more recent years, experiments in converting captured CO_2 and using it for specific purposes such as making diamonds and other durable materials is underway. More ambitious experiments include the use of solar radiation management either through technologically created shading elements, or by maintaining the polar ice caps through large-scale "air conditioning projects." Other efforts include the creation of cirrus clouds to allow for more heat to escape the greenhouse gas layer leading to a reduction global temperatures.

In conclusion, the precautionary approach should be used to determine if geo-engineering is an appropriate strategy given the uncertainty of the long-term global effects of altering the normal processes nature uses to regulate the climate. However, if the climate is already following new patterns as a result of human-induced global warming, then it may not matter.

Additional resources:

Rising CO_2 poses bigger climate threat than warming, study says, www.upi.com/Science_News/2018/06/12/Rising-CO2-poses-bigger-climate-threat-than-warming-study-says/1991528807303/

10.7 Global sustainable design and resilience

Currently, adaptation versus mitigation is one of the major debates among scientists, policy makers, and sustainable designers. Adaption, which is covered in more detail in Chapter 7, is the expression of resilience in the face of extreme climate change. In other words, adaption is an approach to deal with climate change not by fighting against it, but by creating scenarios were humans can still thrive in the face of dramatic effects of climate change. For example, if it becomes more common for major hurricanes to strike coastal communities, how can we better adapt to lessen the damage and recover more quickly? This is already affecting the way realtors and insurance companies are dealing with properties in higher risk areas (Luscomb 2017).

Part of adaptation includes the consideration of migration away from vulnerable areas. According to the European Commission:

> Examples of adaptation measures include: using scarce water resources more efficiently so they will be more available when needed; adapting building codes to future climate conditions and extreme weather events; building flood defences and raising the levels of dykes; developing drought-tolerant crops; choosing tree species and forestry practices less vulnerable to storms and fires; and setting aside land corridors to help species migrate.
>
> (European Commission 2009)

In 2017, a series of storms devastated parts of the world, especially Puerto Rico, which continued to recover eight months after the storm that caused widespread hardship and human loss in the thousands. Evidence is mounting that the pattern of these types of storms are more frequent and more intense. However, there is no definitive proof to this point of view. These weather changes impact heat waves, forest fires, and other climate-related disasters and are, in part, driving the debate about the best approach to address climate change.

Mitigation

The strategy of mitigation, addresses the root causes of climate change and aims to eliminate the threat. Sustainable design is most often associated with mitigation, because its goal is to reduce or eliminate greenhouse gas emissions from buildings, cities, manufacturing plants, products and more. The second half of this book addresses mitigation strategies at all scales.

The "bathtub"

The problem with only pursuing resilience and adaptation strategies stem from the position that the damages to the environment will only get worse, even if we slow down emissions to a steady pace.

The "bathtub" visualization tool, developed by MIT Professor John Sherman, helps to understand this position. Imagine three bathtubs. Water levels in the tubs represent varying levels of CO_2 emissions in the atmosphere as the result of varying levels of mitigation, which have varying degrees of climate impacts. The first tub reflects "business as usual" which means a continuing rise in the emissions of greenhouse gases which leads to a fast and furious overflow scenario where society experiences a point of no return as the *amplification and feedback loops* wreak absolute havoc on the planet's environment. The middle tub features the logical approach of "leveling – off emissions" which is great, but there is already a lot of warming in the pipeline from emissions that have already occurred. Here, the tub still overflows and we are forced to contend with the real consequence of climate change through resilience strategies. The last pathway is to pursue a global effort to dramatically reduce the level of emissions. In this scenario, the tub stays full but never reaches the point where humanity is under a constant threat from human-induced climate change.

The ability for humanity to allow self-interest to gain momentum and block efforts to curb emissions gets more complex when adaption is included in the list of possible options. In an article by the International Council for Science (ICSU) during the Paris Climate talks in December of 2015, is a quote regarding the impact of only pursuing adaptation:

> There is a reckless, almost climate-sceptic, bent to the pro-adaptation argument, some experts argue, in which fossil-fuel companies and industry are let off the hook, and the pressure to reduce emissions is eased. Certainly, many free-market advocates like the American Enterprise Institute have long favoured adaptation over emissions reduction. MIT professor of systems dynamics John Sterman argues that if people believe they are protected from rising seas or more intensive storms, they could be less amenable to supporting mitigation policies.
>
> (Mitigation vs. Adaptation: Which one matters more?
> http://roadtoparis.info/top-list/mitigation-vs-adaptation/)

Given the reality of the direction the climate is heading, The UN established special initiative call the Climate Resiliency Initiative, Anticipate, Absorb, Reshape, or A2R. The mission of A2R, according to the website is "Strategies to manage the unavoidable and to avoid the unmanageable needs to be implemented urgently and at scale" (UN Climate Resilience Initiative A2R 2018). In other words, be prepared to be prepared is the mantra of the initiative.

The motivation of empathy is clearly expressed in the statement by Former UN Secretary General Ban Ki-Moon, when discussing the fact that the world's poor are more likely to be impacted by climate change,

> These are the people who did the least to cause climate change, yet they stand to lose their homes, their jobs, and even their lives because of the growing impacts of climate change. That is why I have asked the UN system to put together a package of initiatives to address this urgent need.
>
> (UN Climate Resilience Initiative A2R 2018)

Lastly, resilience and mitigation are major themes threaded thought the SDGs and are not wrong when pursed in addition to mitigating emissions. SGD #9 is represented below as an example of how resilience has made its way into mainstream thinking about environmental issues.

SDG Goal #9

Build resilient infrastructure, promote inclusive and sustainable industrialization, and foster innovation.

Target 9.1

Develop quality, reliable, sustainable and resilient infrastructure, including regional and transborder infrastructure, to support economic development and human well-being, with a focus on affordable and equitable access for all. **Indicator 9.1.1** – Proportion of the rural population who live within 2 km of an all-season road.

Target 9.4

By 2030, upgrade infrastructure and retrofit industries to make them sustainable, with increased resource-use efficiency and greater adoption of clean and environmentally sound technologies and industrial processes, with all countries taking action in accordance with their respective capabilities. **Indicator 9.4.1** – CO_2 emission per unit of value added.

Additional resources

Bathtub Visualization Tool, www.climateinteractive.org/tools/climate-bathtub-simulation/
UN Climate Resilience Initiative A2R, www.a2rinitiative.org/background/
UN Sustainable Development Goal 9, https://sustainabledevelopment.un.org/sdg9

10.8 Global health and sustainable design

The association of health and well-being with design of the built environmental is well documented in Chapter 8. The individual designer may struggle to apply all the specific requirements at such a broad scale when facing the overwhelming issues of health and well-being at a global scale. And yet, that is exactly what integral sustainable design is about – being able to extend your vision beyond the scale of your profession, and look at the long-term and long-range implications of decisions, or strategies. Everyone's overall health is tied to the state of the global environment. Gro Harlem Brundtland expresses it this way, "More than ever before, there is a global understanding that long-term social, economic, and environmental development would be impossible without healthy families, communities, and countries" (Brundtland Commission 1987). Health + well-being has reached a broader definition, where focus is on prevention and is an inherent part of designing all types of spaces, not just hospitals. The importance of architects, interior designers, planners, and others in creating healthy supportive environments is now well documented.

At the global level, this concept is integrated into the SDGs holistically at every level. The World Health Organization publishes a yearly report: *Monitoring Health for the SDGs*. In the 2017 report, the term "intersectoral" is used reflecting the broadened more inclusive view of health care. For any profession involved in the built environment, there are multiple opportunities to improve the health of occupants at all scales (World Health Statistics 2017, p. 32).

Some examples from the 2017 report are as follows:

Implementation of housing standards and urban design that promote health

Labour sector promotion of occupational standards and workers rights to protect worker health and safety across different industries

Improving product standards, public spaces, and using information and financial incentives, involving the education, agriculture, trade, transport, and urban planning sectors

Intersectoral, or multi-disciplinary projects can leverage greater assets to attack the problem of health equity holistically with solutions that are comprehensive and sustainable over the long term. Health and well-being for all populations and socioeconomic sectors need to be a priority in sustainable design. The principles of social equity and empathy are identified as core values in both quadruple bottom line and integral sustainable design, and should guide our design directives at all levels.

The SDG that deals most directly with health + well-being at the Global Scale is Goal #3 Good Health and Wellbeing.

Overarching Goal: Ensure healthy lives and promote well-being for all at all ages
Targets 3.6: By 2020, halve the number of global deaths and injuries from road traffic accidents
Indicators 3.6.1: Death rate due to road traffic injuries

Target 3.6 addresses design of communities and streets. A possible strategy may be to work towards enacting "complete streets" as the norm or required form of infrastructure. To review, a "complete street" creates safe conditions for drivers, cyclists, runners, and pedestrians, by delineating an area for each type of transportation. It does this with green space and by separating functions. SDG #3 contains many more Targets and Indicators, please review the SDG website for a fuller understanding.

Climate change and global health

By studying cholera outbreaks in and around Bangladesh and India, Rita Colwell from the University of Maryland at College Park has linked cholera outbreaks to climate change.

She was the first to link infectious diseases to climate change in Proceedings of the National Academy of Sciences in 2000. Colwell used geographic information systems to compare cholera outbreaks with the environmental conditions of rising ocean temperatures, water heights, and chlorophyll concentrations. When this information was plotted on a map of the area, the two conditions matched almost identically (Sternberger 2010 270–271). "The approximately ten thousand people who perished as a result of Hurricane Mitch (Honduras, 1998) died not only from the immediate effects of the storm but also from the infectious diseases that flourished in its wake. People living in poverty and in underdeveloped countries are the ones who suffer most" (Sternberger 2010, p. 274).

We can see that the global level affects health at every scale, and what we do at every scale affects the global conditions so vital for our health. Health and well-being will be addressed at every scale as we move through the rest of the book and more detailed information is in Chapter 8.

Additional resources

National Complete Streets Coalition by Smart Growth America, https://smartgrowthamerica.org/program/national-complete-streets-coalition/
Sustainable Development Goal #3 Good Health and Wellbeing, https://sustainable development.un.org/sdg3
World health Statistics, 2017: Monitoring health for the SDBs, World Health Organization, http://apps.who.int/iris/bitstream/handle/10665/255336/9789241565486-eng.pdf;jsessionid=A798712D7D0654B197B098AF260C027D?sequence=1

10.9 Global sustainable design process

Stakeholder engagement and collaborative design has been shown to be a means for truly holistic sustainable design. This is even more true at the global level where there is even more diversity and varied needs to be met with design solutions. The work of the UN in developing the SDGs is an example of an inclusive design process.

The SDGs were established to be inter-governmental, multi-stakeholder sustainability initiatives that help to direct policy and development decisions. They were to be applicable over broad cultural, socioeconomic, and geographic areas to be ready to be adopted at the World Summit on Sustainable Design. The purpose of their approach was based on,

> Partnerships as a means of implementation has never before in the history of international cooperation been more important. A revitalized Global Partnership to deliver on all of the sustainable development goals will facilitate intensive global engagement bringing together governments, civil society, the private sector, and mobilizing and utilizing available resources to transform the world for the better by 2030.
>
> (Division of Sustainable Development 2015)

Their process for achieving the SDGs:

Step 1: Initiate an inclusive and participatory process of SDG localization.
Step 2: Set the local SDG agenda.
Step 3: Plan for SDG implementation.
Step 4: Monitor SDG progress.

Chapter 9 discussed design charrettes and the importance of engaging the stakeholders as a starting point in the design process. Defining parameters and goal setting is a standard and crucial process in participatory design. This makes sure all areas of concern are included and establishes directives to guide the rest of the process. The agenda outlines time tables, expected end results and a path to get there. Implementation is where goals and directives outlined previously are put into action. This action needs to be constantly channeled through the established goals. Monitoring after completion is the feedback loop that will help the project evolve. Constant improvement will benefit the existing project and those in the future. A global project has multiple feedback loops that can influence its implementation or usefulness. Monitoring the effectiveness of our design decisions will help prevent long-term negative effects.

There are times when the design team will decide to have their goals or directives established by a rating system such as LEED. Whatever the source of goals, they should be the driving force. Having established focused goals is critical to integral sustainable design.

Additional resources

Partnership for Sustainable Development Goals: A legacy review towards realizing the 2030 Agenda, https://sustainabledevelopment.un.org/content/documents/2257Partnerships%20for%20SDGs%20-%20a%20review%20web.pdf

10.10 Global sustainable design

To help illustrate the global context of the SDGs, (Figure 10.10.a [next page]) categorizes which goals deal primarily at each scale. The global scale is the largest that we are looking at in this book, but remember that it not only influences, but contains all the other scales. If one area or scale improves it improves all the others. Conversely if there is damage or neglect at one scale or in one area that negatively affects all the others as well.

We see from the diagram below that when the Biosphere, or Eco-system is compromised, all the other systems and scales will feel the effect (Figure 10.10a).

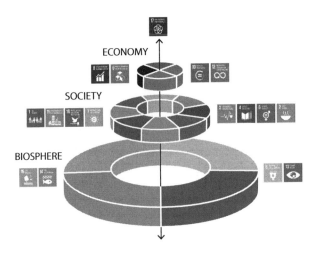

Figure 10.10a Overall structure of the SDGs from a food systems point of view.
Source: J. Lokrantz/Azote.

As we view the diagram above, we can clearly see the categories from the Triple Bottom Line in the diagram developed by Johan Rockstrom and Pavan Sukhdev from the Stockholm Resilience Center. We can also see from the quote, that the goal was to create an integrated relationship between all three categories.

> This model changes our paradigm for development, moving away from the current sectoral approach where social, economic, and ecological development are seen as separate parts. Now, we must transition toward a world logic where the economy serves society so that it evolves within the safe operating space of the planet.
>
> (Stockholm and Beijer 2018)

In this model, the 17 SDGs are organized in way that corresponds with Maslow's hierarchy, though in a different order. Some of the terms for their categories are different like biosphere as opposed to ecology or environment. "Biosphere" implies a global limit or boundary for human activities where ecology and environment are more generalized terms. "Societal goals" are used instead of culture, social equity, or people; "Economic goals" are used for prosperity. The important part is that the SDGs outline a comprehensive integral set of goals with directives and indicators to direct holistic change.

10.11–10.14 Global scale urban, district, building and human scale design

Due to the amount of information and detail provided with each SDG, their targets, and indicators. The readers are encouraged to explore them on their own and apply them to projects. Many of the rating systems and standards to be explored at different scales include or reflect the SDGs. The United States Green Building Council (USGBC) and the Green Building Certification Institute (GBCI) have begun to explore the relationship between the SDGs and the current rating systems. The links to these studies are shared in the resource section below.

Additional resources

Sustainable Development Goals, https://sustainabledevelopment.un.org/sdgs

USGBC (US Green Building Council) / GBCI (Green Building Certification Institute), https://usgbc.org/resources/usgbc-gbci-sdg

10.15 Conclusion

If you went about your daily lives, you probably wouldn't know that there is a worldwide effort to address the most pressing problems. It is difficult, without a conscious effort to think beyond the scale or situation in which you find yourself. It is the hope that by briefly reviewing some aspects of the global scale it will become easier to consider and the long-term effects of any decision will now be placed into better focus.

References

Badger, E., 2017. How Redlining's Racist Effects Lasted for Decades. *The New York Times*. New York.

Cohen, B., 2004. Urban growth in developing countries: a review of current trends and a caution regarding existing forecasts author links open overlay panel. *World Development*, 32(1), pp. 23–51.

Eco-Business, 2014. *Green Mash Up: The Rise of Biophilic Cities*. Eco-Business. www.eco-business.com/news/green-mashup-rise-biophilic-cities/edn.

Frank Lloyd Wright Foundation, September 8, 2017-last update, Revisiting Frank Lloyd Wright's "Broadacre City" [Homepage of Frank Lloyd Wright Foundation], [Online]. Available: https://franklloydwright.org/revisiting-frank-lloyd-wrights-vision-broadacre-city/ [July 25, 2018].

Grimm, N., Faeth, S.H., Golubiewski, N.E., Redman, C.L., Wu, J., Bai, X. and Briggs, J.M., 2008. *Global Change and the Ecology of Cities*. American Association for the Advancement of Science. http://science.sciencemag.org/content/319/5864/756 edn.

Kenzer, M., 1999. Healthy cities: a guide to the literature. *Environment and Urbanization*, 11(1), p. 201.

Lambert, B., 1997. At 50, Levittown Contends With Its Legacy of Bias. *The New York Times*. New York: The New York Times Company.

Litman, T., 2018. *Where We Want To Be Home Location Preferences and Their Implications for Smart Growth*. Victoria Transport Policy Institute. www.vtpi.org/sgcp.pdf edn.

Lucy, M., 2014. *Green Roofs in Washington are Expanding*. GBIG. http://insight.gbig.org/green-roofs-in-washington-dc-are-expanding/edn.

Mostafavi, M., 2010. Why ecological urbanism? Why now? *Harvard Design Magazine*, 1(32). http://www.harvarddesignmagazine.org/issues/32/why-ecological-urbanism-why-now

Newman, P., 2014. *Biophilic Urbanism: A Case Study on Singapore*. London: Taylor & Francis.

Quirk, V., 2014. *The BIG U: BIG's New York City's Vision for "Rebuild by Design."* ArchDaily. www.archdaily.com/493406/the-big-u-big-s-new-york-city-vision-for-rebuild-by-design edn.

Richard, R., 1993. *Ecocity Berkeley*. 1st edn. North Atlantic Books.

Smart Growth Network, *This is Smart Growth*. Smart Work Network. Washington DC.

Smith, M., 2015-last update, Building Cities Like Forests [Homepage of The Modern Ape], [Online]. Available: http://themodernape.com/2015/06/02/building-cities-like-forests-when-biomimicry-meets-urban-design/ [July 26, 2018].

Sternberg, E.M., 2010. *Healing Spaces: The Science of Place and Well-Being*. 1st edn. Cambridge, MA: Belknap.

Stockholm University and The Beijer Institute Ecological Economics, 2018-last update, Stockholm Resilience Centre [Homepage of Ministra], [Online]. Available: http://stockholmresilience.org/ [July 26, 2018].

Sustainable Development Solutions Network (SDSN), 2015. *Getting Started with the Sustainable Development Goal: A guide for stakeholders*. Paris and New York: United Nations.

The Cultural Landscape Foundation, 2018-last update, Ian McHarg – Pioneer Information [Homepage of The Cultural Landscape Foundation], [Online]. Available: https://tclf.org/pioneer/ian-mcharg?destination=search-results [July 25, 2018].

Triman, J., November 4, 2012-last update, Interview with Dr. Stephen Kellert [Homepage of Biophilic Cities], [Online]. Available: http://biophiliccities.org/interview-with-dr-stephen-r-kellert/ [July 26, 2018].

United Nations, 2018-last update, Sustainable Development Goals [Homepage of United Nations], [Online]. Available: https://sustainabledevelopment.un.org/sdgs [July 26, 2018].

Victoria Transport Policy Institute, 2018. Victoria Transport Policy Institute [Homepage of Victoria Transport Policy Institute], [Online]. Available: www.vtpi.org/ [July 26, 2018].

Vidal, J., 2014. *Corporate Stranglehold of Farmland a Risk to World Food Security, Study Says*. Guardian News and Media Limited. www.theguardian.com/environment/2014/may/28/farmland-food-security-small-farmers edn.

Wellington City Council, Open Spaces [Homepage of Wellington City Council], [Online]. Available: https://wellington.govt.nz/services/environment-and-waste/environment/open-spaces [July 26, 2018].

Wikipedia, July 22, 2018-last update, Paolo Soleri [Homepage of Wikipedia], [Online]. Available: https://en.wikipedia.org/wiki/Paolo_Soleri [July 25, 2018].

World Future Council, 2018-last update, World Future Council [Homepage of Future Policy.org], [Online]. Available: www.worldfuturecouncil.org/.

World Health Organization, Types of Healthy Settings [Homepage of World Health Organization], [Online]. Available: www.who.int/healthy_settings/types/cities/en/ [July 26, 2018]. 2018 (that's what the copyright says at the bottom of the page).

Zimmer, L., 2013. *Hundreds of Vacant Lots to Become World's Largest Urban Farm in Detroit*. Inhabit. https://inhabitat.com/hundreds-of-vacant-detroit-lots-to-become-worlds-largest-urban-farm/edn.

11 Urban scale sustainable design

11.0 Introduction

It's hard to imagine a sustainability discussion without a deep look into cities. After all, cities offer so many benefits: dense living, car-free commutes, efficient infrastructure; rich cultural diversity; and moments of extreme beauty. But cities are not, by default, sustainable. They can also be places of extreme pollution, disease, crime, ugliness, and misery. Sustainable designers have the potential to transform cities into thriving, healthy, and equitable places to live and work. The urban scale also includes other types of settlement patterns including suburban areas, small towns, hamlets, and rural areas, each with their advantages and disadvantages when compared with typical cities. As usual, we will begin our look at cities through the lenses of space and time.

11.1 Historical context of urban sustainable design

The Age of the Hunter-Gatherer – 80,000 BCE to 12,000 BCE

The hunter-gatherers roamed vast amounts of space on foot for thousands of years, building temporary settlements that took organic and informal patterns acting more like campsites than actual villages. These early informal settlements were not "designed" by today's standards but they were emergent, rising out of an organic sensibility and the social and functional patterns of living, yet they can impart valuable knowledge about design. There are repetitive strategies, proportions, and spatial arrangements for these early settlement patterns that can offer insight for today's urban designers.

Agriculture – 12,000 BCE to 1750 BCE

It wasn't until the Age of Agriculture, at the end of the last ice age that villages began taking the form that we recognize today. With food sources becoming more predictable and abundant and safely stored in granaries, villages became more permanent. Families grew larger leading to exponential population growth and the transformation of villages into cities.

Towards the end of the Agricultural age in the U.S., the founding of new colonial cities offered the opportunity to rethink European approaches to urban design, especially in the way nature could be integrated into the plan. The first plans for Philadelphia, on land peacefully ceded to William Penn by the Lenape, reflected the desire to build a "green" city (The Cultural Landscape Foundation 2018). Four parks and a center square offered valuable open space and each house included green space. The intention was to combat the dark, congested, dirty, and disease-infected conditions in most European cities of the time caused by the lack of sanitation (Figures 11.1a [next page]).

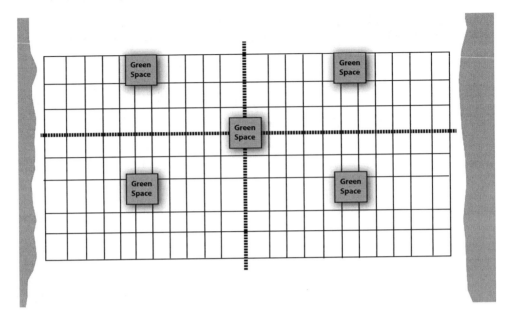

Figure 11.1a William Penn's Philadelphia city plan.

Source: Wikipedia, adapted and redrawn by authors by Thomas Holme – http://xroads.virginia.edu/~cap/ PENN/pnplan.html, Public Domain, https://commons.wikimedia.org/w/index.php?curid=9089244, https:// en.wikipedia.org/wiki/Philadelphia#/media/File:A_Portraiture_of_the_City_of_Philadelphia.JPG

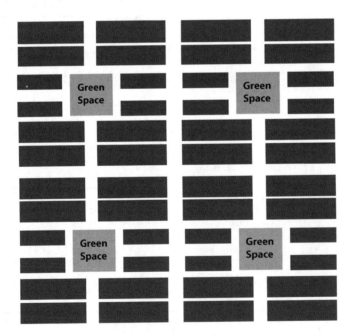

Figure 11.1b Sketch of Savannah's Town plan by Oglethorpe.

Source: Wikipedia, adapted and redrawn by authors. By Fgrammen – Own work, CC BY-SA 3.0, https:// commons.wikimedia.org/w/index.php?curid=19978483, https://en.wikipedia.org/wiki/Oglethorpe_ Plan#/media/File:Savannah-four-wards.png

The plan for Savannah, Georgia, (Figure 11.1.b [previous page]), designed by James Edward Oglethorpe in the early 1700s followed a similar principle. If you look closely, you can see that each green space is surrounded by eight city blocks. These are split into residential areas at the corners, with a small lane providing rear access to each property and the commercial areas are grouped around the square in each sector. It essentially created residential and commercial bands throughout the city broken by green space that created gathering points in the commercial areas. Both Philadelphia and Savannah remain intact 3 centuries later, without changes to the original urban model, and continue to expand using similar patterns, speaking to the value of green space in urban areas.

The Industrial City – 1750 to 1960

The Industrial Worldview, built upon maximizing profit, the exploitation of nature, and the perception of nature as an endless, unchangeable resource, can be seen in the manufacturing-based cities of the 19th and early 20th centuries. Cities became even larger centers of commerce and cultural exchange and the great engine of technical advancement. The Industrial City was filled with factories, and polluted air and water. Accompanying this were the throngs of low paid workers, forced to live in small, poorly built, sometimes dangerous living conditions without sanitation. Living conditions for those in the higher economic spheres were better with some even being beautiful and spacious. Even still, the ills of the city, especially disease, air pollution, congested, and foul streets, affected everyone and by the 19th century, efforts were underway to improve living conditions.

Unlike Philadelphia or Savannah, Manhattan developed without the benefit of a visionary plan. The City's architecture consisted of thousands of tenement buildings with a single stair and very little access to light and air. It wasn't until 1857, on land, partially inhabited by the largest free African-American land-owning neighborhood called Seneca village, that Central Park was founded. Homeowners either lost their homes to eminent domain or received very small sums of money for their properties.

Frederick Olmsted won the competition for Central Park, and his design was executed in 1857. It reflected the principles we now consider essential to sustainable design of increased access to the commons: fresh air, natural light, natural vegetation, and vistas. It also reflects the non-Euclidean, informal, approach to park design, a response in part to rise of the Romantic movement at the time. The design was meant to challenge the rise of technology, mechanization, and industrialization.

In the early 1900s, Howard Ebenezer first proposed the Garden City concept. This was a new settlement pattern that had the potential to be more integrated with nature and to be more self-sufficient than past models of urban design. It was a holistic concept very similar to contemporary sustainable design approaches. His proposals would eventually be adopted in the United Kingdom and in other many other cities around the world.

Frank Lloyd Wright also envisioned city planning in contrast to the grid layout. His Broadacre City integrated areas of industry, untouched nature, and residential areas that included agriculture into each site. Wright based the probability of success of his plan on the advent of the car, greater road infrastructure, and the current technologies of the day that increased connection and communication. Wright demonstrated he was using a systems approach in his thinking, the work "homestead" or individual family was thought of as its own holistic system, embedded in the larger holistic "city," which was part of the geographic area. In later years Wright used the term "Living City" to describe the plan, underscoring his integrated relationship with nature. In that sense, he was already expressing an integrated view of design.

Many blame Wright for the rise of the suburbs and argue that his plan for Broadacre City was fundamentally unsustainable. The reliance on the automobile lies at the center of the criticism, but at that time, the negative effects of pollution from cars were not fully known and the poor health and environmental conditions within the city made escaping an attractive solution. What we do see is that he was looking at cities and urban development "…not [as] an arrangement of roads, buildings and spaces, [but as] a society in action. The city as a process, rather than a form" (Frank Lloyd Wright Foundation 2017).

Postindustrial city

By the 1960s in the U.S., cities experienced mass abandonment, closing industries, and rail lines became defunct. New interstate highways cut through the middle of cities and ended up dividing previously united communities. Interstate 95 in Philadelphia literally cuts through scores of neighborhoods, besides separating communities; there is now noise, air pollution, and the dark underbelly of the highway, ironically a symbol of efficiency, speed, and power. The manufacturing areas at the rivers in many cities were not only abandoned but polluted with toxic chemicals. Creating the basis for the Super Fund established in 1980 to clean up hazardous uncontrolled or abandoned waste sites.

Urban renewal

Originally known as "urban renewal" and currently known as "The Projects," these high rise developments were aimed at alleviating poverty and poor living conditions using the planning principles based on Le Corbusier's stark ideal of the Radiant City, a popular development model of the International Style. In theory, the new residential towers, surrounded by large areas of green space would allow residents' higher quality living conditions with the surrounding areas providing a valuable amenity.

The congested, dirty streetscapes of many urban neighborhoods at the time made this approach seem like a positive step to improve the living conditions of many people. But the approach destroyed neighborhoods, and entire communities were erased. In many cases, land for these new residential towers was only available on the outskirts of the city or in abandoned industrial areas. Many people got newer apartments, but they were now isolated from each other, from shopping areas and the industries where they worked, and from seeing each other on the streets. The large, open green spaces around their new homes turned into dangerous dark areas breeding some of the highest crime rates in the city with no accountability by casual observers or the authorities. Residents crossing these areas had no areas of safety or means of getting immediate help. Residents had to go down by elevators to get outside, and parents or shop owners couldn't watch the neighborhood children play from inside, or gather at the entrances. These problems coupled with a lack of facility maintenance and lack of a sense of ownership, led the "Projects" to quickly fall into disrepair, leading to a lower quality of life for the people meant to be uplifted. Only 50 years after construction, many of these housing projects have been abandoned and blown up, and replaced, by two story houses fronting on urban streets.

Jane Jacobs and Ada Louise Huxtable, among others, led a critical examination of the social qualities that drove urban development from the beginning. In her book, *The Death and Life of Great American Cities,* Jacobs examines cities at every level. She found that streets are really meeting and gathering areas for multiple generations, and that they provide places for children to play under the care of people who know and care about their safety.

The network of streets in a city provides safety because it allows for multiple routes, and businesses provide areas of refuge, accountability, and the equivalent of a crime watch because the employees can identify threats to the neighborhood quickly. She also believed it was the diversity that made the city so appealing and was better at integrating socioeconomic differences. Jacobs and other advocates drew fame and accolades for their advocacy of holistic urban development which decried interstate highways through cities.

During this same time, large urban areas populated by African-American populations were "Redlined" by banks. Redlining refers to the practice of zoning certain areas as a higher risk or ineligible for a bank loan simply because of where it was located in a city or the dominant ethnic background of its residents. It got its name from zoning maps that were traced over with red markers outlining disqualified areas. This made it nearly impossible for people in these areas to have the security of owning their own home or the opportunity to build wealth. It also doomed these neighborhoods to disrepair and apathy because of the social and emotional of segregation. It also resulted in the loss of pride of ownership so important in vibrant, well-maintained neighborhoods. In later years, after the practice of redlining was eliminated, research shows that African-American home buyers still pay higher mortgage rates, on average, than their white counterparts (Badger 2017). Another current problem with urban renewal is gentrification. Gentrification is the practice of renovating and improving a district to appeal to wealthier home buyers and investors. While on the surface it improves an area aesthetically and economically, it leads to the displacement of longtime homeowners, who can no longer afford to live there.

Suburban development

Unlike urban renewal, which was funded by government investment, the suburbs were primarily driven by private development, although the U.S. Transportation Act in 1959 helped to encourage more development by creating more infrastructure. The motivations of developers were mixed between profit and delivering the "American Dream." The American Dream was access to light, air, open space, and more importantly a solely owned and controlled plot of land. These were all primary selling points for new post-World War II housing developments. The early suburban plans had their roots in Garden City concepts developed earlier in the century, but the designs themselves had a decidedly sprawling sensibility and none of the diversity of use or sense of community in the original plans. The negative environmental impacts of the car were not known or even considered. The idea of owning land in the countryside was appealing to millions and seemed like the perfect sustainable solution to raising the quality of life.

From a social equity perspective, the first suburbs reflected the long held story of racism in the United States. Levittown, the first planned suburban community, was designated for "whites only" by the developer. Real estate agents steered minority home buyers away from other suburban areas, and white residents were known to make life miserable for those minorities who were lucky to purchase a house in the area (Lambert 1997).

Suburban ecology

Just as large areas of urban fabric were destroyed to make way for urban renewal, the realization of the suburbs required the widespread removal of farmland and wilderness areas – only to be replaced with green lawns, driveways, and nonindigenous ornamental trees. The net effect of the suburbs from an ecological perspective meant that the local ecosystem services of air purification, water management, and biodiversity were all compromised leaving the suburbs to rely on human-made infrastructure.

The negative effect of one suburban development upon the larger ecosystem is negligible, but when this pattern of development was repeated thousands of times, the effect is magnified. As the suburbs grew and people began living farther away from basic services, depending solely on cars for transportation, roads became congested with more and more traffic. This traffic is contributing to the amount of greenhouse gas emissions and human-induced climate change.

Another issue with suburban development is the isolation that is caused by everyone having to travel longer distances in separate cars to reach shops, activities, business, or their place of work. It is very common for people to live years in the same neighborhood and not know the people who live down the street.

Coming full circle to sustainable design

In the public realm of the postindustrial city, Baltimore Inner Harbor built in 1980 and designed by Wallace Roberts and Todd architects offered the first integrated response to reclaiming the city as culturally vibrant, economically viable, and connected to nature. It wasn't long before other U.S. cities followed Baltimore's example. The Inner Harbor started a revolution of urban design that began to focus on the sense of place so absent in the idealized modernist cities. "Place" has emerged as the major value in planning and urban design. It is the basis for a holistic approach to reimagining and regenerating urban neighborhoods and industrial sites. The activation and reconnection to the water's edge in many urban revitalization projects was the beginnings of what we now refer to as "biophilic design." Human well-being and experience were once again moving to the center of design directives.

Additional resources

12 Secrets of New York's Central Park: Smithsonian, www.smithsonianmag.com/travel/12-secrets-new-yorks-central-park-180957937/#u0YJl1BtKJTKYwhU.99
Seneca Village Site: Central Park Conservancy, http://centralparknyc.org/things-to-see-and-do/attractions/seneca-village-site.html?referrer=https://www.google.com/

11.2 Urban scale sustainable design and the environment

Today, an emerging worldview seeks to integrate the natural world with the city and offers the first glimmers of a new approach to urban design. The death of many cities has offered the opportunity for literally new life – open spaces with flora, fauna, and natural systems. The old views of "protecting nature" or "saving nature" are now replaced with "regenerating natural systems." This has led to the beginning of an urban design revolution. New terms have emerged that underscore the changing mind-set: Landscape urbanism, landscape ecology, green infrastructure, and urban ecology all underscore the new approach to urban design. Mostafavi (2010) writes,

> Ecological design practice does not simply take into account the fragility of the ecosystem and the limits of resources but considers such conditions the essential basis for a new creative imagining.

Ecological urbanism

Urban ecology focuses on uncovering and connecting to the natural systems that once existed underneath the city. Long-buried streambeds and abandoned river frontage offer the opportunity to integrate natural systems back into the city. But to see that potential, designers and stakeholders need to see the city through an ecological lens first. Rather than imagining buildings, parking lots, and streets as the first priority, the site and its natural systems are viewed as a foundation upon which to build all other connections and uses. In cases where dense development has already occurred, urban ecology plays a role in transforming vertical spaces, roof tops, abandoned lots, or other underutilized areas as places to introduce natural elements. These can be in the form of vertical living walls, green roofs, pocket gardens, or urban agriculture, which offer a myriad of benefits, including habitat. Fundamental to all these is the use of a transdisciplinary approach to design through cocreative processes. The key is to get all the team members to see the site through an ecological lens. Viewing urban ecology at its different scales adds in the ideas of watersheds, migratory patterns for birds, butterflies, bees, animal biodiversity along with geological patterns that impact soil and topography, and dark sky opportunities.

Urban regeneration

The concept of regeneration, as discussed earlier goes far beyond the concept of "protecting" nature, to a partnership model of thinking which focuses on restoring and including natural systems in urban environments. This creates cleaner air and water, cooler urban temperatures, enables higher building performance with less energy usage, while increasing positive experiences for urban dwellers. The reason to pursue the more ambitious goal of regenerative design is to restore centuries of degeneration of their natural systems. When Los Angeles county was originally settled 216 years ago all but 1% of its land was used for farming. Regeneration is necessary to reverse the damaging effects of past development and create an integral sustainable solution to growing urban areas.

The future of regenerative urban development should focus as Herbert Girardet, from the World Future Council (2018), writes, "…Initiating comprehensive political, financial and technological strategies for an environmentally enhancing, restorative relationship between cities and the ecosystems, from which they draw resources for their sustenance."

A regenerative approach to cities is needed now more than ever as a greater percentage of the world population is predicted to be moving to cities. Regenerative cities are transformational and reflect dynamic, ambitious, and holistic approaches to creating cities that meet all the needs of human existence along with businesses and manufacturing, the best expression of the emerging Age of Integration.

Additional resources

Journal of Urban Design and Mental Health, www.urbandesignmentalhealth.com/journal-3---la-regenerative.html

Living and Regenerative Cities: World Future Council, www.worldfuturecouncil.org/wp-content/uploads/2016/01/WFC_2010_Regenerative_Cities.pdf

Why Ecological Urbanism? Why Now?: Mohsen Mostafavi, www.harvarddesignmagazine.org/issues/32/why-ecological-urbanism-why-now

11.3 Urban scale sustainable design and motivations

Exploring urban ecology is a critical foundation to the pursuit of sustainable design. In Chapter 3, we explored the relationship between the often-competing motivations of self-interest and empathy. The explosion of population growth in urban centers is occurring not because cities are wonderful places, but because of the opportunity for a higher wages, different types of work, cultural amenities, and in many cases simply survival. In the agricultural zones of many countries, the ability to make a decent living growing food or raising livestock is increasingly more difficult. Climate patterns have led to less water and higher temperatures reducing yields of crops and making conditions for livestock more challenging. In addition to the incessant drumbeat of maximizing profits by large corporations, the smaller farmer is constantly stressed to get more out the land than is logically possible (Vidal 2014). The migration to cities is not just a phenomenon that needs to be dealt with but is a symptom of a much larger global, economic, and cultural challenge. The city does not exist in isolation but is interconnected across scales to natural patterns, global economic pressures, and basic motivation of survival.

> We argue that cities themselves represent the microcosms of the kinds of changes that are happening globally, making them informative test cases for understanding socio-ecological systems dynamics.
>
> (Grimm et al. 2008)

Self-interest motivations and the city

Some believe approximately two-thirds of the population will be living in cities within 40 years (Cohen 2004). The challenges and opportunities to achieve a sustainable future will be focused on cities. With that comes the collision between basic human motivators of self-interest and empathy. The city contains expressions of maximum luxury and maximum poverty – the two coexisting in tension as expressed by radically different built form: Slums for some and luxury towers for others. Motivations for extreme comfort continue to drive the development at the highest strata, but the impacts are mitigated through green design practices.

Don Alexander writes in his book, *The Necessity of Eco-Cities*,

> Obviously, the problems in the cities of the North and South are quite different. In the cities of the North, the problems are largely those of affluence and over-consumption, though many people are still victimized by poverty and a lack of access to adequate services. In the South, many of the urban environmental problems are the result of poverty and overcrowding, though there is also a wealthy stratum whose lifestyles mimic those of people in the rich countries.
>
> (Alexander 2018)

This quote underscores the deep-seated challenges of sustainable development. The motivation of cognitive empathy needs to find its way deeper into the mind-set of developers and city officials when making decisions, because usually the people in the poorest neighborhoods are the most affected.

Slums still account for millions of people living in dense metropolitan areas. These areas often do not have the same resources, opportunities, or quality of infrastructure. Their sociocultural and educational resources are not giving their residents the skills required to be viable and valuable innovators cities need. Slums are also most vulnerable to disasters like drought, earthquakes, mudslides, and tsunamis due to their exclusion from city infrastructure, the temporary nature or poor quality of the structures, and the often "undesirable" qualities of the land they occupy.

Additional resources

Sustainable City | Fully Charged: YouTube, www.youtube.com/watch?v=WCKz8ykyI2E
  The Sustainable City: www.thesustainablecity.ae/

11.4 Urban scale sustainable design and sustainability values

Richard Register (1993) first coined the term "Ecocity" in *Ecocity Berkeley: building cities for a healthy future*. His definition of a sustainable city is one that

> ...can feed itself with minimal reliance on the surrounding countryside, and power itself with renewable sources of energy. The focus of an Ecocity is to create the smallest possible ecological footprint. It will strive to produce the lowest quantity of pollution, use land efficiently, compost, recycle, or convert waste to energy...

> (Register 1993)

But if ecocities are to be truly sustainable, they need to hit all four of the sustainability values of People, Planet, Profit, and Place at high levels. When integrated across scale, time, and difference, they offer the foundations for a truly sustainable future.

11.4.1 Economic prosperity in cities

The wealth disparity in cities has been dissected already and remains a significant problem, but the opportunity in cities to innovate new ideas, technologies, and business models is very high because of physical proximity, cultural vibrancy, and diversity of people. The sharing economy as expressed through business models such as Uber, bike share, and Airbnb are transforming the way cities function. Renting is replacing ownership. At the heart of collaborative consumption is trust and cognitive empathy and the potential exists to leverage those values to build a greater sense of cooperation among citizens.

The principle of a circular economy seeks to optimize all systems by using eco-efficient techniques already built into the urban context. The sharing of walls, floors, and roofs leads to more efficient heating and cooling, the sharing of infrastructure, and shorter travel distances means that the city, compared to suburban and rural models, is more efficient. Consider a simple bus ride with 20 passengers versus 20 cars with one person in each car all traveling to the same destination. Then consider a subway train with 200 riders. The value of efficiency is an inherent aspect of cities and is critical to sustainable design.

11.4.2 Ecological regeneration in cities

The core sustainability value of ecological regeneration is pushing innovation at many levels to achieve a balance between the inherent efficiencies of a city and the natural ecological systems that support human life. The transformation of cities from technically efficient communities to ecologically integrated places is a critical change in the mind-set of urban development. Moving from simply mitigating or reducing impacts to creating eco-efficient design is causing the city to become a lush, vibrant center for ecological health. This leads to greater human health and increased success of other core values.

11.4.3 Social equity in cities

The city is also the place where social inequity is constantly on display simply due to the number and diversity of people living in such close proximity. Inevitably, conflict emerges. Truly sustainable cities mean that access, educational opportunities, and infrastructure are a valued resource and are provided for the entire demographic of a city. Empathy allows for more consideration be given to those with less ability to provide it for themselves. Unfair practices based solely on race, socioeconomic level, or residence location have been previously discussed in Chapters 4 and 5.

11.4.4 Placemaking in cities

The sense of place, a critical sustainability value, and of particular interest to designers, and is the last key to a sustainable city. The fundamental need for beauty and a sense of connection and belonging to a geographical location, established in Chapter 4, urges the designer to integrate existing ecological, historical, and cultural elements into the spaces they design. This is applicable at every scale and creates spaces and places that people want to spend time in, will care for, and create a rich vibrant and diverse urban landscape. The creation of viable open access public space engaged by buildings and informed by ecology is a challenging and rewarding aspect of sustainable design. This can take many forms ranging from the configuration of space, to the relationship with the surrounding buildings or ecological features, to the materials used, or the introduction of natural elements like waterfalls or vegetation.

  While these values may seem obvious to some, the reminder to apply all of these values in a sustainable design project is worth repeating over and over again. Working collaboratively with other disciplines and policy makers adds another level of complexity to any project, but the rewards are great and can influence communities far into the future. When the design team and stakeholders are equipped with a clear set of shared sustainable values at the beginning of the process achieving urban regeneration is easier and more holistic in its solutions. The use of codeveloped guiding principles, covered in Chapter 9, is the key to making the sustainable values visible in a project.

11.5 Urban scale and integral sustainable design

Integral Sustainable Design is an effective framework for applying sustainability values into an urban design project. Figure 11.5a (next page) reflects a holistic set of considerations for an integral urban design project. Notice that the upper left quadrant is focused on "inner health" while the upper right quadrant focuses on "outer health". Outer health can be quantified by health metrics while inner health is more subjective, but can be still be understood through surveys and other engagement practices The goal of culturally vibrant public space suggests that a multi-cultural and inclusive approach to the design of urban space is in order. Such a process will begin the task of moving towards a more socially equitable city by often-marginalized populations an authentic voice in the future of their city. Finally, the lower right quadrant looks

	Subjective	Objective
Individual	*Experience* - **Work towards "inner health" and psychological well-being** - Make the city beautiful, engage all the senses - Apply bio-inspired design to parks, roadways and building exteriors	*Performance* - **Work towards "outer health" Physical health and well-being** - Increase availability of healthy food - Increase walkability - Increase Bicycle friendly urban design - Maximize energy efficiency - Maximize water efficiency
Collective	*Culture* - Create **culturally vibrant urban spaces** and places, Celebrate heritage of historic structure - Embed **social equity** into every decision with transparency and accountability - Build a socially resilient Urban culture	*Systems* - **Life-cycle Systems**: Grow local food, bring back local manufacturing and process waste locally - **Passive Systems**: Organize buildings to maximize potential for solar - **Active Systems**: Use Smart City technology to optimize city systems for resilience - **Living Systems**: Create a regenerative and resilient urban ecology - **Human Systems**: Build a strong economy

Figure 11.5a Sustainable urban design goals within the integral framework.
Source: Created and drawn by authors.

holistically at all types of systems with the orientation of buildings in a masterplan as a key consideration. This will allow better opportunities for net-zero performance of future urban buildings. However, cultural considerations of the urban context require different building orientations that better define the street edge or accentuate an important urban corner.

11.5.1 Performance Perspective and cities

Metrics and standards – The performance of cities is driven in large part by standards and metrics which reflect the values of the community or government. Building codes and zoning ordinances established either by the cities or adopted from larger state, national, and global standards are the key to driving higher energy performance. In the U.S. a number of cities have adopted the CBECS (Commercial Building Energy Consumption Survey) to require buildings over 50,000 square feet to report their energy and resource use, an application of transparency and accountability in design. The reporting does not guarantee that energy will be saved, but it establishes a baseline by which future energy improvements can be measured.

Eco-efficient design is inherent in cities. While cities originally formed for social, economic, and security reasons, the city is an efficient human settlement pattern, especially when compared to suburban sprawl. As stated previously, reduced travel time, shared infrastructure, and shared roofs and walls all provide inherent benefits. They also come with considerable social issues and opportunities, which will be discussed in the Culture Perspective overview below.

Dematerialization is the idea of doing more with less. In this case, it is applied to what is being referred to as "smart cities." Smart cities are urban areas that employ technology to monitor processes, water usage, recycling efforts, and even streetlights with sensors to save energy and increase efficiency. Here are some examples:

- Automatic and remote monitoring reduces the electricity used for streetlamps.
- Hamburg Port Authority reduced operational costs by 70%.
- First responder times are reduced, traffic details can be relayed faster to divert travelers, and carbon dioxide and pollen counts can be better monitored.
- Waste management, recycling, and resource recovery can be more efficient.

A radical response to the suburban ideal

Paolo Soleri, Frank Lloyd Wright's student, used the term "miniaturization" as his way of expressing "doing more with less." He sought to carry Wright's thoughts and ideas of a holistic urban planning system through to their logical end. Soleri designed a city named Arcosanti. It was meant to be a

> …hyper-dense city, designed to: maximize human interaction with ready access to shared, cost-effective infrastructural services; conserve water and reduce sewage; minimize the use of energy, raw materials and land; reduce waste and environmental pollution; increase interaction with the surrounding natural environment.
>
> (Wikipedia 2018)

It was intended as model prototype for desert ecology. Construction was started in 1970 and continues today even after Soleri's death.

Additional resources

Danish Smart Cities: Sustainable Living in an Urban World, www.cleancluster.dk/wp-content/uploads/2017/06/594256e47ab31.pdf

11.5.2 Systems Perspective and cities

The realization of high-performance cities relies upon effective integration of systems. Given the age of cities, their level of complexity, and the sheer number of people involved in the decision-making process, the effective integration of systems thinking is challenging.

Life cycle systems for environmental impact and embodied energy can generally be reduced in cities. Their inherent density shortens travel distances for manufacturing and delivery to the public, and all waste sources are handled more directly. Existing buildings and transportation infrastructure represent additional life cycle advantages. These structures can be renovated or adapted to meet the changing needs of an urban area and lessen new resource extraction. The use of Smart city technology, especially sensors and tracking devices, can help organize the flow of goods and services more efficiently.

Passive systems in cities are controlled by the street grid and building heights. The orientation of the streets, their width, the required setbacks along with building height limits create opportunities or challenges for sun exposure and natural ventilation. Zoning codes determine how far buildings are from the streets which directly affect access to light and air. Many urban areas and communities are becoming aware of this and are placing restriction on building heights so that there is a more equitable distribution of daylight and airflow. When designing new cities or districts, consideration of predominate wind patterns and sun patterns can help to mitigate the local climate. As an example, cities in hot dry climates typically reduce the width of the streets so that the buildings naturally provide shade during most of the day. The heat island effect can also be greatly impacted by passive design strategies. The temperature of the overall city can be reduced if buildings simply use a lighter colored roofing material.

Active systems

Active systems in cities are initiatives underwritten by the governing bodies of the city. They can take many forms from actual technology to policies or incentives that encourage individuals to invest in energy saving technology. Some examples would be stormwater management systems that allow on-site water absorption, sewage treatment plants based on natural purification, clean energy for public transportation, wind or solar powered electric generation plants, policies that reward private investment in Photovoltaic (PV) or high-performance facades. Geographic information system (GIS) can reduce energy use and increase efficiency of services by controlling traffic volume, public transportation, emergency services, streetlights, and many more.

    Living systems in cities will be discussed in more detail later in Section 11.6. These systems also allow for opportunities to reduce the heat island effect with green roofs, shade trees, vertical living walls, and parks.

    Table 11.5.2a (next page) is meant to serve as a useful tool to identify different sustainable strategies for urban design. These are not the only systems that can be applied. Performance, culture, and experience are also critical to a holistic approach. The strategies listed below have direct impact on the other quadrants.

11.5.3 Culture perspective and the city

The sustainable values of social equity, transparency, and accountability are expressed in different cities with varying levels of success. Respect for lower income people in the planning process and including their "voice" in the process is critical. All socioeconomic levels and ethnic populations need to be given the same consideration and access to the natural resources including fresh air, clean water, views, daylight, and quality construction. Homelessness is a complicated issue and requires a holistic examination and solution. There are cities like Portland Oregon who are handling it well because they are addressing it in a comprehensive manner.

    Cities possess high levels of cultural vitality and heritage, key aspects of sustainable design. Experiencing the cultural heritage of an area provides a sense of stability and understanding of the past that is very important in being able to move forward in life. Preservation of historic structures and the adaptive reuse of old factories, for example, allow the residents a deeper quality of life and add layers of aesthetic richness to the urban streetscape. The principle of social responsibility should transform the processes of urban design (see Chapter 9).

Table 11.5.2a Urban systems for sustainable design

	Passive systems	Active systems	Living systems	Human systems
Human health + wellness	Interconnected walking and cycling paths, separation of cars and people	Rideshare, trains, buses, cars, boats	Hiking trails Connect residents to nature Urban organic gardens	Encourage a culture of walking and healthy eating through incentive grant programs
Climate + resilience	Lower heat island effect through light-colored surfaces Reserve areas of the city for storm surges	Pump stations and river Flood management of rivers	Use parks to allow flood waters to enter city and dissipate	Establish socially resilient city agencies to conduct long-term plans to respond to climate events
Food	Create a network of connected urban farms and orchards Eliminate food deserts	Smart city food distribution	Food is energy and can be grown on-site	Policies to encourage large-scale urban food production – eliminate food deserts
Flora + fauna	Locate and connect green space Use trees to lower urban temperatures that attack heat island effect Wildlife bridges over interstates	Smart city technology	Ecological urbanism strategies Connect "patches" of nature	Citywide initiatives to increase and connect green space
Energy + transportation	Interconnected bike and walking trails including high lines Locate solar and wind farms just outside of cities	Smart city tech to optimize energy use Smart city use of GIS and traffic management to improve traffic flows	Use vegetated areas and vertical green walls to save air conditioning costs	Tax and incentives to encourage green energy Congestion fees Policies for use of fossil fuels – carbon neutral goal setting
Earth + materials	Logical location of recycling centers Use local and sustainable road construction materials	Smart city tech to manage waste systems for efficiency Recycling centers that make new products	Citywide composting	Citywide net-zero waste program – recycling and repurpose CBECS Reporting using Energy Star Portfolio
Water	Remove combined sewer systems of using green stormwater infrastructure – direct water to the ground	Smart city – use technology to manage flows of water	Green stormwater Living large-scale infrastructure Daylight streams	Incentives to to collect and process stormwater on all properties
Air	Design new city neighborhoods with wind directions in mind	Smart cities – use sensors to determine health risks	Plant trees Vertical green walls to clean air	Incentives to encourage walking Policies to protect air quality from industry

Source: Created and drawn by authors.

11.5.4 *Experience*

"Space Syntax" is a type of study that defines the relationships between human behavior and city layouts. It uses complex computer modeling to analyze the features of urban areas that encourage walking. Their studies have found that "...more interesting things to do and see in such spaces, the more people will walk, especially if they have many choices and more than one possible path to follow." The study also showed that closer setbacks and mixed-use buildings with retail at the ground level encouraged walking as well (Healing Spaces 266). Aesthetics, a sense of safety, visual interest, along with multisensory experiences are very important in creating a city that is loved and cared for.

11.6 Urban scale sustainable design and bioinspired design

The urban environment is by definition a hardscape, built from concrete, and asphalt. Buildings are also made of hard materials on sites cleared to facilitate construction. Without planning and effort to incorporate natural elements the city is devoid of plant life, earth colors, living creatures, natural light, fresh air, or connection to any type of natural system. The challenge to make cities more habitable, more life enhancing, and more beautiful is starting to be addressed by a wide range of approaches, including the use of biophilia. In Chapter 6, we covered biophilia in depth. E.O. Wilson, the founder of the biophilia movement, states,

> Biophilic cities are different from green cities in which the emphasis is on energy and environmental conservation. In biophilic cities, there is a greater focus on wellbeing and health, celebrating life forms and processes that we as a species have evolved from.
> (Eco-Business 2014)

To review, biophilia is human's inherent desire and need for nature and natural elements. In Chapter 6, we also review some very specific positive changes that can occur with exposure to natural elements. Incorporating nature and natural elements into cities is not new. For many centuries, city parks and greenways have been used to effectively incorporate nature into the city. As cities become denser, more crowded, and the primary mode of settlement, the focus on introducing nature changes from a pleasant diversion to a necessity.

The city of Washington, D.C. now has over 2 million square feet of green roofs (Lucy 2014), and the city of Wellington, New Zealand, has approximately 3,500 hectares of reserved open space lands within the city limits (Wellington City Council 2018). This includes parks and reserves for recreation, wildlife and scenery, children's play areas, walkways, and harbor frontage. Singapore is working hard to develop over 200 km of park connectors that include canopy walks and green walls and rooftops.

Quality is also an issue. Existing parks with a majority of green lawn space lack plant diversity and require constant maintenance and energy from lawn mowers. Biodiversity and wildness in the greening of the city is critical, not just for the measurable benefits of increased species diversity but also for creating even richer and more robust human experience in cities.

Stephen Kellert acknowledged the difficulty in attacking the urban context in an interview before he passed away,

> The big question is how can we live in a compatible and nurturing and positive and beneficial and harmonious relationship to the world beyond ourselves, to nature, in an

increasingly urban and built world. And it can occur in many, many ways, but like I say it is more challenging

(Triman 2012)

Cities that are already built make integration of nature even more difficult, but it is still possible

The city of Detroit, Michigan, is planning to build thousands of acres of urban farms right in the middle of the city (Zimmer 2013). These urban farms will be occupying land that has been abandoned and vacant for years. This will provide resiliency at many levels, local fresh food, opportunities for community involvement, and all the social benefits of renovating vacant property into useful areas. Green areas also collect and process stormwater, clean the air, and create lower urban temperatures.

The city of Singapore is a globally recognized example of a biophilic city. According to Peter Newman (2014), author of *Biophilic urbanism, a case study on Singapore*, "Singapore has bucked the Asian cookie-cutter high-rise tower syndrome through its planning over many years, particularly its recent commitment to biophilic urbanism. It now appears to be a leader in this new approach to city building."

Note that the integration of biophilia can be a multiyear or phased project. Singapore was an established city that chose to incorporate biophilic elements as a way to improve the quality of life for its citizens. The city is now seeing the benefits of this approach and is gaining international attention. The idea of a city in a garden is very different than creating pockets of gardens in a city. The unique aspect of Singapore's approach is to think three dimensional regarding the application of biophilia using roofs, walls, streetscapes, and any surface, thereby transforming the city.

Probably the most iconic is the development of "Supertrees" which are located in the signature urban public space: The Gardens by the Bay. The structures range from 80 to 160 feet high. They collect solar energy used for night-lighting, and they are adorned with over 150,000 living plants. They are connected by skywalks offering views of the bay and the city. This creation of this park is part of an overall concerted effort to build a "Garden City." It demonstrates the power of a set of clear guiding principles, an inclusive planning process through civic engagement, and the power of biophilia itself as a motivating factor for human activity.

Cities that function like forests

Biomimicry is the process of studying nature or natural systems to find solutions to complex human problems. Biomimicry offers another approach to urban design. Cities are complex systems. They have many interconnected systems like sewers, roads, public transportation, electric grids, people, birds, plants, animals, all while functioning and coexisting in a very limited space. The forest is the same, except the forest exists in harmony each supporting the functions of the other. Ilaria Mazzoleni says, "Nature is really a master example of making different things work one to the other and eliminating things that don't fit with the picture." When you observe the nature of a forest you see mutually beneficial processes; systems; and its members, the plants and animals.

The city of Lavas, India, used biomimetic analysis to devise a way for the city to cope with the heavy monsoon rains. Janine Benyus cofounder of the Biomimicry Institute and HOK Architects worked together to evaluate how the local ecosystem handled the monsoon rains. What they found was that the plants and trees stored the excess water in their root system and later released it during the dryer times. From this information, design strategies were formulated: Use permeable paving that allowed water to be absorbed, construct the foundations of buildings to provide stability even when the ground was saturated, and have rooftops act as containment areas with the ability to rerelease the water as vapor and later after the storm surge (Martin 2015).

This is only one example how biomimicry can shape urban design to be more sustainable, resilient, and mutually beneficial. Bioinspired design techniques indicate a shift in worldview and movement towards a more integral way of thinking and perceiving the world around us.

Additional resources

A Skyscraper Made of Bones: How Biomimicry Could Shape the Cities of the Future, www.citylab. com/life/2016/09/a-skyscraper-made-of-bones-how-biomimicry-could-shape-the-cities-of-the-future/497969/

Keynote Address, Janine Benyus, Cities like Forests: www.esri.com/videos/watch?videoid=3158& isLegacy=true

Cities that function like forests: https://biomimicry.org/cities-that-function-like-forests/

Biophilic Cities: http://biophiliccities.org/

Green tower www.world-architects.com/en/woha-singapore/project/oasia-hotel-downtown

11.7 Urban scale sustainable design and resilience

Resilient cities are one of the primary focuses of sustainable design. The specifics of resilient design are covered in Chapter 7, and resources are included at the end of this section for readers who want to learn more about this subject.

New York City was one of a number of cities devastated by Hurricane Sandy in 2012. The city lost power for weeks, the subways were completely flooded, the public transportation system was offline, and some areas where completely cut off from rescue efforts because of flooded streets, with the whole city suffering millions of dollars of property damage. This was the first time anything of this scale had ever happened to the city and was completely unexpected. Because of the crippling and destructive results, and the rise in frequency of extreme weather events, efforts are now being made to protect the city.

BIG Architects is leading a team of international engineers, urbanists, landscape architects, cultural planners, sustainability experts, ecologists, arts and cultural planning specialists, graphic designers, and the School of Constructed Environments at Parsons The New School. The collaborative team will evaluate protecting the city from future flooding along with improving the economic, social, and cultural aspects of the areas effected. The project is called the Big U.

The main goal of the project, according to BIG, is to design a protective ring around the island of Manhattan. The goals for the project as stated in their design brief are as follows:

> ...the Big U protects 10 continuous miles of low-lying geography that comprise an incredibly dense, vibrant, and vulnerable urban area. The proposed system not only shields the city against floods and stormwater; it provides social and environmental benefits to the community, an improved public realm.

> (Quirk 2014)

This shows a dedication to holistic design approach that considers all aspects of integral sustainable design. Their long-term systems approach can be seen with additional statements:

> The resulting designs would not only solve existing problems, but prevent the formation of new ones, proactively enhance the city in many dimensions, and channel its future growth in desirable directions.
>
> (Quirk 2014)

This statement underscores the dramatically different approach afforded by resilient design. The design team has incorporated economic growth, social and cultural richness, financial prosperity as well as the experiential qualities of the project along with deterring physical damage. The principles below further express the desire for deep integration.

The big U principles

- Flood protection and preparation are not a mere line of defense; they must take entire neighborhoods and districts into account.
- The design should be community-driven.
- The system should be compartmentalized and should be able to build incrementally.
- Physical resiliency should be combined with social resiliency.
- The requirements of different sectors (housing/transit/energy/urban development) should be addressed by one solution.
- Flood protection should be tied to community benefits (better open space and better access to housing, jobs and education, lower insurance rates, and possibilities for growth), allowing government investment to be leveraged with local and sectoral funding in a Resilient Community District.

This is just one example of hundreds of new resiliency projects taking a more comprehensive and integral view. The 100 Resilient Cities (100RC) initiated by the Rockefeller Foundation is "dedicated to helping cities around the world become more resilient to the physical, social and economic challenges that are a growing part of the 21st century" (Stockholm and Beijer 2018). 100RC supports the adoption and incorporation of resilience measures that also consider the social and cultural stresses of catastrophic weather events along with the economic and physical damage.

Additional resources

100 Resilient Cities Website: http://stockholmresilience.org
Governing Disaster in Urban Environments: Climate Change Preparation and Adaption after Hurricane Sandy, Julia Nevarez, Lexington Books, 2018
The Big U: www.rebuildbydesign.org/data/files/675.pdf
The BIG U: BIG's New York City's Vision for "Rebuild by Design," www.archdaily.com/493406/the-big-u-big-s-new-york-city-vision-for-rebuild-by-designcc

11.8 Urban scale sustainable design and health + well-being

Solutions to urban health problems require the effective involvement of non-health sectors (e.g. industry, transport, labour, education, commerce/trade, municipal utilities and services, urban planning, etc.), as well as nongovernmental organizations, the private sector, and the community. The overall strategy employed by the Healthy Cities initiatives is to generate intersectoral action and community participation to integrate health protection.

(World Health Organization 2018)

Bioinspired design and resiliency strategies are part of sustainable urban design. They directly impact the health within cities by reducing exposure to pollution and disease, encouraging physical exercise, and increasing access to fresh food with urban agriculture. Some may consider them "preventative health care" because they strive to keep people from needing medical treatment. Citizens who are healthy have a higher quality of life and allow medical resources to be used by those with serious illness or trauma. The social and economic benefits are obvious.

Health care and health equity are subjects at the forefront of both medical professions and sustainable design. Not all cities and health systems have reached the point where comprehensive health care is available to all. "To improve health, it is now believed that projects must address poverty and the deprivations associated with it, not simply those diseases associated with poverty" (Kenzer 1999). The challenges are deep and complicated, and they involve policymakers, social structures, the medical profession, cultural norms, and many other aspects too numerous to fully cover here. "The poor districts of the world's cities still have higher mortality rates than more privileged neighborhoods" (Sternberg 2010, p. 259).

As a small example, New York City has a substantially lower obesity rate than the rest of the state. Further investigation points to the built environment of the city encouraging walking. Congested traffic and fewer parking spaces make walking the better and faster option most of the time. Studies by urban designers have shown that simply having interesting details and finishes to look at can increase the desire to walk. To promote a sense of safety having well lit paths that are adjacent to shops, public areas, and other people is important. Gardens, vegetations, and the proximity to parks also increase the attractiveness of walking.

Place shapes behavior, so as design professionals we need to influence the areas that we have control over. Many times, design professionals can be the leaders influencing clients, city officials, and stakeholders towards a holistic and integral approach by sharing the proven benefits of changes in the built environment. For more specific strategies and their benefits, see Chapters 8 and 6.

Additional resources

Framework urban regions: www.wpro.who.int/health_promotion/documents/regional_guidelines_ for_developing_a_healthy_cities_project
Healthy Cities: A guide to the Literature: http://journals.sagepub.com/doi/pdf/10.1177/095 624789901100103
Healing Spaces, Esther M Sternberg, MD

11.9 Urban sustainable design and integrative process

Collaborative design practices

Architects, designers, landscape architects, planners, urban designers, and civil engineers are discovering the benefits of collaboration and stakeholder engagement early in the design process. This approach not only provides a deeper understanding of the project but fosters community engagement and support. The city of Detroit used this approach when they were establishing their "Detroit Future City: 2012 Detroit Strategic Framework Plan." This plan was to act as a blueprint to transform the city. Because of the decline in the automobile industry, the major employer of the city, population had dropped 24% and the amount of vacant properties increased by 20%.

Planning efforts had been unsuccessful in the past because of public cynicism about planning effectiveness, a sense of immobilization, fatigue stemming from the difficult circumstances in the city, and Detroit's long history of racial tension

The city of Detroit wanted to wage a sustained campaign to engage the community in the visualization of the new Detroit. Their first step was to plan "Increasingly, authentic and inclusive civic engagement…[as] a core element of many planning processes [1]." The examples below are from the Detroit Works plan:

In-person engagement tactics

- Large-scale town hall meetings (community learning);
- Topic-based summits (gathering input, community learning/dialogue, and idea generation);
- Community conversations (gathering input, building trust, and community learning/dialogue);
- The "Roaming Table" (building trust);
- Open houses and drop-in visits to the project's Home Base office (gathering input, building trust, and community learning/dialogue);
- Attending or presenting at existing community meetings (gathering input and community learning/dialogue);
- Street team door-knocking and leafleting (building trust);
- Technical team working sessions (synthesizing input).

In addition to direct strategies to engage stakeholders, virtual engagement tactics were also used by the team. The team also initiated virtual space relationships. This type of interaction is largely an unexplored relationship in the design professions. The Detroit project is making inroads with this type of engagement. Some examples are shown below:

Virtual engagement tactics (including both online and phone)

- Telephone town halls (building trust and community learning);
- "Detroit 24/7" online planning game (gathering input, idea generation, and community learning/dialogue);
- E-newsletters (building trust and community learning);
- Home Base "hotline" calls (community learning);
- Website updates and social media (building trust and community learning);
- "Detroit Stories" video history project (building trust and community learning/dialogue);
- Earned media, such as print, radio, and television media and communications features (building trust and community learning).

These strategies listed above work well for engaging the social, economic, and ecological initiatives. Out of the Detroit urban plan, new projects are beginning to move towards completion. The Fitzgerald Neighborhood Revitalization Plan, designed by Spackman, Mossop and Michaels, is built upon information gathered through the city's engagement plan, and uses many integral and holistic strategies. The project includes interconnected greenways, urban farming, and urban orchards to be interspersed throughout the neighborhood. All these strategies are bringing nature and food production into this neighborhood, increasing it's beauty along with the adjoining properties, and providing a local resource of fresh food and green space, where once there were only abandoned properties.

Additional resources

Detroit Works Long-Term Planning Project: Engagement Strategies for Blending Community and Technical Expertise, www.mdpi.com/2075-5309/4/4/711/pdf
More on the Fitzgerald Neighborhood Revitalization Plan, http://spackmanmossopmichaels.com/project/fitzgerald-neighborhood-revitalisation-plan/

11.9.1 Smart Growth

Starting from the beginning without a road map or a set of guidelines is difficult and overwhelming. Smart Growth is the first of three sustainable planning tools that have been in use for the past few decades. It has much in common with transit-oriented development (TOD) and New Urbanism which will also be covered in this chapter. Smart growth is the most regional of the sustainable urban design approaches. It emerged as a response to the rise of unplanned suburban sprawl and is aimed at providing a more integrated and sustainable model of development. The principles shown below reflect a holistic approach to planning.

Smart Growth Principles from a Triple Bottom Line Perspective are from a report by the Victoria Transport Policy Institute (Litman 2018):

Economic goals: Reducing development cost | Reducing public service costs | Efficient transportation | Clean industries;
Social goals: Improved housing | Community cohesion | Preserving unique cultural resources.

When we include the integral quadrants as a lens for Smart Growth, we can see that the objectives and strategies are holistic. Smart Growth Principles from an Integral Sustainable Design Perspective are as follows (Smart Growth Network 2018):

Mix land uses: This principle aims to create an *integrated system* of land use where each function can strengthen the other leading to more high performing buildings from lower embodied energy costs, lower transportation energy due to walkability, and shared walls decrease heating and cooling costs.
Take advantage of compact building design: Compact building design optimizes function while minimizing material resources in a structure. A smaller more efficient house saves energy and pollutes less, thereby maximizing performance.
Create a range of housing opportunities and choices: A range of housing opportunities and choices is focused on fostering diversity in communities. Low to high rent choices mean that a broader range of people will likely live near each other as opposed to traditional suburban planning which repeats the same house type in a given development (Culture).

Create walkable neighborhoods: Walking is a behavior that improves human health (performance) and increases the likelihood of chance encounters with neighbors (Culture), and if the community is well designed and attractive, it will enhance the experience of the person walking. Distinctive, attractive communities with a strong sense of place are not usually found in suburban developments. This principle is aimed at heightening the experience of residents by using place-based strategies.

Preserve open space, farmland, natural beauty, and critical environmental areas: Environmental *systems and patterns* are critical for human health. Protection of ecosystems will have long-term benefits to residents through cleaner air and water, and will heighten the experience of residents through biophilic connections with nature.

Strengthen and direct development towards existing communities: This principle *maximizes performance* by reducing transportation distances and has the potential to improve the culture of a neighborhood that is underdeveloped thereby increasing economic opportunities and job choices. However, with that comes the risk and rewards of gentrification which, depending on the point of view, can have a devastating effect on families that have lived in the same neighborhood for generations.

Provide a variety of transportation choices: Transport *systems* integration allows for the maximization of transport *performance to save money and reduce pollution.*

Make development decisions predictable, fair, and cost-effective: This principle attempts to bring *equity* and empathy into the center of decision-making processes – a critical initiative when dealing with developers who are attempting to *maximize profit* and will avoid any extra costs that might make a project better for the entire community.

Encourage community and stakeholder collaboration in development decisions and integrated design: Community *culture and equity* is acknowledged by co-creative processes. Equity for residents is improved by inclusion in processes.

Critics assume that consumers are selfish, inflexible, and lazy, and so, once accustomed to sprawl and automobile travel, cannot change. Experience, however, indicates that most people are actually quite generous and creative, enjoy walking, and tend to flourish in Smart Growth communities.

(Litman 2018)

Additional resources

Examples of Smart Growth Communities and Projects, www.epa.gov/smartgrowth/examples-smart-growth-communities-and-projects
Where We Want to Be: Household Location Preferences and Their Implications for Smart Growth: Tod Litman, www.vtpi.org/sgcp.pdf

11.9.2 Transit-oriented development

TOD is very similar to Smart Growth. It zooms into the opportunities for vital communities centered around public transportation, and offers more specifics about placemaking and public space. In fact, a set of specific placemaking principles was developed to help designers make choices that will meet the four perspectives of Integral Sustainable Design. It links the urban, district, site, and building scales in the types of principles it proposes.

Sidewalk cafes and ground floor retail are building scale design approaches that add to the pedestrian scale experience, and enliven the cultural vitality of urban space and an entire urban district. Linking these districts to transit hubs, jumps scale again as the region systems of *transport* are now available. All of these lead to better energy and environmental performance which impacts regional and even global scale health.

A summary of defining principles would be:

- Close proximity to rail station;
- Well-defined public spaces – outdoor rooms;
- Mix of uses to create lively, vibrant places;
- Pedestrian scale – comfortable, safe, enjoyable;
- Active ground-floor retail;
- Sidewalk cafes;
- Tree-lined streets.

TOD, like Smart Growth, seems obvious like high-quality design. It is important to remember that without guidance from design professionals, many developments around the world choose to avoid these basic principles. Cities end up being devoid of cultural vitality, lack heightened experiences, rely on a single car-based transport systems, and as a result end up wasting energy and polluting the environment.

Additional resources

Transit Oriented Development Institute: www.tod.org/placemaking/principles.html

11.9.3 New Urbanism

New Urbanism starts with a slightly different premise than TOD or Smart Growth. New Urbanism begins with placemaking at the center of its approach. It starts with the idea that good neighborhoods have been built for thousands of years and can serve as a model for new developments or urban infill projects.

Examining the principles below, the focus on small-scale neighborhoods is obvious. Of particular note is the principle of mixing all types, costs, styles, and sizes of housing in close proximity. This follows traditional successful development patterns, allows for different housing needs, and promotes a wider economic range of residents, which promotes a naturally created diversity – a core sustainability value. New Urbanism hits many of the sustainability values with its very holistic approach. Beauty, another core value of sustainability, is also included in New Urbanism but is addressed as a preference for historic architectural styles or ornamentation.

Principles of New Urbanism

- Walkability – A ten-minute street friendly walk from home and workplace;
- Connectivity – Interconnected network of streets;
- Mixed use and diversity – Shops, offices, and apartments within the neighborhood;
- Mixed housing – All ranges of sizes and types of houses in close proximity;
- Quality architecture and urban design – Creating a sense of place along with aesthetics and beauty;
- Traditional neighborhood structure – Hierarchy of rural to urban transit. Diversity of land use in terms of public to private spaces;

- Increased density – More residential, office, shops, and other services in close proximity;
- Green transportation – Pedestrian and bicycle friendly along with high-quality trains and other public transport connecting different neighborhoods, towns, and cities;
- Sustainability – Eco-friendly technologies and reduced negative impact on the environment;
- Quality of life – Improve and uplift the quality of life.

The one criticism or area of puzzlement with New Urbanism is the instance on historic architectural precedent. It is true that historic architectural typologies represent a visually unified yet diverse built environment that respects and supports the sense of human scale which heightens *experience* and increases *cultural connection*. But an architect working within the new urbanist system should be given the freedom to creatively meet all the positive aspects of the historical systems while also expressing the cultural aesthetics of the current period. Doing so illustrates the evolving cultural influence which creates a rich experience by the layering of time.

Additional resource

New Urbanism, www.newurbanism.org/newurbanism/principles.html
25 great ideas of the New Urbanism, www.cnu.org/publicsquare/2017/10/31/25-great-ideas-new-urbanism
Smart Growth and New Urbanism: What's the difference? www.cnu.org/publicsquare/smart-growth-and-new-urbanism-what%E2%80%99s-difference

11.9.4 Additional frameworks for urban sustainability

LEED for Cities: https://new.usgbc.org/leed-for-cities
China Urban Sustainable: http://mckinseychina.com/wp-content/uploads/2014/04/china-urban-sustainability-index-2013.pdf
City Blueprints: https://eip-water.eu/City_Blueprints
Global City Indicators: www.globalcitiesinstitute.org/
Indicators for Sustainable: http://sustainablecities.net/
Reference Framework for Sustainable Cities: www.rfsc.eu/
STAR Community Rating: www.starcommunities.org/about/
BREEAM Communities: www.breeam.com/communitiesmanual/
Eco2 Cities: https://siteresources.worldbank.org/INTURBANDEVELOPMENT/Resources/336387-1270074782769/Eco2_Cities_Book.pdf

11.10 Urban scale sustainable design and global scale

The Sustainable Development Goals (SDGs) began as a global scale initiative, and continue to catalyze and shape the sustainable development movement in new and powerful ways. The SDGs directly reference to the design of cities. Chapter 10 covers the SDGs in detail. In this chapter "Goal 11: Sustainable Cities and Communities" is influencing cities to make sustainable design integral to the transformation of cities over the next decades.

Goal 11: Sustainable Cities and Communities
Overall goal: Make cities inclusive, safe, resilient, and sustainable

The SDGs have an underlining theme of resilience and equity, which are important aspects of a sustainable city. The quote below is from the SDG's website:

> The challenges cities face can be overcome in ways that allow them to continue to thrive and grow, while improving resource use and reducing pollution and poverty. The future we want includes cities of opportunities for all, with access to basic services, energy, housing, transportation and more.

(Source SDG website)

The term growth plays a prominent role in the statement, underscoring the belief that sustainable development is as much about private investment as it is about infrastructure. In fact, the public-private partnership model for the development of sustainable cities is gaining favor. Projects that combine public financing and resources with private resources are starting to become more common in large-scale urban projects.

The targets and indicators for this SDGs are too numerous to describe in full here. Below are some selected targets and indicators:

Target 11.1 – By 2030, ensure access for all to adequate, safe and affordable housing and basic services and upgrade slums;

Indicator 11.1.1 – Proportion of urban population living in slums, informal settlements, or inadequate housing.

The first target and indicator directly attacks the fundamental problems of poverty in cities. Maslow's hierarchy of needs is reflected here in assigning basic rights like access to the commons of clean water and air, sanitation, and more. The indicators are a way to measure progress towards meeting the targets. The target below illustrates the desire to increase the amount of participation in the integrated design process.

Target 11.3 – By 2030, enhance inclusive and sustainable urbanization and capacity for participatory, integrated and sustainable human settlement planning and management in all countries;

Indicator 11.3.1 – Ratio of land consumption rate to population growth rate;

Indicator 11.3.2 – Proportion of cities with a direct participation structure of civil society in urban planning and management that operate regularly and democratically.

Additional resources

Goal 11: Make cities inclusive, safe, resilient and sustainable, www.un.org/sustainable development/cities/

Getting started with the SDGs in Cities: https://sdgcities.guide/

Regernerative Cities, www.worldfuturecouncil.org/wp-content/uploads/2016/01/WFC_ 2010_Regenerative_Cities.pdf

Regenerative Cities: Moving Beyond Sustainability, www.urbandesignmentalhealth.com/ journal-3---la-regenerative.html

Regenerative Cities Making Cities work for people and planet Background Paper, www. lowcarbonlivingcrc.com.au/sites/all/files/regenerativecities_backgroundpaper2.pdf

Sustainable Development Goals, United Nations, https://sustainabledevelopment.un.org/sdgs

11.11 Sustainable urban design

Most major cities in the world have well-staffed sustainability departments, a indication that sustainable design is more than just good intentions. Cities are seeing tangible benefits from sustainability efforts in the form of energy savings, costs savings, and environmental improvement. The social goals of sustainable design are helping to build more equity, transparency, and accountability into urban processes. The renewed focus on design, especially the act of placemaking, is making cities more beautiful than ever.

Goal setting

Copenhagen, Denmark is one of the best examples of a sustainable city with ambitious goals to reach carbon neutrality by 2035 and zero waste by 2050. These goals cover Mobility, Water, Energy and Resources, and Strategy. According to the design team the main goal of the Copenhagen plan is to

> …show that it is possible to combine growth, development and increased quality of life with the reduction of CO_2 emissions. It is all about finding solutions that are smarter, greener, healthier and more profitable. And by 2025 we will be able to call ourselves the world's first carbon neutral capital.
>
> (City of Copenhagen 2014)

This is a very broad and ambitious goal that Copenhagen is approaching with smaller integrally focused goals.

Eco-efficiency

The comprehensive consideration of systems thinking across scale led to changes, which over time resulted in dramatic performance improvements. Some of the most recent performance levels are shared below (City of Copenhagen 2014):

- 70% of CO_2 reduction is observed while using seawater-assisted district cooling system as compared to traditional cooling methods.
- 20% of CO_2 reduction was observed from 2005 to 2011, and the target is to be carbon neutral by 2050.
- The reduction of sulfur dioxide in the atmosphere is due to the reduction of the dependency of coal in the city for its heating and cooling.
- The reduction of lead pollution in the atmosphere reduced in the 1990s due to the regulations of banning fuels for transport.
- 98% of the heating comes from district heating grid. The district heating grid is one of the most carbon efficient ways to produce and supply energy. It includes energy from biomass, wind, and geothermal to replace fossil fuels.
- 80% reduction in consumption of electricity when using district cooling as compared to traditional cooling.
- 20% reduction of heat and electricity consumption in buildings compared to 2010.

Copenhagen is using smart city principles as part of its overall sustainable design approach. In 2012, a budget of 40 million euros for intelligent street lighting, transport, and solar energy projects in the city was approved.

One of their priorities is increasing the number of people who choose cycling as their main form of transportation. If the Triple Bottom Line was used to organize the performance, social, and environmental benefits, it would look like this.

Economic

- Cycling provides a low cost form of transport.
- Reduced journey times and traffic congestion increase economic productivity.
- Healthier citizens reduce health-care costs at an estimated rate of 0.77 euros per km cycled.

Social

- 88% of cyclists do it because it's the fastest or most convenient way of getting to work.
- Creation of jobs.
- Improved city life.

Environmental

- Reduced noise.
- Reduced air pollution.
- Reduced carbon dioxide emissions.

Life cycle systems: 2018 move from 45% of household waste recycling to zero waste city by 2050.

Active systems heating and cooling: The use of district heating and cooling for buildings is the key to more energy efficiency, using a combination of biomass, geo-exchange, and fossil fuels to generate district level heating, cooling, and electricity. Waste heat from making electricity is sent to buildings in the winter. Cold seawater is used to help with district cooling.

Active energy systems: 50% of city's electricity needs by 2020 and 100% by 2025 are to be met with wind turbines.

Active transport systems: Copenhagen has made the transition from a car culture to a bike culture – a key strategy in helping to achieve carbon neutrality by 2030. The integration of bike storage with public transportation makes it easier for people to do without their cars. The use of smart city technologies allows for residents to see where buses are and when they will arrive at their destination. The technology also allows for green lights to stay on longer to speed up the movement of buses through the city. The compact layout of the city itself is a prerequisite for the success of bicycles travel because distances are much shorter compared to sprawling cities.

Active water systems: Many cities have combined sewer systems which means that wastewater mixes with stormwater during intense storms. This results in wastewater directly entering water bodies instead of water treatment plants. Copenhagen planners propose to create reservoirs where excess stormwater can be stored until there is capacity to handle the water. They are using a multi-scale approach to water by looking at large-scale infrastructure projects and human scale water meters.

Living systems: The use of green roofs is becoming more prevalent as a means to manage stormwater and increase biodiversity in the city. More green spaces are also introduced as a means to address city resilience.

Human systems: Community ownership of wind turbines and public campaigns to attack negative perceptions of wind turbines are being used to educate and provide incentives.

From the Culture Perspective, Copenhagen is using incentives to encourage property owners to deal with stormwater directly on-site either through cisterns or percolation into the soil – a culture change. Water meters help residents track their water use leading to less water consumption, and the city designates funding to explore ways to protect groundwater supplies.

The culture of Copenhagen is not only using goals and metrics to drive the design of efficient systems, but is using long-term thinking to consider the future of the city in uncertain climatic conditions.

Copenhagen and resilience

Resilience and adaption are parts of Copenhagen's sustainable strategy. Some of the threats identified include a rise in peak of summer temperatures by 2–3 degrees by 2050; precipitation increases of 30%–40% by 2100; and seal level rise of 1 m in the next 100 years. These environmental risks represent substantial threats to the existing city which is only 45 feet above sea level. Design thinking has led a series of strategies to address the threats: Develop infrastructure systems to drain water from city during intense storms; additional green spaces act as retainment facilities and reduce the risk of flooding; passive cooling strategies such as shading, better insulation, and ventilation are proposed to deal with hotter weather and possible interruptions of electricity supplies.

Additional resources

State of Green Website, Denmark https://stateofgreen.com/files/download/1174
Copenhagen Solutions for Sustainable Cities, https://international.kk.dk/sites/international.
kk.dk/files/Copenhagen%20Solutions%20for%20Sustainable%20cities.pdf

11.12–11.14 Urban sustainable design across scales

The impact of cities at the district, site, building, and human scale cannot be underestimated. In the upcoming chapters the role that cities play in shaping the sustainable design approaches at each scale will be explored.

Eco-city initiative

Ten principles of eco-city are as follows:

- Compact, mixed-use urban form that is efficient and protects the natural environment.
- Green space that permeates the cities and provides a major proportion of its own food.
- Road infrastructure to focus on transit, walking, and cycling and minimal use of car and motorcycle.
- To use closed-loop systems for water, energy, and waste management.
- Central city to emphasize circulation by different modes of transit other than cars to bring in more employment and residential growth.
- High-quality public culture, community, and good governance.
- The physical structure and urban design of the city, especially its public environments, are highly legible, permeable, robust, varied, rich, visually appropriate, and personalized for human needs.

- The economic performance is maximized through innovation, culture, and social quality of the city's public environments.
- Planning for the future of the city is a visionary "debate and decide" process, not a "predict and provide" computer-driven process.
- All decision-making is sustainability based, integrating social, economic, environmental, and cultural considerations as well as compact, transit-oriented urban form principles.

(Kenworthy 2006)

References

Alexander, D., 2018. *The Necessity of Eco-Cities: Why Cities are Central to Achieving Sustainability*. www.newcity.ca/Pages/cities_and.html edn.

Badger, E., 2017. How Redlining Racist Effects Lasted for Decades. *The New York Times*. https://www.nytimes.com/2017/08/24/upshot/how-redlinings-racist-effects-lasted-for-decades.html [November 3, 2018].

City of Copenhagen, 2014. *Copenhagen: Solutions for Sustainable Cities*. Copenhagen: Solutions for Sustainable Cities.

Cohen, B., 2004. Urban growth in developing countries: a review of current trends and a caution regarding existing forecasts author links open overlay panel. *World Development*, 32(1), pp. 23–51.

Frank Lloyd Wright Foundation, September 8, 2017-last update, Revisiting Frank Lloyd Wright's "Broadacre City" [Homepage of Frank Lloyd Wright Foundation], [Online]. Available: https://franklloydwright.org/revisiting-frank-lloyd-wrights-vision-broadacre-city/ [July 25, 2018].

Grimm, N., Faeth, S.H., Golubiewski, N.E., Redman, C.L., Wu, J., Bai, X. and Briggs, J.M., 2008. *Global Change and the Ecology of Cities Charles L. Redman3, Jianguo Wu1,3, Xuemei Bai4, John M. Briggs*. American Association for the Advancement of Science. http://science.sciencemag.org/content/319/5864/756 edn.

Kentworthy, Jeffery R., 2006. The eco-city: Ten key transport and planning dimensions for sustainable city development. *Environment & Urbanization. International institute for Environment and Development*, 18(1), pp. 67–85.

Kenzer, M., 1999. Healthy cities: A guide to the literature. *Environment and Urbanization*, 11(1), p. 201.

Lambert, B., 1997. At 50, Levittown Contends With Its Legacy of Bias. *The New York Times*. Available: https://www.nytimes.com/1997/12/28/nyregion/at-50-levittown-contends-with-its-legacy-of-bias.html

Litman, T., 2018. *Where We Want to Be Home Location Preferences and Their Implications for Smart Growth*. Victoria Transport Policy Institute. www.vtpi.org/sgcp.pdf edn.

Lucy, M., 2014. *Green Roofs in Washington are Expanding*. GBIG. http://insight.gbig.org/green-roofs-in-washington-dc-are-expanding/ edn.

Mostafavi, M., 2010. Why ecological urbanism? Why now?. *Harvard Design Magazine*, 1(32). Online resources, no page range given http://www.harvarddesignmagazine.org/issues/32/why-ecological-urbanism-why-now [November 3, 2018].

Newman, P., 2014. *Biophilic Urbanism: A Case Study on Singapore*. Taylor & Francis Online. [Online] Available: https://www.tandfonline.com/doi/full/10.1080/07293682.2013.790832 [November 3, 2018].

Quirk, V., 2014. *The BIG U: BIG's New York City's Vision for "Rebuild by Design"*. ArchDaily. www.archdaily.com/493406/the-big-u-big-s-new-york-city-vision-for-rebuild-by-design edn.

Richard, R., 1993. *Ecocity Berkeley*. 1st edn. Berkeley, CA: North Atlantic Books.

Smart Growth Network, 2018. *This is Smart Growth*. Smart Work Network. Environmental Protection Agency. https://www.epa.gov/sites/production/files/2014-04/documents/this-is-smart-growth.pdf [November 3, 2018].

Sternberg, E.M., 2010. *Healing Spaces: The Science of Place and Well-Being*. 1st edn. Cambridge, MA: Belknap.

Stockholm University and The Beijer Institute Ecological Economics, Stockholm Resilience Centre [Homepage of Ministra], [Online]. Available: http://stockholmresilience.org/ [July 26, 2018].

The Cultural Landscape Foundation, 2018-last update, Ian McHarg – Pioneer Information [Home page of The Cultural Landscape Foundation], [Online]. Available: https://tclf.org/pioneer/ian-mcharg?destination=search-results [July 25, 2018].

Triman, J., November 4, 2012-last update, Interview with Dr. Stephen Kellert [Homepage of Biophilic Cities], [Online]. Available: http://biophiliccities.org/interview-with-dr-stephen-r-kellert/ [July 26, 2018].

Victoria Transport Policy Institute, Victoria Transport Policy Institute [Homepage of Victoria Transport Policy Institute], [Online]. Available: www.vtpi.org/ [July 26, 2018].

Vidal, J., 2014. *Corporate Stranglehold of Farmland a Risk to World Food Security, Study Says.* Guardian News and Media Limited. www.theguardian.com/environment/2014/may/28/farmland-food-security-small-farmers edn.

Wellington City Council, 2018. Open Spaces [Homepage of Wellington City Council], [Online]. Available: https://wellington.govt.nz/services/environment-and-waste/environment/open-spaces [July 26, 2018].

Wikipedia, July 22, 2018-last update, Paolo Soleri [Homepage of Wikipedia], [Online]. Available: https://en.wikipedia.org/wiki/Paolo_Soleri [July 25, 2018].

World Future Council, 2018-last update, World Future Council [Homepage of Future Policy.org], [Online]. Available: www.worldfuturecouncil.org/.

World Health Organization, 2018. Types of Healthy Settings [Homepage of World Health Organization], [Online]. Available: www.who.int/healthy_settings/types/cities/en/ [July 26, 2018].

Zimmer, L., 2013. *Hundreds of Vacant Lots to Become World's Largest Urban Farm in Detroit.* Inhabit. https://inhabitat.com/hundreds-of-vacant-detroit-lots-to-become-worlds-largest-urban-farm/ edn.

12 District and site scale sustainable design

12.0 Terms

Districts are typically defined areas within a city, essentially a subset of a larger urban area. Districts can have themes or sometimes zoning ordinances that dictate what types of activities take place there, like the "shopping" district, "entertainment" district, or "residential district." From a social or economic perspective, districts can also come with negative connotations as in ghettos, favelas, or slums. Sustainable designers are keen to attack the problems in these kinds of districts as a way to fight for environmental, economic, experiential, and social improvements. A neighborhood or community is often contained within a district.

Campuses are similar to districts in that they contain a collection of buildings and shared outdoor spaces but have reasonably defined boundaries as in a college campus or a corporate office park.

Sites are bounded areas where a specific building or project is located. It is bounded by a legally defined property line, which could be a street, river, geological feature, or a line drawn on a map. An urban park surrounded by roads is an example of a site and so is a private plot of land that may contain a building. While the actual "site" has boundaries, it is a smaller part of the district that surrounds it, and the district is part of the urban context.

Infrastructure

Sustainable design at the district and site scale has the potential for integrating nature and sustainable site infrastructure across multiple projects. Shared amenities, stormwater infrastructure, outdoor spaces, and in some cases, renewable resource collection can be shared by multiple owners or the neighborhood. Public space within each district can provide these functions and enhance cultural or social vitality as well.

The district scale, with its smaller governing bodies, can sometimes enact changes quicker than larger urban, state, or federal entities. The officials are closer to the residents of the area and have a more intimate understanding of the area's needs.

This chapter will give the beginning designer a exposure to the possibilities and advantages of working at the site and district level, with the hope to spur creativity and diversity in design directives. For professionals that don't normally work at these scales, this baseline knowledge can serve to expand design concepts to larger scales.

Examining sustainable design strategies at the district scale is meant to provide a contextual backdrop and a rationale for the pursuit of sustainability and promote systems thinking and the interconnectedness of Integral Sustainable Design. For the experienced designer, educator, and expert who may wish for more depth, this chapter is meant as a starting point for individual exploration and discovery.

12.1 A brief history of districts and sites

In the age of industry, the majority of the Western world was finding ways to maximize profit by extracting as much wealth from the land as possible without regard for future impacts. However, there was a thread of sustainability, and in 1905, President Teddy Roosevelt started the process of protecting nature by establishing the national park system. It started with John Muir, advocating for the rights of nature in the Yosemite area, and moved to fighting the construction of dams and educating the public about the importance of nature. It wasn't long after that landscape architecture became an acknowledged profession.

By the 1960s, environmental awareness was growing and Ian McHarg wrote the seminal book, *Design with Nature*, which launched a new design consciousness. The development of Geographic Information System (GIS) accelerated and is now the basis for a new *system-based* ecological perspective. The publication of *Our Common Future* in 1987 and the release of William McDonough's Centennial Sermon in 1993 solidified the concept of sustainability in the minds of built environment leaders around the world.

By 2005, the sustainable design movement had matured and now reflects the emergence of the latest worldview of integration. Today, the designs of districts and sites are heavily influenced by a wide range of new thought patterns and approaches that seek to reintegrate nature back into the city.

12.2 Nature: urban districts and sites

Landscape ecology is a design movement that has reshaped the way landscape architects and scientists think about the built environment. The term itself is a unification and suggests a cross-disciplinary approach that links the science of ecology with the different scales and approaches of landscape architecture, urban design, and planning.

Landscape ecology parallels other linkages in the design world like building science and architecture, or evidence-based design and interior design. Landscape ecology prizes the inclusion of scientific approaches focused on diverse habitats. More specifically, the fragmentation or connection of habitats is a prime concern in ecological health. This approach supports earlier discussions about the health and well-being of ecosystem services and for humanity itself.

GIS technology allows for a much better visualization of landscape patterns, leading to more insights into the intense fragmentation that is occurring because of the built environment. More specifically, landscape ecology calls upon biologists, social scientists, and others to incorporate a research-based approach that includes conservation, management practices, and long-term sustainability. The study of interlinked patterns across the landscape and across scale is of paramount importance in the process of identifying opportunities to make the fragmented landscape whole again.

The use of green infrastructure is central to the achievement of landscape ecology. Like the stepwells in India, the use of green infrastructure, especially for water collection and treatment, is also increasingly used as an amenity for urban dwellers (EPA 2018).

Additional resources

Principles of Ecological Landscape Design, Travis Beck, Island Press
The Evolving Practice of Ecological Landscape Design, https://thefield.asla.org/2017/08/01/
   the-evolving-practice-of-ecological-landscape-design/

12.3 Humanity: urban districts and sites

The first invisible hand, profit, drives the fixation on initial costs and profits, generating tension between altruistic motivations to preserve history, provide connection with natural elements, and leave open space for social interaction. The tension can also be increased by rating systems and codes that dictate higher building standards like Passive House, which require alterations to the facades of old buildings to meet ambitious energy standards.

Other areas where we see conflict between financial interest and altruism, beauty, and cultural interest at the district level are historic preservation, and resource and infrastructure expenditures, which are based on economic or ethnic background of neighborhoods.

If a district is designated as a "historic district," restrictions on new construction or renovations may dictate the preservation of the cultural heritage of existing structures and neighborhood. It has been shown that history, beauty, and culture are very important, but restoration of these structures can be more costly than new construction and is sometimes seen as unnecessary. Real estate values in these areas can become elevated, resulting in less diversity in the people who can afford to live in these areas.

Areas with residents on the lower socioeconomic scale are provided less infrastructure and fewer services than are provided in higher economic areas. Some justifications that are used for this are lower tax-based contributions, no representation, and that the people in control don't have to live in the conditions and so are unaware or don't care; these are all profit-motivated views. The view of empathy realizes that we as a society are only as strong as our weakest link and that when all people prosper the society does better and is stronger. Empathy also understands that no one should suffer simply because of where they live and their ethnic background.

"Branding" or designating a district for specific functions can be profitable and efficient, and can create cultural and social connection as well as be profitable for businesses, as in the "art district" or "theater district." There are also benefits in shared land use and sustainability synergies that can occur in these areas. The problem occurs when branding is used to deny the residents services and opportunities, or to create divisions or segregation. An example would be the "redlining" that occurred in the 1960s to African-American and ethnic neighborhoods spoke of in previous chapters. We can also see it in the quality of the education and facilities of schools in lower income urban areas.

Gentrification also happens at the district scale. It undermines the cultural diversity and displaces long-time residents who can no longer afford to live in their neighborhoods. In most cases, it does not really revitalize an area, but just moves the poverty somewhere else.

Performance relationship

From the performance perspective, the use of green infrastructure to collect, treat, and reuse stormwater is a necessity for a sustainable site. Plants can be used to create microclimates to shade buildings and reduce the overall urban heat island effect. Sustainable sites can help to clean the air, clean the water, support biodiversity, grow food, and simply increase open space. Site design in the urban context, at a basic level, is the act of organizing ecosystem services to make the environment better serve humanity and animals without compromising the ability to be self-sustaining. The inclusion and design of open space in dense urban areas allow access to the commons of light, air, view, and biodiversity. Open space within urban areas also serves migrating species creating resting places, habitat, and greater biologic exposure to human occupants.

Experiential and cultural relationship

Incorporating sustainable site design connects human occupants to the natural world in meaningful ways without leaving the city. The heightened experiences of interacting with nature through the simple act of sitting under a tree or next to a small brook or laying in the sun and hearing birds sing creates well-being, relaxation, and a sense of escape that can be incorporated into daily routines. Cultural and social connections can be more meaningful because of these types of spaces.

12.4 Sustainability values: districts and sites

The sustainability values as organized by the four simple P's – People, Planet, Prosperity, and Place, serve as a foundation upon which sustainable design is built. In the case of districts, neighborhoods, and communities, the values are a unifying element to rally around or expose underlying conflicting beliefs because of the diversity of the various stakeholders. In many cases, the role of the sustainable designer or project leader is as much about facilitation and education as it is about the ability to make beautiful drawings. The rise of the designer as facilitator is covered in more detail in Chapter 9, but it bears repeating here for emphasis. The opportunity for urban designers and landscape architects to educate and promote the core values of sustainability is very high because of the varied clients and broad types of "sites" they have influence over. The team has a large palette of tools and strategies available to them. Open access to public space can be either a great generator of cultural vitality or a death sentence for the district if poorly designed, and a death sentence for the district, as we saw in the urban renewal low-income housing projects we profiled earlier.

Sustainability value: equity

Social equity goals encourage clients and designers to consider open access as a critical part of sustainability, understanding the need to consider larger social concepts as important design directives. Some rating systems like the Living Building Challenge specifically address access and inclusion. For example, they do not allow "gated" communities.

Sustainability values: high-performance and ecological regeneration

The performance value of ecological regeneration can drive design directives to reduce consumption at every level and include higher levels of system integration. The goal is to strive for eco-regeneration or establish new zones of biodiversity that can also handle stormwater and reduce active system loads with passive or renewable strategies.

Sustainable value: beauty

Beauty is always desired but often the first thing eliminated due to budget constraints. Integral sustainable design considers beauty as a real and valuable thing. Transforming and enriching the human experience is one of the main purposes of the built environment and motivates people to preserve and care for their structures.

12.5 Integral Sustainable Design and districts

The application of sustainability values through the action of integral sustainable design at the district and site scales is key to connecting buildings and their occupants to each other, creating a sense of community, ownership and belonging. It is also vital in connecting the building and interior scale to nature and achieving larger scale urban sustainability. The interaction between scales is best perceived at the site or neighborhood scale because districts, sites, and campuses are the places where people interact the most. It is also the easiest scale for designers to observe the holistic characteristics of sustainable infrastructure and open space.

12.5.1 High-performance districts and sites

The use of metrics, goals, standards, and benchmarks to guide the performance of a site or district design project can't be underestimated. Ambitious performance goals encourage the use of dematerialization to eliminate or reduce the footprint of products, services, rooms, and even entire buildings as means to save or reduce energy. Making a building smaller and taller leaves more space for plants and animals to inhabit the site, for stormwater treatment and collection, and for human interaction with the natural world. The smaller building also uses less energy and allows daylight to penetrate deeper into the heart of the building. High-performance design need not be perceived as utilitarian. More information on metrics and frameworks is covered in Section 12.9.1 where ecodistricts are explained.

12.5.2 Systems for districts and sites

When reading the chart on the next page, it becomes clear that the systems in a district and site offer strategies and technologies that can work together in a synergistic way. Shared stormwater can be directed to irrigation in a shared food garden, water feature, or constructed wetland in public open space.

The benefits of these systems to other quadrants of Integral Sustainable Design are obvious. Energy efficiency is greatly improved by system synergies. The experience in a district is deepened by the inclusion of diverse robust green space. Permaculture is a holistic system that seeks to unify human behavior, energy needs environmental impacts, and more. It is its own philosophy that predates sustainable design (Table 12.5.2a [next page]).

12.6 Bioinspired districts and sites

Many sites or districts are devoid of local wildlife and indigenous plants. Imagine the standard shopping mall, an endless sea of asphalt, interrupted by the occasional tree with water systems buried and piped to far away water treatment plants. Indigenous species are important because they are better adapted to survive in the climate zone of the project, require less irrigation, and provide food for indigenous species. There are many beautiful plants that birds, animals, and insects cannot eat because they are not native. Bio-inspired design of districts and sites helps to shift the mindset towards the creation of lush, verdant, and lively spaces for all species.

Table 12.5.2a Systems for districts and sites

	Passive systems	Active systems	Living systems	Human systems
Human health + wellness	Interconnected systems of walking and cycling; separation of car and people on streets	Rideshare, trains, buses, cars, boats	Hiking trails Horse riding	A culture of walking
Climate + resilience	Reserve areas of the site for storm surges	Highly efficient and well-maintained pump systems	Use biomimetic strategies to combats storms	Social resilience activities
Food	Arrange urban gardens for maximum solar access	Food trucks Power, hookups for urban gardens and food trucks	Organic gardening Permaculture	Policies and procedures for efficient gardening Financial systems Maintenance system
Flora + fauna	Locate parks within walking distance of as many residents as possible	Light parks and electrify parks with solar	Transform local parks into ecologically robust open space – connect open spaces for animal movement	Policies and procedures for collecting and sharing solar and wind power
Energy + transportation	Identify and set designate suitable locations for photovoltaic (PV) panels and wind	Carbon neutral districts Shared PV and wind power	Food can be grown on-site	Piezoelectric walkways
Earth + materials	Centrally located recycling and repurposing centers in districts and on-site	Net-zero waste reduces life cycle impacts of a district Highly efficient trucks	Composting within district Use phytoremediation to clean toxic soil in brown fields (move to site)	
Water	Highly efficient and well-maintained pump systems	Interconnected systems of walking and cycling; separation of car and people on streets	Hiking trails Horse riding	A culture of walking
Air	Reserves areas of site for storm surges	Rideshare, trains, buses, cars, boats	Use biomimetic strategies to combat storms	Social resilience activities

12.7 Resilient districts and sites

The relationship between bio-inspired design, resilience, health and wellness, and sustainable design is tightly woven and synergic. The Napa River Flood Protection Project, 1998–2012, is an example of an integrated approach to resilient design. The project addresses a long-standing problem of persistent flooding in the region. Since 1862, 22 major floods have struck the valley, leading to loss of life and property. As a response, a collaborative effort between the government sector, community organizations, and stakeholders developed a set of "Living River Principles" to guide the development of an integrated flood protection plan. Guiding principles and stakeholder engagement were the cornerstones of the sustainable design process. A few selected results of the project expressed though the Triple Bottom Line from the *Landscape Performance Series Case Study* are shown below.

Environment: Restored 75% of the historic wetlands north of Butler Bridge (over 700 acres), which has resulted in 71 species of migratory and resident birds observed on-site, including the peregrine falcon and burrowing owl.

Social: Integrated 2.5 miles of new, paved trail along the east bank of the Napa River into the developing San Francisco Bay Trail network, which will eventually create a continuous, 500-mile recreational corridor around the Bay.

Economic: Created an estimated 1,373 temporary construction jobs and 1,248 permanent retail and administrative jobs at properties developed in expectation of 100-year flood protection along the Napa River.

Notice that social outcomes also key into experiential aspects of the project, making it a true Integral Sustainable Design project. To learn more about this project and others, visit the links shared in the Additional Resources section below.

Additional resources

Napa Valley Case Study, https://landscapeperformance.org/case-study-briefs/napa-river-flood-protection

The Gentilly Resilience District in New Orleans, www.nola.gov/resilience/resilience-projects/gentilly-resilience-district/

Building Resilient Districts, New Orleans, https://medium.com/cities-taking-action/building-resilient-districts-6098ceab2de5

12.8 Health and districts

The potential of districts and communities to attack health problems is quite high. The mixture of buildings, parks, public transport, and infrastructure, if orchestrated properly, can become a catalyst for healthier living. Walkable communities, access to healthy food, fresh air, and clean water are all achievable in a district focused on health. A cognitive shift is needed to change the objective of designing from an artistic expression to include health, well-being, healing, and prevention, for people and the planet. At the district scale, initiatives and strategies are coming from evidence-based design, environmental psychology, the healthcare industry, and design firms or rating systems.

According to the architectural firm of Perkins and Will:

> Health District Planning prepares our healthcare clients for the new healthcare economy, where the investment will be shifting from "brick-and-mortar" facilities to shared community settings where disease can be maintained or prevented at lower costs.
>
> (Perkins + Will 2018)

Here, the shifts in thinking about human well-being, health care, and sustainable design are clearly expressed as being concerned with prevention at the community level. Perkins + Will's proposals, based on research of the changes in healthcare systems below, exhibit cross-disciplinary work that engages stakeholders and thinks across scales. They are proposed as follows:

Building scale: Sustainable healthcare facilities with active design and healthy materials.

Campus scale: Healthcare facilities that are walkable, accessible, and physically integrated into the urban fabric of surrounding communities.

District scale: A mixed-use district that is anchored by a healthcare facility, providing amenities and services that support quality of life such as housing options, access to transit, and access to active recreation, and is supported by resilient infrastructure.

City scale: Guiding, prioritizing, and organizing the healthcare provider's collaborations with other providers, government entities, and community organizations to improve the health of the community.

Regional scale: Identifying opportunities for contributing to the health of the local and regional economy through growth, partnerships, and new investment.

Another example of how a deep focus on health is shaping the city planners is the first WELL Certified District in Tampa, Florida. The mayor of Tampa, Bob Buckhorn, spoke about the project:

> Our city will demonstrate that city design, not just building design, can be healthy and sustainable, and it will position our community as forward thinking … Research shows that people who live in walkable, connected neighborhoods have lower rates of obesity, diabetes, high blood pressure, and heart disease.
>
> (Grauerholz 2016)

The WELL Building Standard, covered in 14.9.2, is a comprehensive rating system that guides the efforts of design teams to approach a project through the lens of health. Categories include air, water, nourishment, light, fitness, comfort, and mind. This is a radical departure from other rating systems covered later in this chapter, which include health but are focused on reducing energy usage and large environmental impacts. The WELL Building Standard is complementary to other standards like Leadership in Energy and Environmental Design (LEED) or Living Building Challenge.

Additional resources

Perkins + Will Research Journal, https://perkinswill.com/sites/default/files/ID%206_PWRJ_Vol0602_05_A%20Vision%20and%20Planning%20Framework%20for%20Health%20Districts.pdf

Living WELL: Tampa Becomes the First City in the World to Introduce a WELL Certified District, www.wellcertified.com/en/articles/living-well-tampa-becomes-first-city-world-introduce-well-certified-district

Avalon Park and Preserve, https://landscapeperformance.org/case-study-briefs/avalon-park-and-preserve#/sustainable-features

12.9 Integrated design process for districts and sites

The Living Well project in Tampa, Florida, consists of a collaborative team between private business, nonprofit organizations, and the district. The WELL Building Standard rating system was key in driving the collective understanding of the project's intent. These specific metrics helped to focus the design team's efforts to innovate.

12.9.1 Rating systems: EcoDistricts

There are a number of holistic, community scale planning methods that are transforming the way designers, residents, and leaders think about their neighborhoods. EcoDistricts, 2030 Districts, Living Community Challenge, and LEED for Neighborhood Development are all major rating systems for guiding the planning, process, and measuring the success of a project. Each approach has differences, but they all aim to achieve sustainability at the district scale.

The EcoDistrict framework is the first comprehensive protocol for the development of sustainable communities. It exemplifies all of the principles, concepts, and approaches covered in this book so far. There are literally hundreds of EcoDistricts scattered around the world.

EcoDistricts are organized around a nested "donut" of "imperatives," "priorities," and "implementation phases." At the center are the imperatives, the core values of an EcoDistrict: Climate, Equity, and Resilience. These are infused in every decision and strategy developed for the project. The middle ring of the donut illustrates the priorities or design principles. The six priority areas are Livability, Prosperity, Health + Wellness, Connectivity, Ecosystems, Stewardship, Climate + Resource Protection. The use of the terms "place" and "prosperity" links directly to the sustainable values covered in Chapter 4. "Resource Restoration" speaks directly to eco-efficiency as shown in the example below, and "connectivity" is related to human systems of transportation and mobility. The outer ring provides four Phase Implementation Models: Formation, Road-map, Action, and Stewardship.

The protocol is organized around a cascading set of topic areas, goals, and objectives. The EcoDistricts Protocol strives to incorporate change by:

1   Putting equity and inclusion at the start and at the heart of solution design.
2   Adopting the most holistic resilience lens to address the mounting climate crisis.
3   Redefining the way we work so collaboration is inescapable.
4   Embacing transparency to propel innovation.
5   Using peer exchange as a catalyst for transformation.

EcoDistrict Protocol

PRIORITY: Resource restoration.
GOAL: Move towards a net positive world.
OBJECTIVES: Increase efficient water use; divert waste from landfills; productively reuse remediated land; and pursue energy efficiency, technology advancements, and clean, renewable energy production that reduces greenhouse gas emissions (EcoDistricts 2018).

In 2002, the first EcoDistrict community went into operation in England. The BedZED project features six hectares of development with 82 units, commercial space, and a nursery. Innovative social concepts were employed such as varied housing options and ownership models, rooftop gardens, and rainwater collection. The project was innovative then and still remains a model for a sustainable urban development. The buildings are 88% more efficient than traditional structures of the same type, and hot water and energy use are also significantly lower. The community itself remains completely occupied with residents reporting they are overall happy.

EcoDistricts are not without criticism. In an article published by Paris Innovation Review, the authors noted the proclivity for politicians to use EcoDistricts as a "greenwashing" opportunity. Others critique EcoDistricts as a tool to enable the gentrification of poor neighborhoods, and others argue the districts are far more socially homogeneous than intended (Paris Innovation Review 2013). Despite the criticisms, EcoDistricts and their relatives shown below offer a framework, a road map, and metrics for those seeking to achieve higher levels of sustainable design. The systems are critical parts of a sustainable design methodology for ecologically design communities.

12.9.2 Rating systems: Sustainable Sites Initiative

Zooming in to the site scale, the Sustainable Sites Initiative (SSI) is offered as a road map for designers to maximize sustainable design opportunities in a project. It was developed as a partnership between the United States Botanic Garden, the Lady Bird Johnson Wildflower Center, and the American Society of Landscape Architects. The website describes its initiative as "an interdisciplinary effort to develop guidelines and a voluntary rating system for sustainable land design, construction, and maintenance" (ASLA 2018).

The SSI was designed as a regenerative guide for ecosystem services themselves. As a reminder, ecosystem services are the benefits that are provided by natural systems that support human survival but are often considered free and not a part of conventional accounting methods. Clean air and water are ecosystem services, but the intrinsic value of such services is difficult to quantify. The list from the SSI is given below:

- Regulate global and local climate | Detoxify and cleanse air, soil, and water;
- Control erosion | Replenish water supply | Decompose, treat, and reuse waste;
- Provide food and raw materials | Provide refuge and pollination services;
- Provide human health and well-being benefits | Cultural, educational, and aesthetic values.

The implication of stating the function of ecosystem services is the possibility that landscape architecture has the potential to replicate the regenerative functions of the services, which is a powerful guide for design.

Create Regenerative Systems and Foster Resiliency

- Ensure future resource supply and mitigate climate change
  This principle encourages the use of eco-efficient design and system integration to reduce negative effects on the environment.
- Enhance human well-being and strengthen community
  This principle taps into social equity and human systems and speaks about the importance that the natural world has on human health.
- Transform the market through the design, development, and maintenance practices.
  This principle ties into life cycle system integration to provide economic benefit in the form of prosperity.

The three principles above are organized by the Triple Bottom Line but at the top, instead of sustainable development, is "Create Regenerative Systems and Foster Resiliency." This takes into account the sustainability value of ecological regeneration discussed earlier in Chapter 4.

The SSI is very ambitious and forward looking and includes the latest examples of thinking about sustainable design. Some have criticized the system as being too ambitious and not achievable. Rating systems are designed to educate design teams and clients and push them beyond their comfort zone into new ways of thinking and practicing. Rating systems live on the fine line of setting the bar too high and turning off potential users or setting the bar too low and missing opportunities for greater impact.

Resources and additional rating systems

EcoDistricts, https://ecodistricts.org/
GBCI Sustainable Sites Initiative, www.sustainablesites.org/
2030 Districts, www.2030districts.org/
Living Community Challenge, https://living-future.org/lcc/
LEED for Neighborhood Development, www.usgbc.org/guide/nd

12.10 Global scale and districts

The United Nations Sustainable Development Goal (SDG) that deals most directly with the district and site scale, sustainability is #11: make cities and human settlements inclusive, safe, resilient, and sustainable.

The targets and indicators for these SDGs are too numerous to describe fully here. The SDG target that deals most directly to site development is **Goal 11: Make cities and human settlements inclusive, safe, resilient, and sustainable** (UN 2018).

Target 11.7 – By 2030, provide universal access to safe, inclusive and accessible, green, and public spaces, in particular for women and children, older persons, and persons with disabilities.

Indicator 11.7.1 – Average share of the built-up area of cities that is open space for public use for all, by sex, age, and persons with disabilities.

Indicator 11.7.2 – Proportion of persons victim to physical or sexual harassment, by sex, age, disability status, and place of occurrence, in the previous 12 months.

The targets are broad to allow for integration into the specific geographic, cultural, and climate of the site. The indicators are also broad, yet measurable, to make them more applicable and adaptable to each specific situation.

Additional resources

Landscape Architects Can Help the World Achieve New Sustainable Development Goals, https://dirt.asla.org/2015/09/23/landscape-architects-can-help-the-world-achieve-new-sustainable-development-goals/
UN Sustainable Development Goals, https://sustainabledevelopment.un.org/sdgs

12.11 Urban scale and districts and sites

Cities are made up of districts. The lines of the districts have been defined by the governing forces of the city and can be based on geographic or geologic features like rivers or topography or by what happens within them like manufacturing or seaports. Districts can be defined by the number of residents or based on the square area of land. How they come to be interrelated defines the overall quality of the city. Streets, greenways, urban parks, interstates, and water frontage are all opportunities for open space and to improve the overall quality of life.

12.11.1 Landscape urbanism

Landscape urbanism is the theory proposing a city be designed around landscape rather than around buildings or traffic patterns. It is not a rating system or a model for sustainable design, but the principles and tenets of the main directives are complementary and consistent with Intergral Sustainable Design.

Landscape urbanism looks to integrate technical and ecological systems, making them visible as a critical part of the aesthetics and experience of the urban context. This approach supports resilience by providing a more flexible landscape system. The landscape is considered a system that spans and connects the whole city at various scales (Gintoff 2016). James Corner (2011) is the author of an essay entitled *Terra Fluxus*. He has identified four general ideas that are important for use in landscape urbanism. The steps are as follows:

1. **Process over time** – "Understanding the fluid or changing nature of any environment and the processes that affect change over time … the idea that our lives intertwine with the environment around us, and we should therefore respect this when creating an urban environment."
2. **Surface extends beyond the horizontal** – "… urban infrastructure sows the seeds of future possibilities, staging the ground for both uncertainty and possibilities."
3. **Operation or working method** – "to conceptualize urban geographies to function across scale and implicate a host of players."
4. **Imagery** – "the collective imagination informed and stimulated by experiences in the material world must continue to be the basis for any creative endeavor."

Corner, towards the end of *Terra Fluxus*, talks about the integral nature that should be part of the relation between the landscape, urban areas, and people; "… engaging real estate developers and engineers alongside highly specialized imaginers and poets of contemporary culture---all these activities and more seem integral to any real and significant proactive design of synthetic urban projects" (Corner 2011).

The most well-known example of landscape urbanism is the High Line in New York City, (Figure 12.11.1a [next page]) an elevated park built on top of an old elevated railway. The project was designed by James Corner Field Operations and Diller Scofidio + Renfro, with Piet Oudolf. The High Line includes amenities like paved paths, benches, and gardens. The park cuts through different districts in New York, linking the city in ways that a building or street could not achieve. The High Line concept started a trend, which other cities around the world following suit.

The shift in urbanism from a focus on buildings to a focus on landscape is a critical step forward for sustainable design because it shifts the focus to ecology and infrastructure. Sustainable design requires a self-sufficient district and site – a microcosm of the larger urban area, the region, and the globe. The biophilic city movement supports these approaches while biomimicry allows different models to be analyzed for possible strategies based on the time-honored success of natural systems.

Making infrastructure visible is one of the ways to engage the public in the process of sustainable design. The Resilience project explored in Section 11.7 by BIG expresses many of the strengths of landscape ecology, choosing to focus on the interconnection of systems and

Figure 12.11.1a The High Line, New York City.
Source: Wikipedia. By Jim.henderson - Own work, CC0, https://commons.wikimedia.org/w/index.php?curid=18390084

the visible expression of infrastructure. Interconnected green spaces link districts through nature walks or through a consistent green experience while moving through the city on foot. Complete streets, greenways, connector parks, highlines, and river walks all serve as important connectors to link city districts together.

12.11.2 Complete streets

The complete streets movement is implemented at the site and district scales but also links to the larger urban context. They can be a "connecting thread" through different districts within a city. The quality of the street itself determines whether people will walk or chose to use a car. Complete streets seek to combine natural elements, green space, walking, biking, pollution and noise control, and cars safely.

Better experiences are created for walking, cycling, and driving. Complete streets also buffer residential areas from the noise and pollution of cars, define functions within a city, create a pleasant atmosphere for shopping and restaurants, and increase exposure to natural elements. When people feel safer and it is more pleasant, they will walk and bike more, saving energy and reducing emissions.

Rivers, highways, train lines, and geological features can all divide cities, yet the principles in complete streets look at these as opportunities to provide higher quality unifying elements between districts and sites. The areas around these features usually go unused, when they should be looked to as increased green space. Smart growth, covered in Section 11.9.1, also features a section on transportation. The design of transportation is changing to become less about speed, efficiency and simplification, to focus on safety, ecology, and mix modes of transportation.

Additional resources

Benefits of Complete Streets, http://old.smartgrowthamerica.org/complete-streets/complete-streets-fundamentals/benefits-of-complete-streets/

Complete Streets Local Policy Handbook, www.smartgrowthamerica.org/app/legacy/documents/cs-local-policy-workbook.pdf

Terra Fluxus, James Corner, http://march1section1.pbworks.com/f/CornerTerraFluxusLR2.pdf

12 Projects that Explain Landscape Urbanism and How It's Changing the Face of Cities, ArchDaily, www.archdaily.com/784842/12-projects-that-show-how-landscape-urbanism-is-changing-the-face-of-cities

12.12.1 Districts

Capitol Hill EcoDistrict case study

The Capitol Hill EcoDistrict, in Seattle, Washington, is one of the most well-documented EcoDistricts in the U.S. A task force of community members and technical experts worked with design professionals and the local university to identify key performance metrics, collect data to establish baselines, and develop a reporting system to share progress over time. This process also included review by the University of Washington's Cities Collaboratory, and feedback from an interdepartmental group of the City of Seattle staff. They used the following criteria to guide decision-making:

- Place specific
- Relevant to real issues
- Measurable
- Scale appropriate
- Congruent with partner organizations
- Evaluated and adjusted over time
- Easy to communicate.

Table 12.12.1a organizes the specific design targets along with the corresponding systems, culture, and experiential approaches to meet the goals. This is a recurring pattern of thinking in sustainable design: developing guiding principles, identifying targets to express the principles, and indentifying design strategies to achieve the targets. This is basically the same approach used by the UN in the Sustainable Development Goals and by many other planning groups seeking to enact change.

SYSTEMS PERSPECTIVE

Notice that policies are shown under the system perspective. This is because a system can be something physical, any kind of workflow, or a set of policies that drive behavior. This is an important aspect of sustainable design which is often achieved as much by words as by physical interventions through typical design processes.

EXPERIENCE + CULTURE

Notice in Table 12.12.1a (next page), that the culture and experience strategies contribute to meeting the performance goals. All of these aspects are considered when designing a

Table 12.12.1a Capitol Hill EcoDistrict performance goals and associate design strategies

Selected performance goals	Possible systems, strategies and policies to achieve goals	Culture + equity strategies and policies to achieve goals	Experience strategies to achieve goals
Reduce building energy use and water use intensity by 50%	**Green building design strategies** (passive, active, and living systems designed to save water)	Foster energy saving culture in the district	
Achieve 70% waste diversion	**Life cycle design strategies** and policies to encourage residents to recycle	Foster a recycling culture in the district	Design attractive recycling bins and trucks
Achieve 21% tree canopy cover	**Living system design strategies** and policies to reward tree planting; maintenance of systems considered	Foster a culture among residents that appreciates nature	
Achieve 100% of district within ¼ mile of the park	**Living system design strategies** Planning approach of adding park space whenever possible	Local leaders develop empathy for local residents who need access to healthy food	Design urban environment to activate the streetscape making it fun to walk
Achieve 100% of district within ½ mile of grocery store	**Passive approach** of adding park grocery stores whenever possible; advertising to encourage locals to shop locally	Local leaders develop empathy for local residents who need access to healthy food	Design urban environment to activate the streetscape making it fun to walk
Double farmer's market shopper count by all incomes	**Passive approach** of making the markets larger; add active systems of night lighting and power for food trucks. Seek external funding	Educate residents regarding benefits of healthy food	Make farmer's markets fun to visit. Add interesting design elements such as benches, tables, and more
Increase pedestrian and bicycle traffic at selected intersections by 33%	**Passive approach** of definition of different circulation paths (passive design)	Educate residents regarding benefits of walking and biking	Design urban environment to activate the streetscape making it fun to walk and bike

Source: Capitol Hill EcoDistrict adapted by the authors.

sustainable district. For example, designing a lively and beautiful streetscape will encourage people to walk more or meet and form friendships with their neighbors. Or, educational programs can help residents see the health benefits of walking more. It all fits together into a comprehensive approach to sustainable design.

Additional resources

Capitol Hill EcoDistrict: A proposal for District Scale Sustainability, https://capitolhill housing.org/downloads/Capitol-Hill_EcoDistrict_Report_2012.pdf
Capitol Hill EcoDistrict Metrics, http://capitolhillecodistrict.org/metrics/

12.12.2 Eco-villages – cohousing model from Europe

Eco-villages are intentional communities based on the goal of achieving the Triple Bottom Line. The term "intentional" is meant to differentiate it from the typical standard intent based on consumption. Eco-villages are an expression of collectivism, cooperation, and altruism. Eco-villages are much smaller than EcoDistricts with approximately 150 residents or less.

The villages are usually built by the residents themselves, which increases the emotional connection and focuses decisions towards quality and longevity. Renewable energy and shared resources such as geothermal or central heating facilities optimize efficiency. Passive systems in construction, resource collection and preservation are used as much as possible.

Residents avoid the use of automobiles as much as possible, with some communities even banning them from being used within the confines of the village. Many times, there is food production on-site, and social areas that are shared like large kitchens, libraries, or event spaces, and residents are usually required to take turns in the regular maintenance activities that benefit the community.

Additional resources

Search for American Eco-Communities, www.greenecocommunities.com/states.php
Global Ecovillage & Sustainable Community Network, www.globalecovillages.org/
Auroville, https://auroville.org
Auroville (2017) is an example of an intentional community that falls somewhere between an ecovillage and an ecodistrict. It is located in Pondicherry, India and was founded by Mirra Alfassa in 1968. With a total population of approximately 2,200, it is about 2.5 km in diameter and is surrounded by a "greenbelt" forest.

12.12.3 Sustainable site design

The Phipps Center for Sustainable Landscapes is an example of the kinds of projects that express the emerging world view of integration. The project designers sought to create a project that unifies with nature rather than the more typical approach of domination or exploitation. The Center for Sustainable Landscapes (CSL) at Phipps Conservatory and Botanical Gardens was designed to be the first project in the world to simultaneously achieve LEED Platinum, Sustainable Sites four-star certification, and the Living Building Challenge. Built on a previously paved-over city maintenance yard and documented brownfield, the nearly three-acre site supports a new education, research, and administrative building. The CSL is open to the public, and its building and landscape performance is being extensively researched and monitored to inform the design and construction of similar projects (Figure 12.12.3 [next page]). The project achieves the following:

Performance

- Manages all sanitary waste on-site
- Manages a ten-year storm event
- Introduces 150 native plant species
- Net positive energy and net zero water.

Figure 12.12.3 Water capture and treatment at the Center for Sustainable Landscapes.
Source: Wikipedia, By Dllu – Own work, CC BY-SA 4.0, https://commons.wikimedia.org/w/index.
php?curid=44448369, https://en.wikipedia.org/wiki/Phipps_Conservatory_and_Botanical_Gardens#/
media/File:Phipps_Conservatory_23.jpg

Systems

The level of integration of systems in the Phipps project is remarkable, especially in the landscape where the plant selection, water collection, water treatment, and circulation paths all work together to create a rich human experience and highly performative set of systems.

Life cycle systems: The building meets the Materials Petal of the Living Building Challenge which means the "red list" of forbidden materials is adhered to – a very challenging accomplishment given the lack of product manufacturers willing to disclose their ingredients. With the help of the builder, the team was able to comply with the strict requirements. The team decided to extend the Red List Imperative to furniture within the building, even though it isn't a requirement for Living Building Challenge 1.3 Certification (International Living Future Institute 2018). Eighty-two percent of the construction materials for the project were acquired within 500 miles of the site, and reclaimed, salvaged, and recycled materials were used throughout the project, thereby saving energy and reducing associated air pollution.

Passive systems: Orientation of the building and integration of the building with the sloped site led to a more integrated project. The linking of the main ground floor space with the gardens outside is an important strategy to encourage building users to spend time outdoors.

Active systems: The combination of passive, active, and living strategies to collect and treat wastewater and stormwater is the most remarkable part of the project. The site plays a key role not just from a utilitarian approach but because the green infrastructure doubles as experiences of beauty and nature.

Living systems: The design of the landscape was completed by Andropogon, a landscape architecture firm located in Philadelphia. It featured an ambitious approach that includes a living roof and living landscape.

Human systems: The use of the integrative design process was essential to the successful design and implementation of the project. The human systems of projects, especially at the beginning of the design process, are so critical, because the design team needs to work around sets of shared values and agreed upon performance measures. At the heart of successful projects is an enlightened client who is willing to invest the time and resources to bring the project to fruition. Richard Piacentini, executive director of Phipps, says, "It took an integrative design process with great partners and a talented design team led by The Design Alliance Architects, to create a place that would meet the highest standards related to health, buildings, people and the environment." The commitment to maintaining the net-zero status of the building involves attention to human behaviors in space to make sure that energy is saved in the ways intended.

There is so much more to learn from the Phipps project. The links are given in the Resources section below.

Additional resources

The Design Alliance Architects, http://tda-architects.com/phipps.html

International Living Future Institute, Phipps Center for Sustainable Landscapes, https://living-future.org/lbc/case-studies/phipps-center-for-sustainable-landscapes/

AIA COTE 2017, Center for Sustainable Landscapes, http://aiatopten.org/node/507

Whole Building Design Guide, Center for Sustainable Landscapes, https://wbdg.org/additional-resources/case-studies/center-sustainable-landscapes

12.13 Conclusion: integrative buildings, interiors, and materials

Integrating landscape architecture with buildings and interiors is more than simply "connecting" inside to outside. Offering viable ground floor access between public space inside and outside the building is critical to creating an authentic and long-lasting connection. Furthermore, an entirely new subdiscipline of interior landscape architecture and indoor ecology is emerging as a means to wrestle with the harsh technical demands of interior living walls. Matt Gindlesparger, an associate professor in the Architecture Program at Jefferson University, has been developing a living wall that integrates with the mechanical systems of a building to clean the air (Urban Omnibus 2015). These systems reflect the technical requirements to really understand the impacts and interactions of living systems in space, and underscore the need for mainstream designers to fully understand the technical complexity of living technologies along with the specialized experts they hire.

References

ASLA, 2018-last update, The Sustainable Sites Initiative: Sites [Homepage of The American Society of Landscape Architects], [Online]. Available: www.asla.org/sites/ [July 28, 2018].

Auroville, 2017-last update, The Auroville Charter: a new vision of power and promise for people choosing another way of life [Homepage of Auroville], [Online]. Available: www.auroville.org/contents/1 [July 28, 2018].

Capitol Hill EcoDistrict, 2014-last update, EcoDistrict Index [Homepage of Capitol Hill Housing], [Online]. Available: http://capitolhillecodistrict.org/metrics/ [July 28, 2018].

Corner, J., 2011. *Terra Fluxus*. PB Works. http://march1section1.pbworks.com/f/CornerTerraFluxu-sLR2.pdf edn.

EcoDistricts, 2018-last update, EcoDistrict Protocol Principles [Homepage of EcoDistrict.org], [Online]. Available: https://ecodistricts.org/get-started/the-ecodistricts-protocol/priorities/ [July 28, 2018].

EPA, 2018-last update, Stormwater Management and Green Infrastructure Research [Homepage of United States Environmental Protection Agency], [Online]. Available: www.epa.gov/water-research/stormwater-management-and-green-infrastructure-research [July 28, 2018].

Gintoff, V., 2016. *12 Projects that Explain Landscape Urbanism and How It's Changing the Face of Cities.* "12 Projects That Show How Landscape Urbanism is Changing the Face of Cities," ArchDaily. www.archdaily.com/784842/12-projects-that-show-how-landscape-urbanism-is-changing-the-face-of-cities edn.

Grauerholz, M., 2016. *Living WELL: Tampa Becomes the First City in the World to Introduce a WELL Certified District.* International WELL Building Institute. www.wellcertified.com/en/articles/living-well-tampa-becomes-first-city-world-introduce-well-certified-district edn.

International Living Future Institute, 2018-last update, Certified Living: Phipps Center for Sustainable Landscapes [Homepage of International Living Future Institute], [Online]. Available: https://living-future.org/lbc/case-studies/phipps-center-for-sustainable-landscapes/#materials [July 28, 2018].

Paris Innovation Review, 2013. *EcoDistricts: A Sustainable Utopia?* Paris Innovation Review. http://parisinnovationreview.com/articles-en/ecodistrict-a-sustainable-utopia edn.

Perkins + Will, 2018-last update, Health District Planning [Homepage of Perkins + Will Architecture], [Online]. Available: https://perkinswill.com/type/health-district-planning [July 28, 2018].

United Nations, 2018-last update, Sustainable Development Goals [Homepage of United Nations], [Online]. Available: https://sustainabledevelopment.un.org/sdgs [July 26, 2018].

Urban Omnibus, 2015. *Ventilation Goes Vegetal*. Architectural League of New York. https://urbanomnibus.net/2015/11/ventilation-goes-vegetal-cases-plant-based-air-filtration-system/edn.

13 Building scale sustainable design

13.0 Introduction

Why does the word "sustainable" need to be placed in front of the word "architecture"? How does that change the equation for what constitutes good design? The idea that somehow sustainable design is just part of "good" design has dominated the mindset of architects for the last 20 years or more. The term we use for this is "casual greening" – the process of designing in a normative way and then trying to "add" sustainable elements in a desperate attempt to mitigate the destructiveness of a built project. In this book, we are looking for something better, something deeper, and something that gets to the core of goodness, a holistic conception of architecture that can transcend its time to persist for generations.

Earlier in this book, we argued that great architecture was an integral part of sustainable design. Sustainability is the overarching goal for all of humanity seeking to turn the momentum towards a sustainable future. In this way, we can share a common understanding of what sustainable architecture really is and how we might move forward. In this chapter, we will explore sustainable design at the building and interior scale through time, space, and the four perspectives to understand the richness, depth, and possibility of sustainable architecture.

Architecture and interiors were combined in this chapter in the same way that districts and sites were combined in the previous chapter. One of the main premises of this book is the blending of professions across scale which is a critical element to successful sustainable design. Interior designers can and should help to shape architectural projects. Architects are already heavily involved in interior design. Each field has its own set of specialties, and its own educational focus. The importance of focusing on the interiors scale is underscored by the fact that most people in the U.S. spend 90% of their time indoors along with the rise of interest in health and the built environment. The use of evidence-based design as a means to guide design decisions is considered a radical approach for some, but that is exactly what needs to happen if healthy interiors are to become a critical focus in the profession. The Well Building Standard, covered in Chapter 8 and later in this chapter, is a good place for interior designers to start to engage evidence-based design in a tangible way.

13.1 Sustainable architecture across time

Indigenous architecture

Hunter-gatherers emerged from over a million years of evolution to become the human beings that we might recognize today, and with them came the origins of architecture. Their worldview is one of direct connections to nature. Residential architecture came out of necessity, born of the materials that were nearby. Forms were derived from what the materials could easily accomplish. Shapes were made to be strong, and therefore, there was an absence of straight walls and roofs. Everything had a sense of biomimetic design in that the maximum volume was made with the minimum amount of structure and material – hence the primitive hut. The word "primitive" is often thought of as a negative term, but as we move further into the worldview of integration, we are rediscovering indigenous architecture, and we are beginning to see that, perhaps, these were the first sustainable architects and builders. Local materials, orientation towards the sun, and openings facing prevailing breezes or away from cold winds are common features in early architecture, suggesting an instinctual approach that is more sophisticated than it appears. Strategies were passed down through direct teaching in real time on real projects – not that dissimilar to the design build courses today. We would not see such high-performance architecture again till the early 21st century – thousands of years later.

At the same time, the *cultural* goal to embed meaning in design drove indigenous cultures to perform heroic feats of engineering. The *systems* needed to build a ceremonial structure such as Stonehenge required a deep knowledge of physics and a *performance*-based process. Imagine moving a one-ton boulder four miles and then placing it in an exact location within a circle in order to reveal and celebrate the path of the moon and sun. Architecture was a way for early humans to bring more meaning and spirituality to their rituals. The alignment with sun angles to deepen *experiences* in spaces and places is as strong in Stonehenge as any modern structure.

Architecture in the Age of Agriculture 12,000 BCE to 1750

Vernacular architecture has always been associated with sustainable design, because the buildings are sited in the best locations and respond to the local climate through variations in building shape, roof slope, and location/size of windows. Local materials are used by local craftspeople. The colonial farm houses nested in rolling hills and lush farmland of Bucks County, Pennsylvania are an excellent example of vernacular architects at work. The shape of the building is long and thin allowing the south side of the house to absorb heat in the thermal mass of the brick walls, while the west wall typically has few windows to avoid the heat gain. The kitchens were often located in a separate building to keep the main house cool and safe. Fireplaces play a prominent role in the architectural expression as there was complete dependence on this type of energy to create comfort and prepare food. The performance of the building was a direct result of passive design and the effectiveness of the fireplace design, which needed to project heat into spaces without allowing smoke to enter the space.

Architecture in the Age of Industry 1750–1960

With the advent of efficient mechanical systems to heat and cool spaces, the demand to make buildings passively respond to climatic forces was replaced by the international style – a form of architecture disconnected from place, context, and climate. In contrast, Frank Lloyd Wright's "organic" architecture prized the use of local materials, integrated site and building, and the expression of the complex unity of nature in his forms and motifs. As Wright moved later into his career and as the worldview began to shift, he built the hemisphere house, which was designed to harvest heat from the sun in order to avoid the use of traditional heating systems. Although the house was not as effective as intended, the gesture towards a new conscious form of sustainable design was expressed.

Architecture in the Age of Information 1960–2005

Ian McHarg's (1971) seminal work, "Design with Nature," not just impacted planners and landscape architects but also influenced architects. McHarg's clear approach in documenting patterns in nature and human development led to a rise in architects seeking to better connect their projects to the ecological context. In 1973, the OPEC Oil Embargo greatly reduced access to oil and gasoline in the U.S., bringing an awareness that fossil fuels were not infinite and that energy itself was an issue in the design of buildings. This led to the birth of energy-efficient design in buildings, a precursor to the more holistic sustainable design that is practiced today.

James Wines of Site Architecture explored quite a different avenue of expression, an emerging cultural attraction to nature. His work predates the rise of biophilia, which would come later in the current worldview. Wines' underlying theme in design was not necessarily the integration of building with nature but rather "nature's revenge," as if nature itself was finally getting retribution for centuries of destruction and neglect. In his approach, greenery found its way up and over buildings suggesting the idea that architecture was really as much a ruin as it was a vital space for human activity.

This time period also marked a return to the understanding that humans and nature were once again the focus for the design of buildings and interiors. Postmodernist architecture emerged as a clear departure from the often stark and brutalist modern architecture and shifted to make historical references in forms and ornamentation, and to reconnect humans to architecture through human scale, recognizable forms, and culturally meaningful symbolism. In thinking about the Quadruple Bottom Line of sustainability which includes a focus on people as a core value, the rise of postmodernism opened the door to a more inclusive, less abstract expression of built form. The works of Denise Scott-Brown and Robert Venturi were well known for this design approach which not only explored traditional forms and iconography but also sought to celebrate architecture as a form of pop culture. This can be seen as the precursor to inclusivity and stakeholder engagement so prevalent in today's sustainable design approaches.

Historic preservation became supported by local and federal governments, acknowledging the value of history and heritage in people's societal connections. Interior design finally reached maturity, not just in the design of large office floors but in the desire to adaptively reuse old buildings. These approaches represent some of the most foundational aspects of sustainable design: saving energy by reusing structures and preserving culture to meet intrinsic human needs.

Architecture in the Age of Integration 2005

We've now entered the Age of Integration, a paradigm shift in the way humanity views nature, uses energy, and communicates with each other. With that come significant changes to the fields of architecture and interior design. Sustainable design is the expression of the new worldview in the design process which is becoming more inclusive and transparent, and by the shape, form, materiality, and performance of buildings. This chapter will explore these changes in more detail.

13.2 Sustainable architecture and nature

In the Age of Integration, sustainable design has matured as a movement and produced an array of striking examples of design projects that directly relate to nature, not just visually or thematically, but literally connecting for a direct benefit. In earlier chapters, ecosystem services were discussed in some detail. Now, landscape architects, architects, and interior designers are beginning to understand that the site itself is not a blank slate but a rich tapestry of ecological interactions all capable of helping to provide lighting, cooling, power, and water to a project. This profound shift in the relationship between building and site is the ultimate reflection of an integral worldview where humanity and nature do not just coexist but work in partnership to meet basic human needs. This is an exciting time, indeed, a new frontier of architecture expression for building and interiors with biophilic design emerging as a major driver of sustainable design. The role of the scientist, especially in the case of McDonough and Braungart with the Cradle to Cradle™ Design Framework, exemplifies the new sense of collaboration across disciplines to achieve higher levels of environmental performance and health.

However, despite the examples discussed in the brief history of sustainable architecture, most building projects designed today damage the natural world by wasting resources, destroying local ecologies, and polluting the atmosphere. The full transition to an integrated model for design and nature has yet to truly hit the mainstream.

13.3 Sustainable architecture and humanity

In Chapter 3, we studied how the competing motivations of self-interest and empathy play out in the design world. For architecture, the drive of self-interest on the part of clients has led to architecture becoming a commodity, just a container of square footage to house functions aimed at maximizing profit and/or minimizing the first cost of construction. The short-term view of architecture held by many clients is one of the most destructive forces to the built and natural environments.

Meanwhile, architects are often consumed with the need for aesthetics and an overriding drive for beauty. On the one hand, this is so necessary, but on the other hand, there is little consideration for the long-term environmental impacts or effective cost management of a design project. The intrinsic need for beauty was clearly established by studying Maslow's expanded hierarchy of needs where aesthetics were identified as an important growth need. The question is: How do you pursue beauty and meaning while still reducing resources, consumption, and ecological damage?

To meet all the goals of sustainable design, the entire *sustainable* design team needs to be dedicated to the project, using empathy to meet the needs of all the stakeholders – end users, community members, and the environment itself. Evidence-based design, stakeholder-driven design, and post-occupancy evaluations are all practices that help to achieve that. These processes are often pursued through a host of rating systems, standards, and codes, which are covered later in this chapter. Rating systems help design teams to organize their thoughts and activities to effectively deliver buildings that are less damaging to the planet. Rating systems can provide a starting point or basis for design, but they don't always reach the deeper more holistic needs. It is hoped that deeper, more authentic motivation for the environment is emerging as a part of an overall shift in human consciousness.

13.4 Architecture and sustainability

The Quadruple Bottom Line was introduced in Chapter 4 as a means to better integrate designers into the core matrix of sustainability and as a way to place the economic, social, and environmental values of a project on equal footing with a more typical emphasis on design expression. Architects have long struggled to develop a holistic approach to design, partly due to the pressure from clients or by being fixated on the image of the building. Environmental or social concerns are often downplayed or not acknowledged. Aesthetics was identified by Maslow as a legitimate human need, so it's not surprising that designers and clients desire it. Sustainable design does not require the sacrifice of high-quality aesthetics or beauty to achieve better performance. In fact, the opposite is true, as we saw in an earlier example of the Sidwell Friends School and we will see later in the case study on the Kern Center (13.13). Both projects underscored the power of uniting aesthetic expression with environmental functionality.

As a response to the conflict between architects and budget, and between architects and nature, Quadruple Bottom Line sustainability offers a pathway for architects to regain balance in practice. The four bottom lines – People, Profit, Planet, and Place – offer an integrative and holistic sustainability framework to pursue practice.

One of the core sustainable values discussed in Chapter 4.4.1 attempts to shift thinking about the cost of green buildings from short term to long term. But even in the short term, the investment in time, energy, and resources pays off. In an exhaustive and carefully compiled study on the economic benefits of green buildings, it was found that an increase in market value and a decrease in operating expenses were "statistically significant changes from non-green average buildings." In a study by the Institute of Building Efficiency in 2007, the cost of green buildings varies within a range with a LEED Gold Building being one of the least expensive examples. Design teams, including the builder, with deep experience in green building are able to accomplish the task of meeting sustainability goals in a cost-effective manner. However, as explored in depth in Chapter 3, the constant focus on first cost continues to block efforts and hinders the growth of the green building movement.

Additional resources

The Well Building Standard: Introducing the WELL Certified Office, https://higherlogicdownload.s3.amazonaws.com/CORENETGLOBAL/f2d6ddf2-66f5-4042-b2ea-71e165781e82/UploadedImages/Events/CoreNet%20Lunch&Learn_Delos.pdf

Does a Healthy Workplace Improve the Bottom Line? www.gensler.com/research-insight/gensler-research-institute/toward-a-wellness-based-workplace

Costs and Benefits of Green Buildings, https://thinkprogress.org/costs-and-benefits-of-green-buildings-ceef267baf06/

Economics of Green Building, http://buildingefficiencyinitiative.org/articles/economics-green-building

13.5 Integral sustainable architecture

While the Quadruple Bottom Line is a useful framework to be used in discussions with most clients and stakeholders, deeper, more powerful meta-frameworks are needed to guide the design process towards more profound levels of sustainability and to better meet the regenerative goals of authentic sustainable design. In Chapter 5, we studied Integral Sustainable Design as an effective meta-framework for sustainable design, because it includes the qualitative and quantitative aspects of design, thereby creating a holistic approach to design.

The building scale offers the opportunity to express those principles shown in Figure 13.5a in very direct and visible ways. Architecture is more than "frozen music;" it is the built manifestation of philosophy and motivation, which also expresses social and cultural concepts forming and shaping human experience and perception. The integral worldview shifts

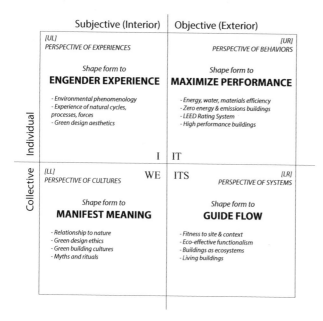

Figure 13.5a Integral sustainable design by Mark DeKay.
Source: R. Fleming (2013).

towards a more united relationship with nature and a more sophisticated approach that moves away from short-term thinking including the goal of long-term prosperity and sustainable design. In the following sections, we share a brief overview of some fundamental aspects of sustainable design. They are meant to start the conversation, spur creativity, and begin the process of integration across the four perspectives of Integral Sustainable Design. Reviewing Chapter 5 at this point is recommended to gain the full understanding of how Integral Sustainable Design operates.

13.5.1 Maximizing performance of buildings

Eco-efficient design and goal setting

The principle of eco-efficiency is an integral part of all the rating systems in a variety of ways and with different levels of ambition. The development of guiding principles and goal setting is part of the sustainable design process and was covered in Chapter 9. The use of rating systems and standards to drive performance was also covered in Chapter 9, and some specific examples are covered in Section 13.9 of this chapter.

Carbon neutrality and net-zero ready buildings are two standards that are critical metrics for the buildings and interior design disciplines. A net zero energy building (NZEB) is typically a highly energy-efficient design coupled with a renewable energy system such as solar panels mounted on its roof. The total amount of energy produced by the solar array is equal to or above the total amount of energy used in the building. Oftentimes, the term "ready" is included in the designation of a NZEB because the standard relies on the proper use of the building by residents. There are also net-positive energy buildings, which reach a certain percentage above the energy they use. Currently, the Living Building Challenge (LBC) (13.12) requires projects to generate 110% of the total energy use of the building, thereby generating a 10% surplus.

Carbon neutral or zero net carbon (ZNC) projects are similar in some respects to NZEBs, except that the measurement of carbon includes some useful additional factors. In this case, the total net carbon emitted due to operations of a building is "offset" by renewable energy. Decarbonization is also a term that is used to describe the process of designing systems that emit less carbon. This is typically achieved through eco-efficient design and dematerialization. Metrics and standards like *zero net carbon* drive design teams to maximize performance.

Dematerialization

Imagine a building that provides for all the needs of the occupants but is 20% smaller due to the use of more efficient equipment and digital technologies. The most obvious example is the change from large CRT monitors that took up space and generated heat, to the flat-screen monitors and laptops that reduce the need for space and reduce air-conditioning loads. The amount of paper and books is much lower than only ten years ago, which also reduces the need for physical space. In Chapter 9, the integrative design process was discussed including the use of programming to find opportunities to reduce building size and by default, reduce energy use and environmental impacts. Calculating the necessary square footage of a building, along with specific furniture and equipment requirements, is the key to starting the sustainable design process.

The small and tiny house movement

Smaller residential structures equal less capital output, less time and money for maintenance, and in some instances greater mobility. In 1997, Sarah Susanka started the small house movement that emphasizes a smaller footprint with greater attention to detail and higher-quality construction. This was in response to the phenomenon of "McMansions" – large-scale, mass-produced track housing based on the idea that bigger is better, sacrificing quality and design for size. She has gone on to write about the lifestyle changes that value quality over quantity in every area. Small houses can be 1,000–2,000 s.f. The tiny house movement is said to be popularized by Jay Shafer who lived in a 96 s.f. house for a year. Tiny homes are generally under 500 s.f. and are often mobile. They have been adopted by people with active lifestyles as a second home, those conscious of environment and resource use, as well as for providing alternatives to homelessness. The energy savings and reduced environmental footprint of a tiny house are remarkable. The ability to live happily in such tight quarters is a cultural and experiential question. For many, the trade-off of space is worth the benefits of low-cost and low-impact living. The tiny home movement has matured with tiny home eco-villages coming into existence that include a large shared space to be made available for group functions and events.

Resources for maximizing performance of buildings

The Whole Building Design Guide, National Institute of Building Science, https://wbdg. org/resources/net-zero-energy-buildings

Zero Energy Buildings: A Critical Look at the Definition by National Renewability Energy Laboratory, https://nrel.gov/docs/fy06osti/39833.pdf

13.5.2 *Thinking in systems for buildings and interiors*

Mark DeKay's use of the term "guide flow" is an apt way to begin to talk about systems in the design of a sustainable building. So often, we become fascinated by the form and expression of the design that we often relegate important functional issues to later in the design process, after the primary design aesthetics have been decided. This is the classic problem of the traditional design approach which tends to favor the experiential aspects of the project first, thereby missing opportunities for significant cost and energy savings. The shape and form of buildings not only impact performance levels, but they can also help to connect occupants to the best views and to open it up to the natural world. As stated earlier, a building with light, air, and views increases the productivity and satisfaction of workers. These are considered "passive" systems. They are built once, and they generally remain fixed for the lifetime of the building, requiring no additional energy source to function. Other types of systems also play a role in sustainable design. Lifecycle systems have a huge impact on the embodied energy and carbon footprints of a project.

Active systems such as HVAC and artificial lighting have a direct impact on energy costs. Finally, the use of living systems in buildings reflects the emerging worldview of integration, where humans and nature are once again united in a common purpose. Green roofs and living walls are some examples of this approach.

Bioclimatic design affects all systems choices in very direct ways. In the U.S., buildings constructed in Florida and Alaska have very different climate conditions which require different responses. The sustainable designer is responsible to evaluate and determine which systems are best suited for the project's specific climate. The building's use also dictates which strategies and systems are the best choices. There are entire books written about bioclimatic design; resources are included at the end of this chapter.

Systems are strategies

These kinds of systems are expressed in very specific ways at the building scale: life cycle, passive, active, living, and human. These systems deliver benefits such as higher-energy performance as described above, but they also facilitate human comfort and the quality of experience and interactions within the spaces. Systems are often invisible, but they serve as the platform for a functioning building. In sustainable building design, the systems are most often discussed in terms of strategies. A strategy is a means to meet a specific performance goal. For example, if the end goal is to realize a net zero ready building, then the use of photovoltaic (PV) panels is a *strategy* to achieve that goal. If the goal is to have clean drinking water, then water filtration is a *strategy* to achieve that goal. Technologies are also a system. If the goal of a building project is to reduce energy, then an LED lighting technology is a *strategy* to achieve that goal. Technology and strategy are often used interchangeably.

Life cycle systems in buildings (embodied energy)

Life cycle systems for buildings and interiors are important, because the energy profile of a building over its life cycle can be very large. Table 13.5.2a outlines some of the strategies used to optimize the life cycle of a building. Notice that different steps are involved in bringing a building into existence. Each step carries opportunities for eco-efficiency (reduction of energy use and environmental impacts).

Table 13.5.2a Life cycle for buildings and interiors

Life cycle strategies	Mining and harvesting	Transport	Manufacture	Transport	Use	End of life
	• Use sustainable mining techniques • Specify materials that are extracted or harvested locally	• Reduce distances between extraction points and manufacturing plants • Use efficient transport modes	• Eco-efficient manufacturing • Specific materials made by local manufacturers	• Specify efficient transport methods	• Specify materials that are durable	• Recycled • Biodegradable • Reuse • Repurpose • Design for disassembly or reuse

Source: Created and drawn by the authors.

Table 13.5.2b, shown on the next two pages, begins to identify the rest of the sustainable systems, organized by the elements of nature, as ecosystem services. It's best to start at the bottom of the chart and work your way up. Each level of strategies and technologies requires more cost and more energy and is usually more complex to install and maintain.

The systems are shown in isolation in the Table below. But in reality, they interact at all times in synergistic relationships or trade-offs. For example, the effective use of passive strategies such as daylighting and natural ventilation will reduce the need for active systems such as artificial lighting and air conditioning, thereby saving energy and money. The living systems of a green roof reduce the amount of storm water leaving a building, thereby reducing the size of infrastructure to process water on-site.

Trade-offs in buildings occur where one strategy has a huge energy-saving benefit but results in the reduction of the effectiveness of another technology. For example, the "trade-off" of using PV panels on a roof is that there is less space for a green roof, thereby reducing the stormwater benefits. Sustainable designers make a lot of effort to optimize all the technical strategies to achieve the highest level of energy performance and the least amount of environmental damage. The guiding principles and goals of the project determine which systems are selected for a project. The sets of strategies shown below are the primary approaches to sustainable building and interior design, but there are many, many more options. As a note, rainwater collection is illegal in many places in the world, because in areas prone to drought, water needs to find its way into the ground to recharge aquifers.

13.5.3 Culture + equity

The inclusion of culture and equity into the makeup of sustainable design considerations for a building and interiors project is supported by discussions around the Quadruple Bottom Line and Integral Sustainable Design. Social equity in the design of buildings is not a "feel good strategy." It goes beyond empathy to deliver real economic and environmental benefits. In Chapter 3, we explored how access to light, air, and view in buildings is part of the expression of equity in design. For example, in Germany, the building code requires access to light, air, and view. The synergy of open access to the elements of nature leads to higher

Table 13.5.2b Systems for buildings and interiors

	Passive systems	Active systems	Living systems	Human systems
Human Health + Wellness	Orient views/access to nature; Encourage the use of stairs	User control of lighting and HVAC	Engage building occupants with food gardens and maintenance of green roofs and vertical living walls	Education about sustainability; Train occupants to open and close windows
Climate + Resilience	Avoid flood plains, Find high ground; Raise building off the ground in flood prone areas	Use redundant and distributed active systems	Food gardens; Site design to account for storm surges, heat waves, drought	Social resilience activities
Food	Locate building to not interrupt solar access to crops; Design roof to support food gardens		Bee colony; Herb gardens indoors and out	Incentives for employee participation in on-site food production
Flora + Fauna	Avoid areas of habitat and animal paths; Allow plants and animals up and over building	Artificial lighting to support indoor ecology	Indoor ecology; living walls	Maintenance of living systems
Energy + Transportation	Clear solar access to roof and best slope/orientation of roof for solar; Maximize passive systems to minimize need for PV panels	PV panels + Small wind turbines	Food is energy and can be grown on-site	Turn off lights and computers; Bike racks and showers; Green energy purchasing and other policies
Light	Orient building to collect light; Optimal window-to-wall ratio with roof monitors and light shelves; Interior layout to bounce light deeper into spaces; Light-colored interiors	Efficient artificial lighting; Daylight harvesting; Automatic shade systems	"Natural" light spectrums for humans and plants	Task lighting; Opening and closing of blinds; Furniture position; Retrain to accept wider range of light

	Passive systems	Active systems	Living systems	Human systems
Heat	Orient building to collect heat Efficient building envelope – high 'R' value, low 'U' value Vestibules and buffer spaces	Energy-efficient heating systems Building automation system Geo-exchange heating	Landscape buffers and berms	Thermostat setbacks Proper dress Open and close windows Expand temperature range for comfort
Cooling	Orient building to avoid heat gain or to collect breezes Efficient building envelope Thermal mass White roof color Vestibules Earth tubes Shading	Efficient cooling Chilled beams Building automation systems Geo-exchange cooling	Shade from deciduous trees Vertical green walls Earth berming Green roofs	Thermostat control Proper dress Open and close windows Expand temperature range for comfort
Earth + Materials	Recycling space near loading dock		On-site composting Composting toilets Biodegradable materials	Reusable cups, plates, etc. Dishwasher
Water	Avoid existing water ways Shape roofs to collect water Collect water in cisterns and rain gardens Collect rain water for urban gardens Make sure to keep water moving to prevent contamination	On-site water purification Gray water recycling systems Dual flush toilets	Composting toilets Living machine for water purification Green roof Compost toilets	Education for occupants on water savings Water saving competitions
Air	Collect favorable breezes Block winter winds Accelerate breezes through building Cross-ventilation Venturi effect Low VOC nontoxic materials	More air changes Building flushing ERV/HRV wheels	Air purification through plants	Open and close windows

Source: Created and drawn by the authors.

productivity from employees, which yields not only more profit but also a greater sense of community.

13.5.4 Experience

Deepening experiences in buildings requires a complex choreography of design methods to engage the senses in direct and indirect ways. The use of proportion, color, geometry, rhythm, hierarchy, sequence, and much more are all strategies that are used to deliver a positive physical and psychological experience in space. The use of biophilic principles in space adds to the visceral experience, deepening the emotional connection with nature and the space. In Chapter 4, Maslow's requirements for a self-actualized person include access to beauty, uniqueness, playfulness, and more. The role of the designer is to develop spaces and places that meet these needs. In Chapter 4, we also saw the inclusion of beauty as a sustainable value, making the experience quadrant of sustainable design equal to other quadrants already discussed here.

Coming full circle

All four of the quadrants converge into a grand synthesis – a framework of frameworks that constitutes an integral sustainable building. Now we see the potential for buildings to have high-performance, integrated systems, social equity and deep beauty – the ultimate expression of humanity's highest and best purposes.

13.6 Bio-Inspired architecture

The use of biomimicry and especially biophilia reflects the changing mindsets of clients, design teams, and end users. The benefits and details associated with this approach are detailed in Chapter 6. Bio-Inspired architecture intentionally uses natural elements and exposure to nature that goes beyond visual or aesthetic experience to influence the psychological, physiological, and performative aspects of building occupants. The Khoo Teck Puat (KTP) Hospital in Yishun, Singapore, built in 2010, offers an excellent example of how biophilic design can transform a place and its people. It is the first project to win the Stephen R. Kellert Biophilic Design Award.

The project transformed the existing "Alexandra" hospital, which, according to CEO Liak Teng Lit, was drab, boring, and clinical (Newman 2014). As a response, he appointed a "chief gardener," Rosalind Tan, and an occupational therapist to work with volunteers to landscape the hospital. The landscape included a medical garden, a fragrance garden, and water features. As a result, over a hundred species of butterflies returned to the site. The result was transformative both from an experiential point of view and more so in how it was able to garner volunteers to help with the project. This underscores the relationship between living systems and human systems. The two are needed together for success.

The success of the landscaping project inspired the hospital's leadership and design teams to explore biophilic design throughout the complex, making it a healing place for people and nature. The concept was to design "a hospital in a garden" and have it be a place "where one's blood pressure lowers when he or she enters the hospital grounds" (Newman 2014). Liak Teng Lit, the former CEO, remained involved in the project. Jerry Ong from the firm CPG was the lead architect, who was new to the principles of biophilic design, but it wasn't long before the project began to take shape.

The hospital's goals for incorporating nature were to

1   Help patients forget their pain and improve their rate of recovery by immersing them in a natural healing environment
2   Create an invigorating parklike ambiance for caregivers and the general public
3   Enhance views and access to nature to create a conducive working environment for staff

The buildings were constructed in a "V" configuration, open to the north to catch the dominant breezes which came in over the retaining pond. The exterior envelope took into consideration shading, natural air movement, and permeability. These features allow patients to have exterior views, access to daylight, and natural ventilation without heat gain and glare.

Plants are available at every level of the building, including fragrant plantings on the balconies of the upper-floor hospital rooms. Areas of open space, light, and view of the sky penetrate all the way to the basement levels. As the landscaping has grown, it truly appears that the hospital is part of a growing forest.

Even though KTPH is in a dense urban area, with their creativity and planning, the hospital has

> ...managed to achieve a green plot ratio of 3.92; [meaning...the total surface area of horizontal and vertical greenery is almost four times the size of the land that the hospital sits on. In addition, 18% of the hospital's floor area account for blue-green spaces and 40% of all such spaces are publicly accessible.
>
> (Living Future.org)

This project exhibits the principles of biophilic design holistically applied. For a full understanding of the multiple strategies used, see the Resources section at the end of this section and Chapter 6. As a quick review, applying biophilic principles at the building scale covers the following areas:

DIRECT EXPERIENCE OF NATURE: involves direct visual or physical contact, and/or brief momentary experiences. These need to be meaningful, diverse, and include movement. This covers elements like light, air, water, plants, and animals.
INDIRECT EXPERIENCE OF NATURE: involves representations of nature or natural forms, processes, systems, or other aspects of the evolving natural world. Elements in this category are light + space, shadows, interior and exterior spaces, natural shapes and forms, materials, sensory variability, transitional spaces, and the integration of parts into wholes.
EXPERIENCE OF SPACE + PLACE: elements or attributes of the built environment that inspire an emotional, intellectual sense of attachment or belonging to a physical location. Elements in this category include evolving human relationships, prospect + refuge theory, information + cognition, cultural + ecological attachment to place, and spirit of place.

Additional resources

A New Urban Forest Rises in Milan, Bosco Verticale, http://global.ctbuh.org/resources/papers/download/2099-a-new-urban-forest-rises-in-milan.pdf

Biomimicry Case Study: Eastgate Building Zimbabwe, https://inhabitat.com/building-modelled-on-termites-eastgate-centre-in-zimbabwe/

Biophilic urbanism: a case study on Singapore, Peter Newman, https://www.tandfonline.com/doi/pdf/10.1080/07293682.2013.790832

Healing Through Nature: Koo Teck Paut Hospital, http://greeninfuture.com/pdf/greenPulse_Jan2018.pdf

Healing Through Nature, Khoo Teck Puat Hospital, International Living Future Institute, https://living-future.org/biophilic/case-studies/award-winner-khoo-teck-puat-hospital/

Whole Building Design Guide, www.wbdg.org/resources/whole-building-design

13.7 Resilient architecture

A resilient building is different from a green or sustainable building. The added requirement to design a building that survives through a myriad of unknown future events, and persist 100–200 years or more from now as a useful contributing member of the built environment makes the challenge greater. Events such as floods, earthquakes, and hurricanes are "natural" threats, and there are human-generated events like power outages, economic depressions, and terrorism which also exert forces on buildings and communities. Resilient buildings are designed to respond to these and other conditions through careful planning. This is discussed in detail in Chapter 7.

The Brock Environmental Center design by SmithGroupJJR in 2015 is an example of how resilience changes the equation for a sustainable project. The project serves as an example of contemporary resilient design approaches, and it has reached "fully certified status" for the Living Building Challenge (LBC) (Figure 13.7a).

Performance

The use of ambitious performance standards in the LBC requires the design and construction team to reach net positive energy, meaning that it provides more power on-site over a year than it needs to operate. The building is also net zero water, a very difficult goal to

Figure 13.7a Brock Center.

Source: Wikipedia. By Jwallace72 – Own work, CC BY-SA 4.0, https://commons.wikimedia.org/w/index.php?curid=38048661.

reach that requires deep technical knowledge and the application of inventive water management systems. Eco-efficient design is integral at every level, and every system is optimized for maximum performance.

Passive systems + durability

Designing with passive systems includes evaluating the building's site location, which in this case is situated on a coastal site. The design anticipates sea level rise and coastal impacts. Brock is set 200 feet back from the shore and is raised 14 feet above sea level, anticipating future storm surge impacts associated with sea level rise – even a 500-year storm and winds of up to 130 miles per hour (Urban Land Institute 2016). The interiors are designed for natural ventilation and daylight, meaning that when there is a power outage, the building remains functional. The high 'R' values of roofs, walls, and the use of high-performance windows mean that the building is not only energy efficient but, because of Virginia's mild winters, it can also remain functional during winter if the heat is interrupted. The shape of the roof is designed to reduce the impact of high winds by "guiding the flow" over the building as opposed to hitting the walls.

Active systems

The active systems include battery backup for emergency lighting, and there are plans to add a full battery backup so that the building can stay functional in prolonged power outages. Since the building collects and treats its own water, it can provide clean water to the building's users even if the municipal water supply is interrupted for as long as six weeks.

Living systems

To comply with the LBC requirement of food being grown on-site, indigenous plants including blackberries were used to support local biodiversity, and a pollinator garden was also included to promote the health of bees and butterflies. Oyster beds were introduced into the river bed as a way to clean the water and provide educational opportunities about healthy ecosystems to visitors. Composting toilets, which use no water, were installed to save water.

Human systems

Due to the use of composting toilets and other technologies, the maintenance requirements are higher than a typical building. Living systems are a great way to meet sustainable goals, but they do come with additional maintenance requirements. Human systems include maintenance.

Culture

Due to the elimination of toxic materials and the introduction of natural ventilation and ample daylight, employees reported feeling "better" in the new space than in the old space (Wilson 2015). The shape and form of the building is driven in large part in response to the need to be resilient in the face of high winds and storms. The form with its many windows brings in light and air to all spaces. The materiality of the metal skin reflects the sky and so the color of the roof/walls is always changing. The question as to whether the project is beautiful in the traditional sense is an interesting one. One theory argues that the act of aligning the shape and form of buildings to maximize performance will yield some form of beauty, while others prefer traditional forms.

Additional resources

The Brock Environmental Center: A Pinnacle of Sustainability—and Resilience, www.resilientdesign. org/the-brock-environmental-center-a-pinnacle-of-sustainability-and-resilience/

Whole Building Design Guide, Brock Environmental Center, www.wbdg.org/additional-resources/case-studies/brock-environmental-center

International Living Future Institute, Brock Center, https://living-future.org/lbc/case-studies/the-chesapeake-bay-brock-environmental-center/

13.8 Architecture and health

The relationship between architecture and health is becoming of great importance in sustainable design. This shift changes the focus of the architect and interior designer from pure aesthetics to buildings that can help heal. Many of the details, research, and strategies are outlined in Chapter 8.

With the shift in focus, multidisciplinary design, which includes all the stakeholders, becomes more important than ever. This shift also draws on many different professions along with the end user, science, neuroscience, evidence-based design, and environmental psychology to create supportive environments. The Well Building Standard, which is explored in Chapter 8, is an excellent starting point. It was developed by a multidisciplinary group, supported by scientific evidence, and is focused on the health and well-being of the built environment.

The American Society of Interior Designers (ASID) headquarters in Washington, DC was recently named the greenest and healthiest office on the planet. Perkins + Will designed the groundbreaking interior, which has received both LEED and WELL Platinum Certification. By receiving these honors, the project encourages designers to keep the highest certification standards in mind. "We began this project with a clear goal of showcasing the many ways design can positively affect the health and wellbeing of employees while boosting resource efficiency," Fiser says. "We believe in research-based results in design and placed an emphasis on third-party validation of the space" (Nieminen 2017).

Performance

Randy Fiser, CEO of ASID, stated,

> So we set a lofty goal for ourselves to ensure that the environment we created for our employees would improve those elements, and that we'd measure and create those documents to see whether or not we were successful in doing that.
>
> (Nieminen 2017)

The project is an example of evidence-based design. Instead of picking materials for their aesthetic and durability requirements, every product also conformed to a third-party standard, like Declare, Cradle to Cradle, and Bifma, that makes sure that air quality and other measures were met in the project. The effort to source these materials requires more time but is easier and less time consuming with each successive project. As firms use these rating systems more and more often, it gets easier to accomplish a highly rated sustainable interior.

To underscore the role of research in design, ASID staff participated in the design process by wearing sensors that measured speech patterns and body movement when they interacted with each other. These sensor readings were compared to show how their interactions changed as a result of the new office design. A sustainable project is a living project, and the effects of the design change over time. Surveys are a tool used post-construction to determine if the strategies are effective.

Systems

To meet the requirements of WELL Standards for circadian lighting, a combination of automatic window shades activated by the movement of the sun to deliver optimal light levels while reducing glare was used. This allows natural daylight into the building, while not increasing heat during the summer or glare responsible for eye strain. It serves to override the natural tendency of occupants to leave the shades down once they have been lowered. Air quality and acoustics were also addressed.

Culture

Employees have no assigned seats, but rather select from a variety of workplace environments based on what best supports their specific tasks that day. Spaces range from highly collaborative teaming areas to more heads-down, focused sections. Employees can reserve spaces for a few hours, or a day, through a room reservation system that is integrated into ASID's software. ASID's mission is to promote the importance of interior design and the built environment's effect on occupants. They are using the highest standards for their own office building which is also an expression of their culture and seeks to educate others in the importance of high building standards.

Experience

Biophilic elements are used in the space, not just living plants but patterns and colors that help to connect people to nature. The space provides daylighting and view as well as places to gather. As part of the research, employees wore monitors to record their emotions and neurological reactions towards space as part of the design to increase the beauty of and satisfaction with work spaces. The project has only been recently completed, so the metrics are not reported yet, but the link between research, evidence, and design approaches is becoming more pronounced and is transforming the design professions to think more holistically about the importance of experience.

Additional resources

AISD Headquarters, Interiors + Sources Magazine, https://interiorsandsources.com/article-details/articleid/21250/title/asid-headquarters

Perkins + Will Creates Workplace of the Future for Leading Interior Design Professional Society, https://perkinswill.com/news/perkinswill-creates-workplace-future-leading-interior-design-professional-society

Well Building Standard V2, https://v2.wellcertified.com/

A picture of health – Well Building Standard at Cundall, https://cibsejournal.com/case-studies/a-picture-of-health-well-building-standard-at-cundall/

ASID (American Society of Interior Designers), www.asid.org/

13.9 Sustainable design process for buildings and interiors

Design charrettes for buildings and interiors

In Chapter 9, an in-depth discussion on integrative design practices explored an emerging set of design methods that is dramatically reshaping traditional design practice. The integrative design processes rely upon stakeholder engagement to create guiding principles, targeted performance goals, predesign research and analysis, and the actual collaborative design process itself. These participatory design approaches are typically reserved for urban design or public park projects where large segments of population invest time and energy in visualizing how their communities can be transformed (Abdallah 2018).

In the case of a building or interiors project, the same process can be used in very effective ways, with the stakeholders including, but not limited to, end users, clients, employees, maintenance people, community leaders, neighbors, and all the design disciplines. It's important to note that only the guiding principles, major site plan, floor plan, and sometimes the section are developed collaboratively, leaving the normative design process to flesh out the details and ensure functionality and code compliance. The social equity gained from a collaborative process has benefits beyond informing the design itself, as end users begin to take pride in the design and sustainability of their building, leading to a greater of sense long-term stewardship.

The intentionality of integratively designed buildings and interiors is uplifting to all participants, and it adds valuable information to direct the design process. However, skilled designers/facilitators are required to organize and oversee these intense work sessions. An unfacilitated event or one where the input from stakeholders is not respected can actually increase frustration, resistance, or apathy around a project or its purpose. To learn more about the design process, see Chapter 9.

Mini case study for integrative design

Alice Ferguson foundation, Hard Bargain Farm

Hard Bargain Farm is an LBC project designed by Re:Vision Architecture and Andropogon. A six-day design charrette was used to pursue deep collaboration and high levels of integration. The building program included overnight accommodations for camp students and expanded the dining services.

Guiding principles

The guiding principles, also called touchstones, "stretched" the team's thinking in order to unlock hidden potential in the project. The touchstones are meant to make sure that the project is understood holistically by all stakeholders. They are listed below:

- Make kids the priority – everything is child centered
- Protect open space
- Protect the viewshed to Mount Vernon
- Preserve and protect the rural and historical character of Hard Bargain Farm
- Be comfortable and connected to nature

- Be healthy, with natural light and clean fresh air
- Be adaptable and use flexible spaces that can change over time
- Turn green building strategies into teaching tools
- "Walk the talk" and inspire by example! LEED, climate neutral, regenerative
- Be a trash-free facility

Designing from principles is a key to sustainable design. Aesthetically driven design concepts can fit within the principles and help to call attention to the key aspects of the project. For example, "Turning green building strategies into teaching tools" is a great opportunity to express creativity and bring "delight" to the project.

Multidisciplinary design

The design team for the charrette was diverse with all disciplines including the builder present for the entire experience. Staff, administration, end users, and neighbors were all invited to parts of the charrette. The use of a facilitated, inclusive process meant that the process took longer but created transparency, accountability, social engagement, and support. The multidisciplinary team meant that the design professions were in constant contact and started with a full set of design directives.

One of the great challenges of the project was the location of the building and its impact on the open space of the farm. While technically there was an abundance of space, the site inventory and analysis indicated a specific location for the building to perform at its best. This meant the loss of some open space typically used for soccer and other athletic functions. By having camp staff participate in the charrette, these concerns were voiced leading to the building location being "pushed and pulled" until the group was satisfied that all requirements were met.

Systems thinking

Building up, not out became one of the passive approaches to the building design which lightens the impact of the project on the land. Instead of spreading cabins around the site, the decision was to build "cabins in the sky" where children would experience the tree canopy and be closer to the stars. This is an example of where the integral perspective of *systems* was married to the integral perspective of *experience* to create a synergy between the objective and subjective aspects of the design project – a key victory in sustainable design projects. It was also decided to move the dining function down into a pavilion alongside the main green space of the camp with overnight and cabin functions up on the hill (Figures 13.9a and 13.9b).

An additional challenge was that the building on the hill was always in shade and very wet, while the proposed building in the meadow received lots of sun but needed additional on-site water to meet its needs. A bio-inspired idea came to the team during the charrette to have the "meadow" building act as "grass" and collect sun for both buildings and the building on the hill act as "moss" absorbing and collecting water for both buildings. The charrette culture allowed for such creative ideas to not be rejected out of hand. The design of the buildings changed significantly after the charrette due to budget concerns and other factors, but the principles, concepts, and social cohesion gained in the process helped to buttress the design team in their pursuit of LBC.

"Grass"   "Moss"

Figure 13.9a "Moss" and "grass" images for the Hard Bargain Farm project by Re:Vision
Architecture.
Source: Re:Vision architecture.

Figure 13.9b The "grass" and "moss" buildings from the Hard Bargain Farm by Re:Vision
architecture.
Source: Re:Vision architecture.

Additional resources

Living Building Challenge Case Study Website, https://living-future.org/lbc/case-studies/
morris-gwendolyn-cafritz-foundation-environmental-center/

Re:Vision Architecture website, https://revisionarch.com/projects/potomac-watershed-
complex/

Architecture's Evolving Role: How Community-Engaged Design Can Encourage Social Change,
www.archdaily.com/890691/architectures-evolving-role-how-community-engaged-
design-can-encourage-social-change/

13.10 LEED green building rating systems

The LEED rating systems were founded by the United Green Building Council in 2000, ten years after BREEAM was established in the U.K. BREEAM stands for Building Research Establishment Environmental Assessment Method. In the U.S., it was the first national-scale rating system for green buildings. Eco-efficient design is the focus of LEED, which stands for Leadership in Energy and Environmental Design. LEED and BREEAM organize all the pertinent environmental aspects of a green building project into easy-to-understand categories.

Nine categories of the LEED rating systems are as follows:

- Location and transportation
- Sustainable sites
- Water efficiency
- Energy and atmosphere
- Material and resources
- Indoor environmental quality
- Innovation
- Regional priority.

Specific strategies within the categories are organized by a point system that reflects different "weights." In the LEED system, energy efficiency is worth a higher number of points, while the use of bike racks and showers receives the minimal amount of points. The system is third party validated and based on scientifically vetted criteria. The establishment of LEED and BREEAM was revolutionary, because prior to that, clients, architects, and engineers had no agreed upon definition of green building. This proved to be a critical tool in the rapid rise of green building activity in the first decade of the 21st century. It was transformative, because it gave non-built-environment stakeholders a vehicle to understand and measure the positive impacts of green building. Furthermore, from a design process perspective, the formation of LEED relied on a process that included multiple disciplines, and it was, and continues to be, a transparent process (U.S. Green Building Council 2018) that holds design teams accountable. In short, LEED and BREEAM met the goal of transforming the marketplace and educating clients on the different challenges and opportunities associated with green building. It is the "platform of success" upon which a wide range of new systems have recently come into vogue, such as Passive House and LBC.

Criticism

LEED has faced a series of challenges mainly due to perceptions of the systems being "expensive" and complaints about an overly prescriptive system that some say is too strict. Some architects, for example, complain about the system, because it added cost to their project. Frank Gehry, a high-profile architect, criticized the LEED rating system as adding cost to the project and being overly concerned with "political" issues in the prominent on-line journal *Inhabit* (Singh 2010).

Those with concern for the ecosystem, emissions, and the built environments immediately saw the benefits of LEED and BREEAM and adopted their use for a wide range of projects. Today, these and other rating systems are very useful tools, because they direct attention to many of the core values and principles for sustainable design, which are often

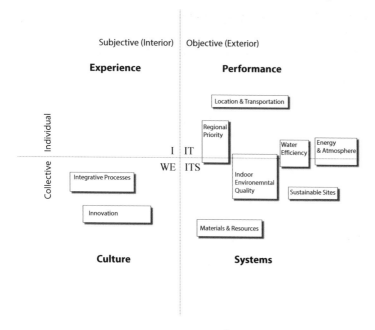

Subjective (Interior) | Objective (Exterior)

Experience | **Performance**

Location & Transportation

Regional Priority

Water Efficiency | Energy & Atmosphere

I IT

WE ITS

Integrative Processes

Indoor Environemntal Quality

Sustainable Sites

Innovation

Materials & Resources

Culture | **Systems**

Individual

Collective

Figure 13.10 The categories of the LEED® rating system organized by Integral Sustainable Design.

Source: Created and drawn by the authors.

overwhelming or too complex to grasp without specialized knowledge. Figure 13.10 indicates that the LEED credits are not holistically distributed among the four quadrants of Integral Sustainable Design. In fairness, the LEED system is meant to be focused on energy and environment and was never designed to attack the full range of sustainable design opportunities.

Transparency, accountability, and ecological regeneration are all reflected in the systems as well as the principles of performative design and the integration of systems. For other values that should be addressed like social equity and beauty, other rating systems will be shown.

Additional resources

To learn more about the LEED Rating Systems, visit: https://new.usgbc.org/leed

13.11 Passive House

Passive House is a demanding performance standard for the eco-efficient design of buildings. It is an *absolute standard* in that there is no baseline to measure against but rather requires meeting stringent energy performance requirements. Passive House or Passivhaus was founded in Germany in the 1990s. Passive House uses all the best passive strategies, especially a tight building envelope and proper orientation, to "max out" the performance of the building prior to the use of active systems for heating and cooling. According to the Passive House Institute in the U.S., buildings can achieve up to a 90% reduction in energy costs by using this system (Passive House Alliance 2018). The system is widely popular in the European Union (EU) and gaining popularity in the U.S. and around the world. Some clients prefer the straightforward nature of Passive House over the more holistic multi-attribute rating systems, because it addresses energy usage directly, while leaving issues like water savings, green materials selection, and site issues to be pursued at the client's discretion without the imposition of standards beyond the local code. The primary strategies to achieve Passive House are shared below in a bit more detail.

Super insulation

Imagine a scenario where you are sleeping at night and you get very cold. You have the choice to wake up and get another blanket or turn the heat up. The first choice is a passive system. It requires no energy for use, but it helps to keep you warm. The second choice of turning up the heat is an active system that requires the use of outside energy – which typically requires fossil fuels. Passive House relies on the first option by super insulating the building, thereby reducing the need for heat. Insulation is a one-time passive strategy that bears energy savings for the life of the building. It is a very effective strategy with low initial costs.

High-performance windows

Now, imagine that you have added your blankets to stay warm, but there are gaps where heat escapes, making you colder. We lose a lot of heat through windows and doors of the home. High-performance windows, often triple pane, prevent heat loss and gain. Thermal breaks in the structure of a window prevent thermal bridging and loss of heat – giving these new triple-pane windows an R-value of ten.

Air tightness

Passive Houses are required to be super insulated and airtight. A Blower Door test is performed after construction to measure the airtightness of the structure, which reflects the potential cold air infiltration and loss of heat. The airtightness requirement for certification requires a secure air barrier making sure there are only 0.6 air changes per hour at a 50 Pascal pressure difference (about a 25 MPH outdoor wind). To put that into context, most buildings in the U.S. vary from 7 to 14 air changes per hour!

Thermal bridging

Thermal bridging occurs when cold moves through the actual materials of the house rather than through the air. Insulation is great to use, but 15% or more of a building's envelope is comprised of structural elements, in the form of studs, columns, and roof joists. These elements can transfer temperature into and out of the house. Stopping thermal bridging is a subtle but important technique in design, and it can be achieved by adding a layer of exterior insulation or foam to interrupt the thermal conductivity through the structure.

Air exchange

Because of the extreme airtightness of a Passive House, the *active system* of circulation/ventilation of fresh air must be employed in order to keep the occupants healthy. Heat Recovery Ventilator (HRV) units are used in the U.S. giving a required 0.4 air changes per hour and absorbing potentially wasted heat for reuse in the building. Using the example of the warm blankets at night, imagine if you cover your head as well as your body. You would be even warmer, because there are no gaps around your neck, but you will need an influx of fresh air to be able to continue to breathe. Because the buildings are designed to such high levels of air tightness, air exchange is a critical feature of the Passive House System.

Criticism

The most widely held criticisms of the Passive House System are its lack of a holistic approach to sustainable design and the need for a mechanical unit to move fresh air into and out of the house. Passive House can be used in tandem with other rating systems where it can help to accomplish the energy requirements. The LBC covered next relies on a well-designed and well-built building envelope to achieve many of its goals.

Conclusion of Passive House

This rating system helps us design based on balancing many systems and metrics to generate a super high-performance building. Because of an increasing awareness of the need to protect common resources and extend our global energy reserves, more Passive Houses are being built every day. Looking at this in terms of Integral Theory, (Figure 13.11) we can see that behaviors/performance of systems are very well represented. The culture quadrant is indirectly addressed through the ethical motivations to save energy. There is a perception that high-performance buildings would be ugly, but that has not been the case. In the Resource section, there are links to see case studies of Passive House buildings.

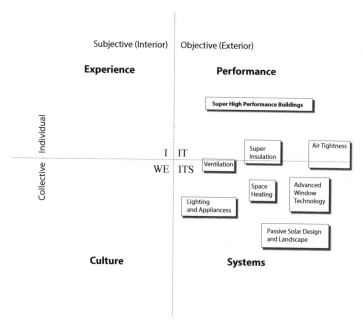

Figure 13.11 Integral analysis of the Passive House rating system.
Source: Created and drawn by the authors.

Passive House affordable housing design by Onion Flats

A set of three affordable row homes in Philadelphia were designed and built by Onion Flats, a design-build firm specializing in Passive House projects. The intention of this project, according the firm's website is "for it to be a model of affordable and sustainable housing for the City of Philadelphia." The design of the homes met the Passive House Standards at an affordable price and included solar panels. The project used the process of off-site construction, where the units were built in a factory and shipped to Philadelphia for final assembly. The use of factory-built architecture is an expression of the principle of dematerialization discussed in Chapter 5.1. Even with the impact of transportation, the process still has a smaller footprint than conventional construction, costs less, and leaves far less waste from the process.

Additional resources

Onion Flats Website, http://onionflats.com/projects/residential/belfield-townhomes.php

13.12 Introduction to the Living Building Challenge (LBC)

The LBC is a third-party rating system created by the Institute for Living Futures (ILFI) which was designed to go beyond LEED to drive the design and construction companies to push beyond the limits of possibility to achieve ambitious groundbreaking levels of eco-efficiency and ecological integration.

ILFI is working to transform how we think about design at the philosophical level, with a worldview that relies on a paradigm shift. This "unified tool for transformative thought" is a call to action as we enter into the age of peak oil, peak water, peak phosphorus, and a rapidly growing population which asks us to reinvent our relationship with the natural world and see ourselves as part of it. If any rating system expresses the new worldview of integration, it's the LBC.

The seven performance areas for the LBC are called "Petals." They are place, water, energy, health/happiness, materials, equity, and beauty. This is a holistic system that successfully integrates the four perspectives of Integral Sustainable Design and incorporates all the sustainability values covered in Chapters 4 and 5. Each of the petals is further subdivided into 20 imperatives, which are applied to projects at a range of scales and locations.

To understand the LBC, there are two simple rules. First, all imperatives are mandatory. There are "Petal Handbooks" which contain updated rules and exceptions according to fluctuations in the market. Second, LBC projects must be operational for at least 12 consecutive months prior to the evaluation process required for certification. It is the actual building performance which is reviewed, not what is "expected" or "designed for."

Two other important aspects that set this rating system apart from others are its requirements for strict vetting of materials and owner education in reference to ongoing building operations/performance. Designs that emerge from the LBC framework are socially just, culturally rich, and ecologically restorative. A brief overview of each petal is shared below.

Water petal

The intent of this petal in the LBC is to redefine "waste" in the built environment so that people see water as a precious resource. In the Eco-Sense Residence, the architects were able to design a system that uses less water than standard systems in addition to using composting toilets. Less water in equals less water out. The building is hard-wired for water conservation. Water cisterns are used for the irrigation of gardens. Groundwater is used as drinking water – a thoughtful consideration based on the filtration necessary to turn rain water into potable water. The house has been used as an example for educating the community on responsible water use since its construction. You can read more about the policy obstacles which had to be overcome during design and construction at the link below.

Energy petal

The intent of this petal is to create a future in which a decentralized power grid, powered entirely by renewable energy, supplies super-efficient buildings and infrastructure without the negative impacts of combustion.

Health and happiness petal

The intent of this petal is to address the experience quadrant of the Integral Theory diagram by optimizing physical and psychological health and well-being. At the Bullitt Center in Seattle, Washington, building occupants are enticed to walk the "Irresistible Stair" rather than use an elevator. Incredible views of the city skyline layered with the Olympic Mountains beyond reward stair climbers as well as individuals at their workstations throughout the building, allowing tenants to remain connected to the world outside its walls. Materials were also carefully chosen to provide pristine indoor air quality. This intent can be compromised by the building owner who doesn't manage the facility, its occupants' habits, and the use of building systems over time.

Materials petal

The intent of this petal is to support the LBC vision for the future in which all building materials are regenerative and have no negative impact on human and ecosystem health. During construction, materials should be recycled or reclaimed and new materials should be natural, nontoxic, and locally sourced whenever possible. The LBC has identified a list of materials shown to negatively affect health, which is called the Red List. All finishes, products, and components used within the project need to be examined to make sure they do not contain any of the listed items. This has a positive benefit on the indoor air quality of the project.

Equity petal

The intent of this petal is to encourage the development of communities that allow equitable access and positive treatment of all people, regardless of their physical abilities, age, or socioeconomic status. There are many limitations to reaching this idea, which stem from ingrained cultural attitudes about the rights associated with private ownership, the varying rights of people, and overzealous zoning standards in building codes. This petal supports healthy, diverse communities that encourage multiple functions and are organized in a way that protects the health of people and the environment. The Smith College's Bechtel Environmental Classroom has achieved this petal by creating a space that enriches a wide range of the College's courses. Massachusetts also has the Architectural Access regulations, which were a precursor to the Americans with Disabilities Act code that is now a national standard of equitability.

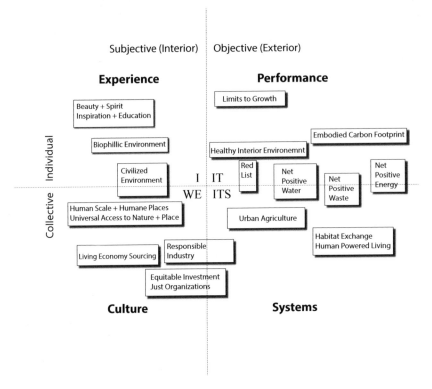

Figure 13.12 Integral analysis of the LBC.
Source: Created and drawn by the authors.

Beauty petal

The architects and owners worked closely together on the design of The David & Lucile Packard Foundation building in which "design stimulates the senses and enriches people's ability to work together on solutions for critical problems facing our world." An operable facade blurs the line between indoor and outdoor spaces. Ultimately, the building structure and landscape created enriched the lives of staff and community who come in contact with them. This is the intent of the beauty petal. It is impossible to mandate or judge specific attributes of aesthetic beauty, but the LBC requires an effort be made to enrich people's lives with each square meter, recognizing beauty is a precursor to caring enough to preserve. The diagram analyzing LBC against Integral Theory is shown above in Figure 13.12.

Additional resources

International Living Future Institute: https://living-future.org/

Declare. https://living-future.org/declare
A transparency platform and product database website. Manufactures are encouraged to submit information to the database to aid in the process of achieving the materials petal. Each project in the database is given a label that profiles components including

Red List items or EPA's contaminants of concern (COC), end-of-life options, location of extraction, VOC information, and if the product is LBC compliant with Red List.

Just. https://living-future.org/just

A voluntary disclosure tool for organizations or businesses to disclose their operations, including how they treat their employees and where they make financial and community investments. The organization is given a score of 1–3 in six categories related to social justice: worker benefits, safety, stewardship, diversity, equity, and local benefits.

Reveal. https://living-future.org/reveal

A verification of the energy performance of buildings, including three separate metrics: energy use intensity, the zero energy performance index, and reduction in energy use from baseline.

13.13 Other green building and interiors rating systems

The WELL Building Standard. https://wellcertified.com/en/explore-standard

This standard is explained in Chapter 8. The principles of the standard are shared below:

Equitable: Provides the greatest benefit to the greatest number of people, inclusive of all demographic and economic groups and with special consideration for groups of the least advantage or vulnerable populations.

Global: Proposes interventions that are feasible, achievable, and relevant across many applications throughout the world.

Evidence-based: Undergirded by strong, validated research yielding conclusions that can reasonably be expected to receive acceptance by the scientific community.

Technically robust: Draws upon industry best practices and proven strategies, offering consistency in findings across the relevant field or discipline.

Customer-focused: Defines program requirements through a dynamic process, with multiple opportunities for stakeholder engagement, and by tapping the expertise of established leaders in science, medicine, business, design, and operations.

Resilient: Responds to advances in scientific knowledge and technology, continuously adapting and integrating new findings in the field.

Architecture 2030

Founded in 2006, The 2030 Challenge is a rating system for new construction that seeks to reduce the carbon footprint of buildings. http://architecture2030.org/

Zero Net carbon (ZNC). http://architecture2030.org/zero-net-carbon-a-new-definition/

13.14 Global architecture

The United Nations Sustainable Design Goals and the Paris Agreement on Climate Change are drivers of architecture and design. The goals are reducing CO_2 emission and reversing pollution. All of the pressure on designers to pursue sustainable design are coming from global-scale initiatives that shape entire industries. In the EU, architects are required to design houses that meet stringent performance standards. California is pursuing a new ordinance that requires solar on all roofs of new housing (Penn 2018), and a green roof or solar is required for new construction in Paris (France-Presse 2015). Rating systems for green buildings and interiors have transcended LEED and LBC and are used all over the world. At the scale of a single architect, the global influence of the worldview itself can greatly motivate the inevitable evolution to higher levels of consciousness and drive individual decisions.

13.15 Architecture and the city

In Chapter 12, we looked in detail at the large and integral role that landscape plays in eco-system restoration, experience, and the social and cultural influences of a city. Architecture is what makes up the urban fabric. The buildings create and shape the interior and exterior spaces within a city. From the systems perspective, buildings are a smaller system that is part of the interconnected urban fabric, which also house the smaller systems of residence, businesses, shops, services, and medical facilities to name a few. In a city, the built environment provides for all our most basic physical needs up to those required for self-actualization.

Buildings and interiors are responsible for shaping our interactions with each other. They can promote social interaction, curiosity, learning, and healing. Going back to what we looked at in Chapter 5 with experience, we saw that "although we fundamentally shape our surroundings, ultimately place exists independently of human life, in turn shaping us" (Trigg 2012, p. 2). There are people who rarely leave the city, and it is the only experience of the world they have so the built environment we are creating is the only one that is shaping their perceptions of both the natural and man made world.

The built environment accounts for approximately 40% of carbon emissions and energy use. New buildings should be using the best strategies to reduce consumption and improve the quality of life. Existing buildings should be renovated to do the same. The city of Copenhagen in their *Solution for Sustainable Cities* realizes the role that buildings play in energy performance and has established a set of guidelines for new and existing buildings. Building tall is another strategy that has been mentioned before to create efficiency and provide more open space for light and air at the ground level. Resilience and adaptation are covered in detail in Chapter 7.

Additional resources

The Role of Tall Buildings in Sustainable Cities, www.witpress.com/Secure/elibrary/ papers/SC08/SC08033FU1.pdf
Copenhagen Solutions for Sustainable Cities, file:///C:/Users/shrob/Downloads/ 1353_58936BnEKE.pdf
Architects versus Economists: The Battle for the Future of Urbanism, From Honduras to Up-state New York www.blouinartinfo.com/news/story/804804/architects-versus-economists-the-battle-for-the-future-of

13.16 Sustainable architecture in communities and sites

The continuing theme of scale jump is key, especially when it comes to buildings in their site and their neighborhoods. Normally, designers are commissioned to design only one building or one small part of a community or district. This sometimes causes a disconnect if special effort isn't given to evaluating what is already there and the possibilities for the future, or even the hopes and aspirations of the stakeholders. It is very important to remember that when a designer is working on a project, it is really a very small part of a whole system.

Performance – One of the things that is learned quite quickly in the design of carbon neutral or net zero energy buildings is that there is often a limited amount of roof space for PV panels, leaving projects short in their performance goals. By increasing the scope of the project to include surrounding structures and parking lots, a shared PV array system between buildings is a likely scenario to increase overall performance of projects. The Bullitt Center in Seattle is a rare example of a building project being able to generate all its own power. It's more likely that buildings work in concert with large sites and ecodistricts to reach the desired performance level.

Culture – The culture of a community can be greatly affected by the buildings that shape public space. Thinking about culture on a larger scale is imperative to successful buildings and successful communities.

Experience – A sensory experience for a building design has already been established as a critical aspect of sustainable design. The relationships of a building to its community and its site are critical components of any community. Providing a sense of safety increases well-being and makes people more willing to walk or bike. Christopher Alexander, in his book *A Pattern Language*, goes into great detail about designing the built environment to create feelings we don't normally think about like privacy, wonder, providing long views for "prospect," and places of shelter for a feeling of "refuge." His development patterns also focus on creating social interactions as well which contributes to a sense of belonging.

Sustainable architecture and site

While many landscape architects have an ecological perspective, architects struggle to shift their goals away from buildings as isolated objects sitting disconnected from the landscape to a more integrated vision of buildings as participants in the landscape and as shapers of space. Early collaboration between architects and landscape architects is critical if the building is to play a meaningful role in the landscape and the community.

The Ray and Joan Kroc Center, built in Philadelphia and designed by Andropogon Landscape Architects and MGA Architects, is an example where the building and site-work are in a symbiotic relationship to support their mutual performance goals. The building benefits from careful placement of trees to shade harsh summer sun, while the landscape benefits from the collection of water via the roofs of the building. The connection of indoor space to outdoor space is critical to break down the typical separation between the two. The four perspectives of Integral Sustainable Design are reflected throughout the integrated use of systems: the meeting of performance standards (LEED Gold Rating); the attention to beauty through the use of proportion, scale, and color; and finally, the culture of the client being reflected by the inclusion of religious images and symbols throughout the project. There is more information and examples on sustainable site design in Chapter 12.

13.17 Sustainable design of buildings and interiors

The R.W. Kern center

The R.W. Kern Center, designed by Bruner/Cott Architects, was completed in 2017. It is located on the campus of Hampshire College in Amherst, Massachusetts. The 17,000-square-foot project exemplifies the application of sustainability values through innovative and thoughtful sustainable design practices. The multipurpose center project located in the heart of the campus (Figure 13.17b) provides all the typical functions and activities of any student center, and so much more for the college from a cultural and experiential perspective (Figure 13.17a).

Project overview

The project is an ambitious design effort that explores the depths of sustainable design from all four perspectives of Integral Sustainable Design (Figure 13.17c [next pages]).

Figure 13.17a The R.W. Kern Center's aerial view.
Source: Courtesy Bruner | Cott & Associates.

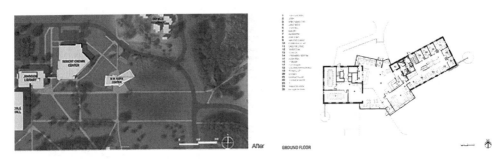

Figure 13.17b The R.W. Kern Center left: site plan Right: first-floor plan.
Source: Courtesy Bruner | Cott & Associates.

Performance Perspective

The Kern Center maximizes eco-efficiency by using ambitious performance goals to drive the entire design team to create an extremely efficient building. The performance goals of net positive energy as per the LBC are actually more demanding than net zero energy with a requirement to generate 110% of the building's energy requirements. The net zero water requirement compelled the design team to work with water harvested on-site, rather than the standard approach of piping potable water into the project. This is by far the most radically difficult performance goal to achieve, especially because local zoning codes typically don't allow for water to be purified on-site and used for drinking.

Systems Perspective

The systems integration is noteworthy on this project especially when it comes to water and energy. Figure 13.17d (next page) reflects an integrated approach that unites architecture and engineering to reach high levels of performance. The energy and water diagrams illustrate the complexity of high levels of eco-efficiency. Furthermore, the integration of these systems into a beautiful approach to architecture reflects a holistic and comprehensive design solution.

	Subjective	Objective
Individual	**Experience** - Local stone and timber - Access to daylight, fresh air and views - Biophilic experiences with plants, materials and wood structure - Beauty of restored site	**Performance** - Living Building Challenge Certified - Net-positive energy - Net-zero water - Eco-efficiency - Red-list materials - Minimalist use of structure and materials (dematerialization)
Collective	**Culture** - Supports college mission and values - Gathering area for college - Collaborative design process - Stakeholder engaged process - Post occupancy surveys - The project also inspired new courses at the college	**Systems** - **Life Cycle**: Using only local, sustainable, and non-toxic materials Red list – free materials - **Passive System**: High insulation, Solar orientation, Natural Ventilation, Storm Water collection - **Active Systems**: Solar panels, Heat recovery wheel, Gray water recovery systems, Inverter-driven heat pump - **Living Systems** Rain gardens, Gray water reclamation, Composting toilets, Pollinator plants - **Human Systems:** Operable windows, Maintenance

Figure 13.17c The R.W. Kern Center highlights within the Integral Sustainable Design framework.
Source: Created and drawn by the authors.

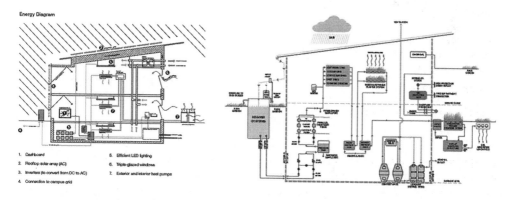

Energy Diagram

1. Dashboard
2. Rooftop solar array (AC)
3. Inverters (to convert from DC to AC)
4. Connection to campus grid
5. Efficient LED lighting
6. Triple-glazed windows
7. Exterior and interior heat pumps

Figure 13.17d Left: building section with highlighted energy systems; right: building section with highlighting water systems.
Source: Courtesy Bruner | Cott & Associates.

Culture Perspective

The Kern Center used the integrated charrette process including students, professionals, community members, contractors, and other project stakeholders. It demonstrates the college's commitment to sustainable design and environmental restoration. Jason Jewhurst, one of the lead architects on the project, stated:

The inclusive design process was critical to the project's success. In the early stages of the project, we led monthly design workshops and open discussion campus forums. Each time, we engaged in rich and insightful dialogue about building design, building systems approach, and even detailing of how various assemblies would come together. The result is a building that embodies the college's commitment to sustainability, community-led action, and hands-on learning.

This project, like many of the mini case studies contained in this book, reflects an effective use of the integrative design process, which is covered in detail in Chapter 9. Furthermore, the project itself achieves important cultural goals of the client.

Experience Perspective

The design of the Kern Center employed local, bio-inspired, natural materials of stone and wood that connect the occupants to nature and the surrounding mountains. The natural materials and colors are soothing, and the daylighting, views to nature, and natural ventilation make it a place that people find a sense of wellness – perfect for an educational facility. Of particular note is the attention to detail, especially when it came to the design of the structural connections (Figure 13.17e). Jason Jewhurst states:

> The Kern Center was designed for people. The natural light, inside-outside spaces, and local building materials offer an innate visual connection to the local vernacular and place. Visitors have expressed that when they are inside the Kern Center they feel deeply connected to nature and the campus surroundings; it is calming, makes them feel safe, and has a positive impact on their wellbeing.

The Kern Center project was recognized as an AIA COTE 2017 winner along with many other accolades and media attention. The project proves that beautiful, high-level sustainable design is possible, especially when enlightened clients are involved. The role of the sustainable designer is as much advocate and facilitator as it is designer in the traditional sense. This is a dramatic change for the professions, but instead of lamenting the loss of the focus on the experiential, the inclusion of social equity and engagement means that architecture and design can play a new, more important role in society.

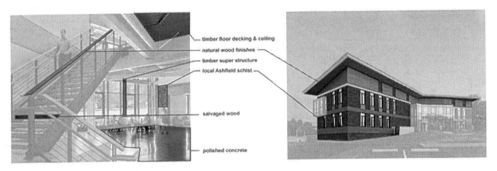

DESIGN WITH A HEALTHY PALETTE

Figure 13.17e Materials for the Kern Center.
Source: Courtesy Bruner | Cott & Associates.

Additional resources

AIA COTE 2017 Top Ten Awards, Kern Center, www.aia.org/showcases/76921-rw-kern-center

Architecture Magazine article on the Kern Center, https://architectmagazine.com/technology/detail/the-rw-kern-centers-minimalist-timber-structure_o

13.18 Conclusion

Sustainable design at the building and interiors scale offers opportunities to tie into the larger historical movement towards sustainable design and to attack the pressing environmental and social challenges of our time. The sustainability values of People, Prosperity, Planet, and Place are consistently expressed throughout all the examples in this chapter, and the Integral Sustainable Design principles were clearly expressed in the last example of the Kern Center. Bio-inspired design, resilience, and wellness have emerged as key drivers of sustainable design, and they offer deeper motivators than energy efficiency or fighting climate change. On the one hand, this is great as clients have more access points to engage sustainability, but this is also disconcerting considering that climate change will ultimately compromise everyone's health and well-being. Thinking across scale and time, core skill sets for sustainable design demand that design teams maintain a focus on regenerative design even when the clients are more focused on short-term localized goals. After all, like doctors, built environment professionals hold a special responsibility to protect the future of generations to come. The short case study of the Kern Center proves that all of the wonderful benefits and possibilities are not only possible but probable.

References

Abdallah, C., 2018. *Architecture's Evolving Role: How Community-Engaged Design Can Encourage Social Change.* ArchDaily. www.archdaily.com/890691/architectures-evolving-role-how-community-engaged-design-can-encourage-social-change/edn.

France-Presse, A., 2015. *France decrees new rooftops must be covered in plants or solar panels.* The Guardian News and Media Limited. www.theguardian.com/world/2015/mar/20/france-decrees-new-rooftops-must-be-covered-in-plants-or-solar-panels edn.

McHarg, I., 1971. *Designing with Nature.* 1st edn. New York, NY: Natural History Press.

Newman, P., 2014. *Biophilic Urbanism: A Case Study on Singapore.* London: Taylor & Francis.

Nieminen, R., 2017. *ASID Headquarters.* Stamats Communications, Inc. www.interiorsandsources.com/article-details/articleid/21250/title/asid-headquarters edn.

Passive House Alliance, 2018-last update, Passive House Institute [Homepage of Passive House Alliance], [Online]. Available: www.phius.org/home-page [July 28, 2018].

Penn, I., 2018. *California Will Require Solar Power for New Homes.* The New York Times. www.nytimes.com/2018/05/09/business/energy-environment/california-solar-power.html edn.

Singh, T., 2010. *Frank Gehry Slams LEED, Calls Sustainable Design "Political".* Inhabit. https://inhabitat.com/frank-gehry-calls-sustainable-design-political/ edn.

Trigg, D., 2012. *The Memory of Place: A Phenomenology of the Uncanny.* Athens: Ohio University Press.

U.S. Green Building Council, 2018-last update, About USGBC [Homepage of U.S. Green Building Council], [Online]. Available: https://new.usgbc.org/about [July 28, 2018].

Urban Land Institute, 2016-last update, Brock Environmental Center [Homepage of Urban Land Institute], [Online]. Available: http://returnsonresilience.uli.org/case/brock-environmental-center/ [July 28, 2018].

Wilson, A., 2015-last update, The Brock Environmental Center: A Pinnacle of Sustainability—and Resilience [Homepage of Resilient Design Institute], [Online]. Available: www.resilientdesign.org/the-brock-environmental-center-a-pinnacle-of-sustainability-and-resilience/ [July 28, 2018].

14 Human scale sustainable design

14.0 Introduction – materials + products

Materials are our first and sometimes only impressions of a project, place, product, or person. It is what we immediately see and often use to make judgments. What we see and touch affects our perception of stability, integrity, and level of professional excellence of a business or person. Materials also give us clues to the socioeconomic level of a neighborhood, greatly affect the aesthetics and beauty of an area, and confer or reflect self-worth. Materials and finishes can make us feel secure or exposed. They affect insulation values, energy use, the longevity of a building, and its ability to withstand seismic events or extreme weather events. In medical facilities, it can promote sanitation and give nonverbal clues to the level of care we will be receiving.

Materials include more than just the finished surfaces that we see in a building or what is visible in the products that we use or the clothing we wear. The materials and finishes we choose can deeply impact the environment and human health; it includes resource extraction, the manufacturing processes, all transportation, the longevity of the material, the health effects, and what happens to it at the end of its useful life.

14.1 Historic context for materials + products

Hunter-Gatherer Age – Local materials that were readily available for use without extensive changes to its form or the need for transportation. Materials were used in their natural state and were therefore biodegradable when the end of useful life was reached. The form of the structures was directly related to the material and its properties that were being used, the specific site and its relation to climatic factors.

Agricultural Age – Natural materials were still the norm, but they began to be manipulated by cutting stones to have straight edges to make assembly easier and less natural. Structures became more ornate and reflected human influence over natural elements as a sign of control. Concrete was invented around this time in the regions of present day Syria and Jordan and allowed large scale and more elaborate structures than had been previously possible. Rome's Emperor Hadrian completed the Pantheon in 125 AD, which has the largest unreinforced concrete dome ever built.

Industrial Age – In this period, the perception of the earth as possessing unlimited energy resources and materials gained momentum. Creation of synthetic materials, many based on the use of petrochemicals to create plastics and synthetic binders started to appear.

These seemed like great improvements over traditional materials and appeared to increase the resource base for our expanding consumerism. Because there was no long-term research or projections, contamination of the natural world eventually resulted.

Age of Information – In this period, experimentation with synthetic materials led to great advancements in structural capability, cost reduction, and ease of installation. However, these products created other problems like mold, moisture deterioration, and sick building syndrome. The growing awareness of resource scarcity and environmental pollution meant that the search for more sustainable materials and products would soon begin.

Age of Integration – Today, a more holistic approach to material selection is now becoming the norm. Looking at longevity of usage and the long-term effects of the material on the humans and animals it comes into contact with is a requirement. A holistic approach also deals with the changes to the ecosystem as a result of the resource extraction and disposal at the end of life. There is also a trend to look to materials that have already been manufactured and reusing them. For example, the Whitney Museum in New York City recently renovated a portion of its facility. The new flooring in many of these areas was reclaimed heart pine floors that were harvested from historic North American structures like barns, factories, farmhouses, and decommissioned New York City rooftop water towers.

14.2 Nature and materials + products

Impact of materials on the environment is one of the main reasons to pursue sustainable materials and products. The impacts to water, air, earth, and to our energy supplies are significant and are mainly centered around three broad areas; resource extraction and manufacturing, end of life disposal, and pollution.

The manufacturing of products starts with resource extraction. Depending on the raw material and methods of harvesting or extraction, there can be great disturbances to the ecosystem. By selecting materials that are extracted or harvested in a sustainable or restorative way, we are effecting greater levels of change. An example would be the Forest Stewardship Council (FSC) which has established forest management standards that also address social and economic benefits. The council oversees the growth and use of land, tracks and certifies chain of custody of the harvested wood, and assists smaller family forests in meeting standards and being economically viable.

The amount of waste in society is creating significant problems for the ecosystem because of the space occupied by landfills, chemicals leaching into water supplies, and plastic ingested by local species. Even worse is the garbage that doesn't make it to landfills and is floating in our rivers, streams, and ocean. The Great Pacific Garbage Patch is located in the Pacific Ocean between Japan and California. It is comprised of two main areas that have come to be formed because of circular ocean currents called Gyres. These patches are mainly plastic items that have found their way into the waterways and are dumped into the ocean. The problem with plastics is that they don't ever biodegrade. They only become smaller and smaller pieces of plastic, which end up being ingested by birds and sea life that either die or, as part of the food chain, pass the plastic on to humans. Efforts are being made to clean up these areas, but understanding disposal and designing ways to reuse, recycle, or dispose of the materials we specify in a way that benefits the ecosystem is crucial.

The pollution of water includes the amount of plastics floating in the ocean which are entering our food supply, see Chapter 2 for a deeper look. In the U.S., government agencies have laws that dictate minimum standards to reduce pollution. Other developing countries have nothing. Changes need to occur everywhere for our practices to become restorative in their manufacturing and disposal. As already mentioned, Sweden is developing a mesh based on cell proteins that has the potential to filter out all contaminants from manufacturing processes as well as the salt in seawater to increase our freshwater supply.

14.3 Materials + products and motivations

The motivations for material selection are shaped by self-interest and empathy. Self-interest has to do with public or cultural perceptions of affluence, comfort, ease, conformity, profit for manufacturers + retailers, and lower pricing for consumer. Empathy takes a longer view and looks at environmental conditions, long-term effects of practices, long-term conditions of the earth, and the ability for future generations to sustain their life. The motivation for self-interest plays a strong role in hindering widespread, systemic change. However, there are plenty of examples where companies have found a way to not only fall in line with the goals but also use them as a competitive advantage to differentiate their product lines.The Living Products Challenge initiated by the International Living Futures Institute takes a long-term view of materials to ensure that they are life enhancing, nontoxic, and resilient.

14.4 Sustainability values and materials + products

The Quadruple Bottom Line values of people, profit, planet, and place form the backbone of a sustainable approach to materials. Harvesting and manufacturing of materials and products can be unethical due to low pay, unsafe working conditions, and exposure to toxic chemicals. Environmental impacts can be quite severe depending on the type of extraction process that is used and material being extracted, which also affects the sense of place. The amount of profit companies derive from the manufacturing process is directly tied to extraction and production methods and how employees are treated. There needs to be a balance between all aspects of the Quadruple Bottom Line.

As a designer, specifier, manufacturer, or retailer, there is a chance to effect change in how the employees are treated and paid, how the manufacturing processes and waste disposal is handled, and have a hand in restoring ecosystems while still making beautiful products and being financially sound. The sports clothing manufacturer Patagonia is a great example of a company that produces high quality products, from environmentally sound resources, while requiring ethical treatment, fair wages, and environmentally responsible manufacturing practices even in countries without environmental regulations.

14.5 Integral sustainable design and materials + products

The materials and finishes we choose can deeply impact the environment, human health, performance, and the cultural standing of a building, product, or article of clothing. As stated earlier, it is the primary thing we see and use to make our subconscious judgments.

Below are some very general integral theory considerations. Your actual choices should explore the four quadrants for your specific project, climate, purpose, and geographic areas. The items should then go on to direct the guiding principles, project goals and objectives, and design decisions. Having a holistic understanding of the all the stakeholders allows for solutions with deeper meaning (Figure 14.5a [next page]).

14.6 Bio-inspired design and materials + products

Bio-inspired products and materials look to natural systems for their inspiration. In Chapter 6, a new material called Sharklet was discussed. It features microscopic-raised diamond shapes that prohibit bacteria, viruses, or germs from sticking to its surface and is based on the patterns of sharkskin. Material researchers and engineers in Japan are using

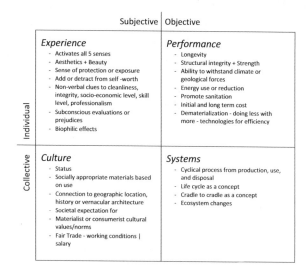

Figure 14.5a Materials and products within the integral sustainable design framework.
Source: Created and drawn by the authors.

the design of mosquito mouths to create needles that are more efficient and less painful. Medical researchers in the U.K. are using similar designs for neurosurgery that require less pressure from the surgeon so that there is less damage to the surrounding tissue (Biomimicry Institute 2016).

Biophilic materials are natural materials that are found in nature and provide very beneficial psychological and physiological benefits when in any environment. Natural materials can have a longer life span, be more resilient, and convey a sense of permanence with their beauty.

14.7 Resilience + adaptation and materials + products

Considering historic structures and vernacular architecture is one of the best ways to assess what materials will best withstand the local climate and weather events. These are the structures that have withstood the test of time and can give us valuable insight. The materials for the structure and exterior envelope will take the most force and should be as durable as possible. Interior finishes matter as well because they can provide flexibility in use for tenants without needing to be replaced. For example, solid wood flooring can easily and inexpensively be resanded to remove any damage from one tenant to another, or even from the building being used as a factory to an apartment, or restaurant. This would not be the case with carpet.

For clothing or products, looking for the most durable and highest quality fiber or base materials will allow a longer product life. It may involve a higher initial cost, but will end up having a lower long-term cost. Products should also be manufactured with accessible parts so that they can be repaired and not immediately disposed of if a component fails.

For buildings, products, or clothing, this requires a shift in consumerism and social concepts about usable life and the value of the "new." It will require a quality over quantity approach.

14.8 Health + well-being and materials + products

In Chapter 8, we saw the great impact that materials can have on the indoor built environment where we all spend approximately 90% of our time. We also saw in Chapter 6 how deeply the materials that we select can affect our mental, emotional, and physical health. There are many sources to help evaluate the chemical composition and health effects. One of the many resources available is Perkins + Will's Materials Transparency website with case studies, government involved information, and databases for searching materials and products.

14.9 Design process for materials + products

Decisions for materials and products used can occur all through the design process. As you research and understand the specific requirements for your project, focused clear objectives will arise. Your decisions should be based on your precedent studies, integral theory client and stakeholder analysis, performance requirements, and resilience and adaptation goals. There is no silver bullet for sustainable materials, but there are some general rules that can apply to any type of project. Please feel free to add any that your research reveals to be important to your specific situation. For more detailed information on the integral sustainable design process, see Chapter 9.

14.9.1 General principles + guidelines for materials + product selection

Some general principles that apply to material and product selection are:

- Consider using recycled concrete for work below grade
- Choose materials, products, and furniture with the highest recycled content
- Specify biodegradable release agents, inks, or cleaning solutions whenever possible
- Avoid products or materials that are not recyclable or reusable at the end of life
- Specify Forest Stewardship Council (FSC) wood for construction, millwork, and furniture
- Avoid endangered species wood
- Find the highest efficient window and doors
- Specify formaldehyde-free products – plywood, particle board, insulation, wallcovering, and carpet
- Use low or no Volatile Organic Compounds (VOC) finishes – varnishes, paints, stains, adhesives, and sealers
- Consider engineered products that use less raw materials
- Consider using installation techniques the reduce the use of adhesives
- Consider using natural materials – wood, tile, stone, and natural fibers
- Consider refurbishing equipment and furniture
- Select leathers with environmental regulation – leather is low maintenance and has the least environmental impact over time
- Consider the highest efficient lighting sources and fixtures
- Consider salvaged materials for interior and exterior

Integral sustainable design does not require a rating system, but you or your client may choose to comply with one.

14.9.2 *Applicable rating systems*

Below are the current major rating systems that are applicable to materials, finishes and the built environment at the human scale. These rating systems are constantly being revised with the current best practices and codes. It's important to review the latest changes before each project. The brief description will help you to decide which system is best for each project.

Leadership Energy and Environmental Design (LEED): https://new.usgbc.org/leed.

The LEED system was established by the U.S. Green Building Council (USGBC). It is a certification process based on third party verification of a comprehensive list of green building strategies. It is a point-based system that allows buildings to achieve a different level of rating depending on the amount of green building strategies used. Each of the rating systems has a credit category that covers materials and resources, water efficiency, indoor environmental quality, and sustainable Sites. Within each of the categories there are specific goals. Each goal reviews the intent along with any requirements. Project members can choose the goals that are most applicable for each project.

Living Building Challenge (LBC) – https://living-future.org/lbc/

A holistic, integrative system that has seven performance categories called "Petals," which are place, water, energy, health/happiness, materials, equity, and beauty. Buildings trying to achieve full Living Building Certification must be Net Zero Energy, Net Zero Waste, and Net Zero Water over a minimum of 12-month occupancy.

Well Building: https://v2.wellcertified.com/v2.1/en/overview

WELL v2 is founded on the following principles: equitable, global, evidence-based, technically robust, customer-focused, and resilient.

Comprehensive Procurement Guidelines (CPG) Program: www.epa.gov/smm/comprehensive-procurement-guideline-cpg-program

Environmental Protection Agency's (EPA) Sustainable Materials Management initiative promotes a system approach for reducing materials use and the associated environmental impacts over the materials' entire life cycle. It promotes the use of materials recovery and reuse in the manufacturing of new products.

Environmental Product Declarations (EPD): www.environdec.com/

EPDs are standardized, third-party verified documents that communicate the results of a product's Life Cycle Assessment (LCA), including all relevant performance information. EPDs also include information on mechanical, safety, human health, and any other issues that are of particular importance for that product.

Forest Stewardship Council (FSC): https://us.fsc.org/en-us

It regulates the growing, harvesting, and certification of forests. FSC certifies the chain of resources to assure that wood used in construction, furniture, and finishes is grown in an ecologically responsible way.

GreenBlue: http://greenblue.org/

It is a nonprofit corporation whose mission is to foster the creation of a resilient system of commerce based on the principles of sustainable material management. They look at packaging, chemical composition, large scale composting, recycling, and forest management.

Green Guard: http://greenguard.org/en/index.aspx

A third party certification party that is a division of Underwriters Laboratory (UL) which assesses and certifies building materials, furniture, and furnishings; electronic equipment; cleaning and maintenance products; and medical devices for breathing gas pathways. Certification is based on a set of standards to evaluate the chemical composition, off-gassing, and leaching of chemicals during use. Certified products will have a Greenguard label.

Green Seal: www.greenseal.org/

It is a nonprofit environmental standard development and certification organization is used by product manufacturers and services providers. Certification is based on a list of Green Seal standards covering performance, health, and sustainability criteria.

National Institution for Occupational Safety and Health (NIOSH): www.cdc.gov/niosh/index.htm

It is a part of the Center for Disease Control (CDC) and helps to ensure safe and healthful working conditions by providing research, information, education, and training in the field of occupational safety and health. NIOSH provides national and world leadership to prevent work-related illness, injury, disability, and death by gathering information, conducting scientific research, and translating the knowledge gained into products and services.

Occupational Safety and Health Administration (OSHA): www.osha.gov/law-regs.html

It creates laws, regulations, and codes with detailed requirements for materials, chemical exposure, and work environments. The intention is to provide safe working conditions. The regulations cover construction, maritime, agriculture industries, and other general business types along with state specific mandates for worker safety.

Passive House: www.passivehouse.com/

The Passive House Institute (PHI) is an independent research institute that has played an especially crucial role in the development of the Passive House concept. Passive house building standards rely on high insulation and airtight envelope with conditioned air being introduced through mechanical means. The type of materials used and their chemical composition are extremely important.

United States Environmental Protection Agency (EPA): www.epa.gov/

Agency of the U.S. Federal Government established with the stated mission of protecting human health and the environment by creating policies about regulating the extraction and use of natural resources, economic growth, energy, transportation, agriculture, industry, and international trade and areas that affect human health.

14.10 Global scale and materials + products

The United Nations (UN) Sustainable Development Goal (SDG) #12 – Ensure sustainable consumption and production patterns. Some of the UN targets associated with this goal address reducing consumption through national policy, efficient use of resources, the reduction of waste (especially food), the use of life cycle assessment, and reduction of pollution. The UN's targets also support development of scientific and technological advancement in developing countries, education and monitoring systems.

All of these goals can be supported by the design, construction, and manufacturing industries through careful evaluation of sources of materials and products. In the textile industry, for example, fabrics that are currently made at the small industry scale outside the United States or Europe are not regulated for environmental impact. It is difficult to get individual companies to upgrade their processes without government sanctions or codes, because their ability to compete is compromised. There are also companies like Patagonia, a clothing retailer who sources their production globally. They look at sustainability in a holistic way and apply systems thinking to how they produce their merchandise. They require that any of their offshore manufacturing plants comply with the most stringent U.S. pollution standards, that the workers are paid a living wage, and that all their materials are organic, fair trade, or from recycled content.

Additional resources

Sustainable Development Goals, www.sdgfund.org/goal-12-responsible-consumption-production.

14.11 Urban scale and materials + products

The materials that we use at the urban scale need to support the principles of longevity, resilience, and integral sustainable design. There are millions of dollars spent on the built environment of an urban area. If after a short time they are structurally compromised, they will need to be replaced, wasting all the resources and energy as well as increasing the volume of landfills.

Vernacular materials play a very large part in creating a sense of place that is tied to the geographic location of a city. Creating a sense of place enhances the quality of life, which promotes care and maintenance. Vernacular materials also have the benefit of time-tested durability within a particular climate. Biophilic materials help connect people to the natural world and have many psychological and physiological benefits. They also help mitigate storm water runoff and reduce the heat island effect.

Any products produced in an urban area have the benefit of providing valuable employment without long commutes and the associated transportation emissions. Industry contributes to waste disposal and pollution. These affect the ecosystem and therefore the quality of life for all.

14.12 District + site scale and materials + products

"What is required is a shift in attitude from architects, planners, and developers alike to find ways to develop buildings as communities and not as competition to one another." (SOM 2015, p. 21)

One of the most important goals is to create a sense of community, connection, cohesion, and caring for each other and the built environment. Materials can help with this. Many of the same benefits and cautions that we see in materials and products at the urban scale apply at the district or site scale. The negative aspects like pollution can be seen directly at the site and district scale, but so can the positive benefits. The rise of farm to table restaurants and urban agriculture are great examples of people's desire to know where their food is coming from and feel connected to the whole process. Vernacular materials are more resilient and reduce transportation emissions along with the biophilic benefits mentioned above. Local materials and locally manufactured products create jobs and employment.

14.13 Building + interiors scale and materials + products

The building and interior scale is where we see, touch and sense material and product choices the most. This is also where we can be exposed to any toxicity or health risks most intensely because we spend about 90% of our time indoors. The wrong choices can bring about a whole host of negative physical reactions to the individual occupants of a building.

We also saw in Chapters 6 and 8 that our choices can just as easily bring about positive psychological and physiological reactions. Material choices can increase healing, reduce pain, lower our blood pressure, and increase cognitive function.

Designers of the built environment should take a very active role in the research, selection, and specification of materials and products that will be in their projects. Rating systems and classification databases can be a great tool in this process. In Sections 14.9 and 14.14.5, lists of the current rating systems and resource databases are available.

14.14. Products and materials

Products and materials are at the basis of anything that we do with the built environment, product design, or the fashion industry. As we have seen throughout the book, they affect every level and scale of integral sustainable design. There are thousands of materials out there to choose from and thousands more reasons; we will be giving an overview of how to choose the best material or product for your specific project type, client, use, and geographic area.

14.14.1 Performance

How a material performs over time makes a huge difference in the sustainability of a project. Diverting material from landfills is also part of integral sustainable design at the site, district, urban, or global scale. Choosing the right material for the right project may take a little extra thought and research but will pay large dividends over its life. Here are a few broad-range issues to consider related to the performance of your materials and products.

Toxicity

We looked at Toxicity and off-gassing of VOCs in Chapter 8. The most important aspect to look at and understand regarding toxicity is the composition of each material or product that is being used as well as any treatment that is applied to the product during manufacturing. The Living Building Challenge Materials Petal requires that each component of each product or system be examined to assure it is free from a list of 14 chemicals known to have adverse health effects. This is accomplished by requesting a Materials Safety Data Sheet (MSDS) from the manufacture or looking it up in an online database if available. Many of the current rating systems require that you address the potential for lowering the Indoor Air Quality (IAQ) of a space. Even if you are not using a rating system, it is still important to understand the long-term effects of the individual materials you are specifying. Specifying all natural materials when possible is an easy way to assure there will be no toxicity issues in your project.

Longevity

Longevity refers to the expected usable life of a material, product, or built environment. Before the advent of the industrial age and mass-produced materials and products, the items that we used everyday and the buildings we inhabited were considered to be permanent or to last a very long time. Until recently, the cost or work to produce or purchase items was relatively large in proportion to the average person's resources. The main selling points were the longevity of a product and that it could be repaired when and if it broke.

The current social structure has changed dramatically in a way that demands constant change to the latest model for products, clothing, cars, and buildings. Manufactures are deliberately designing their products to fail so that new sales may be made; this is called planned obsolescence. This view is detrimental to sustainable design because we are constantly requiring more resources and filling up the landfills with discarded products. These products represent raw materials, labor, transport costs, and large amount of CO_2 emissions as a result of all those processes.

Longevity looks to reverse this process and choose materials, processes, and products that will allow a product, or building to last 50, 100, or 200 years. Infrastructure, buildings, and products can be designed to be renovated, changed, repaired, and then disassembled and reused or easily returned to raw material state to be made into something else. This is possible with many natural products like wood, stone, metal, glass, or wool. Using the highest quality materials, manufacturing or building standards available or affordable goes a long way towards promoting longevity. We can see from all the vintage cars in Cuba that still run and work very well that it is possible to produce products that will last 50–75 years. We can look at many European countries and find buildings that are hundreds of years old and still being used.

A company presently doing this is The Mohawk Group with its carpet ReCover Program that will take carpet from any manufacturer and recycle it into nylon and polypropylene pellets to be used by the automotive and furniture industries. It also recycles plastic bottles to be used in its Everstrand carpet line. Mohawk makes high-quality durable carpet but also considers how it can be reused after it requires replacement.

Durability

The usable life span of a material is extremely important in your material selection process. Many natural materials have an incredibly long life span. Solid wood or stone floors can last and be beautiful for hundreds of years. There are manufactured products that can achieve the same durability. Looking to vernacular architecture, buildings that reflect a local form and aesthetic can give important clues in choosing materials or products that will last. These materials and their forms have lasted over time and were created in response to locally available and abundant materials and climate conditions. The key is selecting the best material for the use, amount of user traffic, and selecting colors and finishes that will not easily go out of style.

Emerging sustainable materials

New materials are now being used as high performance building solutions. These emerging materials provide a great opportunity for sustainable structures because of their enhanced physical properties. We discuss some examples below.

Fiber-Reinforced Composites – Technically known as fiber-reinforced polymers (FRP), they consist of two components: a reinforcement fiber and a polymer binder. They are engineered to specific performance characteristics. The advantages are durability, lightweight, corrosion resistance, high strength, and low maintenance requirements. A well-known example would be the Epcot Center at Disney World in Florida.

PTFE Fabrics – Polytetrafluoroethylene (PTFE) is a Teflon coat applied to woven fiberglass fabric membrane. This combination produces an extremely durable, thermally stable surface that is light and weather resistant. It reflects 73% of the solar energy while allowing 13% of daylight to be transmitted. A well-known example is the Denver Airport in Colorado.

Carbon Fiber Honeycombs – Carbon fiber reinforced polymer honeycombs mimic the properties of balsa wood for strength. Its chemical enables 3D extrusion processes to control the alignment of fibers within the honeycomb structure, optimizing structural strength. This technology could also have implications for conductive composites as well. One of the immediate uses for this technology is the expanding size of wind turbine blades, now reaching 246 feet (75 meters) long.

Translucent Concrete – Translucent concrete combines the strength of concrete with the light transmission properties of optical fibers. Developed in 2001 by Hungarian architect Aron Losonzi, the mixture of fine concrete with approximately 5% optic fibers allows the finish material to transmit light, and weigh less.

Cross Laminated Timber – Cross-laminated timber is made of alternating layers of wood which are then laminated with non-VOC binder or a wooden peg system to create pre-manufactured building panels or structural beams. This reduces construction time, allows sculpture forms, and has superior strength. They also have noise cancelling and insulative properties.

Organic Form Structural Systems – New design and modeling software and available building materials have caused a return to organic form. They have given us the ability to build what we could previously only imagine in a cost effective way. The Birds Nest designed for the Beijing Olympics is a building of outstanding beauty, which employs an organic structural system of superior strength while using less material than a conventional structural grid.

14.14.2 Systems for products + materials

When we apply systems thinking to materials and products, we realize that all of our choices are interrelated. Our choice of materials affects the location and ecosystem of resource extraction, the fuel sources to transport and manufacture the final product, the CO_2 and other emissions from all the delivery and transportation, and the impacts of the landfill disposal.

Cradle to Cradle™ was developed by William McDonough and Michael Braungart in 2002. It uses the model of a natural ecosystem as an approach to materials, goods, and manufacturing to help companies understand the continuous industrial flows of materials, energy, and water within our global community. Their idea is to focus on doing more good, rather than looking to do less bad. The underlying goal for creating the Cradle to Cradle™ approach was "...a delightfully diverse, safe, healthy, and just world with clean air, soil, water, and power – Economically, equitably, ecologically, and elegantly enjoyed" (McDonough and Braungart 2002).

Cradle to Cradle™ is based on five executive vision statements:

1   Value materials for safe and continuous cycling (through the manufacturing and consumption process)
2   Maintain continuous flows of biological and technical nutrients
3   Power all operations with 100% renewable energy
4   Regard water as a precious resource
5   Celebrate all people and natural systems

C2C-Centre website profiles all the currently certified Cradle to Cradle™ products, projects and companies.

Additional resources

Cradle to Cradle, William McDonough and Michael Braungart
Cradle to Cradle Design, TED Talk, www.ted.com/talks/william_mcdonough_on_cradle_
   to_cradle_design

14.14.3 *Culture and equity*

The materials used in the built environment can tell us many things in regards to the culture of the organization or geographical location. They are able to convey status, cultural standing or socioeconomic conditions/levels/positions. As a society, we are very sensitive to materials and their implications. Think about a car with a leather interior versus a cloth interior, or a blouse made from pure silk to one made from synthetic fibers. We can see the difference and automatically have made judgments. Thinking about what is culturally correct is just as important as considering the performance or aesthetics.

Materials used in the built environment help to locate us geographically. Each area has typical building forms made from typical materials this is known as vernacular architecture. These buildings and materials have evolved over time to mitigate climate forces and are highly associated with the sense of place and cultural heritage of an area. Connecting and or acknowledging the cultural and historical aspects of a community are extremely important. Using local materials allows new structures to correspond and fit in with existing structures, making them part of the culture. Because we get so many clues from the materials used in the built environment, we can also influence the culture of an organization with the materials that we choose. If we choose beautiful natural materials, we are conveying a sense of worth to the occupants and organization and can change the way people act towards others. Using natural materials can help connect us to nature and natural cycles and, as we saw in Chapter 6, even people's actions towards each other.

14.14.4 *Experience*

Materials used in the built environment, products, or clothing activate all of our senses, our sight and sense of touch being the greatest.

We also explored in Chapter 6 how the materials used either as finishes, structural elements, or furnishings could cause psychological and physiological changes. According to FP Innovations in their studies regarding the use of wood in medical, office, and school settings, some of the benefits that are associated with exposure to wood are as follows:

Reduced stress levels, measured with cortisol, which is a chemical produced by the adrenal gland and is elevated in response to physical or psychological stress. Long-term elevation of cortisol levels can cause many different health risks from heart problems, to weight gain to insomnia.
Lower blood pressure increases the ability to concentrate, lowers pain perception, or speed recovery times. It actually changes our alpha brain waves improving the ability for focused, creative work.
Melatonin production was found to be higher in a bedroom setting when there was light reflected off its surface. This implies that it would be better for sleeping than other materials.

14.14.5 *How materials + products are related to integral sustainable design*

In the building, clothing, and product industry, everyone claims to be using recycled or sustainable materials. We need to not only investigate the claims in ads and labels; we need to go all the way back to the sourcing of raw materials. We need to then calculate all the transportation, manufacturing processes and the raw materials they use, the delivery and installation or manufacturing to come up with a final product, and then look at how long it will last and what will happen when that usable life ends. The name for this process is Life Cycle Analysis (LCA). LCA measures embodied energy, which are all the things previously mentioned.

14.14.5.1 *Embodied energy |life cycle analysis| carbon emissions*

To fully understand which type of material is the most sustainable in each specific situation, it is vital to look at the whole process of creating each material. The LCA process examines each step of manufacturing from resource extraction through final product delivery including transportation, construction, and building maintenance. It also considers life expectancy, end of life disassembly, reuse or disposal. It gives a fuller understanding of each material or product on the ecosystem and human health (Figure 14.14.5a [below]).

LCA is used to analyze the wide-ranging environmental footprint of a building – including aspects such as energy use, global warming potential, habitat destruction, resource depletion, and toxic emissions. We also need to examine how the community and surrounding ecosystem will be affected aesthetically, physically, and socially even though there may not be facts or numerical figures to put on a spreadsheet.

LCA was invented by Coca-Cola, and it is the first company to use LCA to increase profits by tracking and analyzing every step of production for efficiencies. Today, within the architectural and sustainable design fields, LCA is used to analyze the wide-ranging environmental footprint of a building – including aspects such as energy use, global warming potential, habitat destruction, resource depletion, and toxic emissions. LCA can be performed at the material, product, building, or industry level.

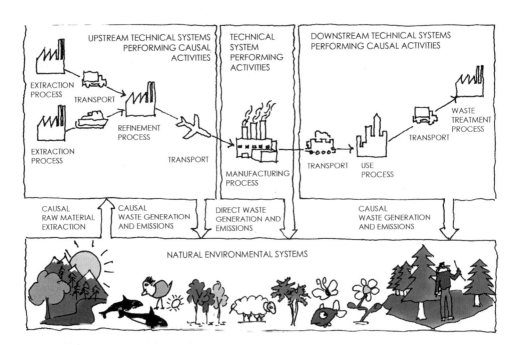

Figure 14.14.5a A graphic representation of life cycle analysis.
Source: Structural special topics, courtesy Routledge.

LCA examines in explicit detail all of the following areas:

Embodied Energy – Energy consumed by the product during all phases of manufacturing and entire life of the product or building. It would include all the processes listed below.

Material Extraction – This includes removing the raw materials from the earth and the energy in transportation to the manufacturing location.

The Ecosystem – The impacts and any reparations required because of material extraction would also fall into this area.

Manufacturing and Formation – The manufacture of finished or intermediate materials, building product fabrication, packaging and distribution of building products all effect the surrounding ecosystem and community and need to be considered as well. Does the facility release warmer water to damage the fish population or change the soil, air or water composition in any way? How is the community affected by this physically and socially?

Construction or Installation – All activities related to the actual building project. Including but not limited to material transportation costs, energy to construct, and maintain the job site, or even fuel for heavy equipment used for site work.

Building Operation and Maintenance – Considers both short- and long-term energy consumption, water usage, environmental waste generation, repair, and replacement of building assemblies and systems, and the transport and equipment use for repair and replacement.

Life Expectancy – Here we look at the typical life of the all the components of a building and take into consideration the durability and typical replacement timeline for each feature. This can help influence our choices towards ones that will have a longer life span even though they may be more expensive at initial installation.

End of Life – This area looks at what happens after products or building features are replaced. Can it be reused in a different function? Can it be disassembled and recycled or does it have to go to a Landfill. This category also looks at the energy used and waste generated during demolition and disposal.

All the energy, raw materials, emissions water, and soil disruption for each step are recorded and combined into a flow chart. This allows us to see the true cost of producing any material, product, or building. Just as important as recording and calculating all the figures is a consideration for any negative impact on the soil, air, and water quality. We also need to examine how the community and surrounding ecosystem has been affected aesthetically, physically, and socially by all the steps above even though there may not be facts or numerical figures to put on a spreadsheet.

LCA is one tool to aid in understanding the complexity of sustainability and shape our design decisions and is best done during the pre-design phase. It should be an integrative analysis and design process and include all the professions involved. Evaluating LCA can have dramatic impacts on the environment as seen in the research SOM Architects did when they analyzed the embodied energy for a typical 60 story concrete residential building to a hybrid building that used wood columns, walls, and floors (SOM 2015, p. 18). Their comparison showed an 80% reduction in carbon coming from reducing the amount of concrete and steel, and considering the carbon sequestering that wood is responsible for.

Other benefits of utilizing LCA, as cited by an American Institute of Architect's study, are:

- Choosing between building design or configuration options
- Choosing between building structural systems, assemblies and products
- Reducing environmental impacts throughout a building's life cycle
- Improving the energy performance of entire buildings
- Mitigating impacts targeted at specific environmental issues

As an emerging designer you can introduce this new process. There are many software packages that now help you track and calculate LCA. Some of the most popular ones are Tally, Athena, GaBi, and Simapro. The NREL U.S. Life Cycle Inventory Database (NREL 2013) is also a great resource for information. Review each software's capabilities compared to your needs to find the best one.

14.14.6 Green product directories

Products, manufacturers, and rating systems are always evolving, so it is best to access the latest version. Below are just a few resources to help with specifying materials for the built environment that reduce embodied energy, toxicity, off gassing, and are manufactured in a responsible way.

Cradle to Cradle™ (C2C – Centre): www.c2c-centre.com/home
Website with a directory of products, companies, and buildings that comply with the C2C Centre rating system and are produced.

Stanford Library: http://library.stanford.edu/guides/green-building-resources
Green Building Resources – Books and Journals on Green Building.

Comprehensive Procurement Guide (CPG) a division of the EPA: www.epa. gov/smm/comprehensive-procurement-guideline-cpg-program

GreenSpec: www.greenspec.co.uk/

International Living Future Institute: https://living-future.org/

Declare. https://living-future.org/declare/
A transparency platform and product database website. Manufactures are encouraged to submit information to the database to aid in the process of achieving the Materials Petal. Each project in the database is given a label that profiles; components including Red List Items or EPA's Contaminants of Concern (COC), End-of-life options; Location of extraction; VOC Information and if the product is LBC compliant with the Red List.

Just. https://living-future.org/just/
A voluntary disclosure tool for organizations or businesses to disclose their operations, including how they treat their employees and where they make financial and community investments. The organization is given a score of 1–3 in six categories related to social justice: Worker Benefits, Safety, Stewardship, Diversity, Equity, and Local Benefits.

Reveal. https://living-future.org/reveal/
A verification of the energy performance of buildings, including three separate metrics: energy use intensity, the zero energy performance index, and reduction in energy use from baseline.

MSDS Online Database: www.msdsonline.com/msds-search/

Important for Living Building Materials Petal Research. The chemical composition of any product or systems component can be searched with the CAS or Chemical Abstracts Service number. A CAS# is a unique identifier assigned to every chemical substance in open scientific literature. All MSDS are required to list the CAS#s for all their components or ingredients.

Oikos Green Building Library: www.eerl.org/index.php?P=FullRecord&ID=1308

Environmental energy resource library

Perkins + Will Material Transparency Website: https://transparency.perkinswill.com/about#usingthelist

Resource for Health Product Declarations (HPDs), database of generic products, and product database using filter for certifications and other "transparency documentation" including HPDs and more. The website also includes case studies for various project types and information on government agencies.

Perkins + Will Precautionary List: https://transparency.perkinswill.com/lists/precautionary-list

Information on materials or products of concern searchable by Product or Project type, CSI specification, or Hazard.

Well Building: https://v2.wellcertified.com/v2.1/en/overview

WELL v2 is founded on the following principles: equitable, global, evidence-based, technically robust, customer-focused, and resilient.

References

Biomimicry Institute, 2016-last update, Ask Nature: Needle-like Structure Inserts Painlessly [Homepage of Biomimicry Institute], [Online]. Available: https://asknature.org/strategy/needle-like-structure-inserts-painlessly/#.W1233dJKiU1 [July 28, 2018].

McDonough, W. and Braungart, M., 2002. *Cradle to Cradle: Remaking the Way We Make Things.* 1st edn. New York: North Point Press.

NREL, November 18, 2013-last update, U.S. Lifecycle Inventory Database [Homepage of National Renewable Energy Laboratory], [Online]. Available: www.nrel.gov/lci/ [July 29, 2018].

Skidmore, Owings and Merrill, Inc. 2015. *Intelligent Densities | Vertical Communities.* London, NLA Breakfast Talk July 2015: NLA London's Center for the Built Environment.

15 Moving sustainable design forward

15.0 Introduction

By now the reader must be anxious to see what's next. After all, the full range of sustainable design principles, strategies, and practices has been introduced. Even with so much covered, the surface has only been scratched, and the conversation has barely started. Sustainable design is a lifelong pursuit, an ever-widening sea-change in thought and aspiration. Maybe, somewhere in one of the chapters, the reader found a foothold, the first rung in a ladder of lifelong learning and definitive action towards a sustainable future. The diagram (Figure 15.0a [below]) reflects the strange nature of change. One can wait for the razor's edge to reach you forcing a painful reaction. Others jump over the razor to become an early adopter or even a pioneer; Either way, change is coming. This last chapter of the book takes an abbreviated look at how change for sustainability is achieved.

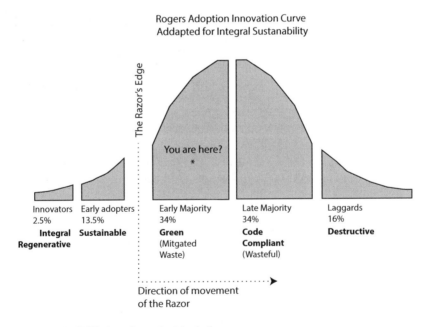

Figure 15.0a Diffusion of sustainable design.
Source: Created and drawn by authors.

15.1 Sustainable design across space and time

It's hard to believe, but we have come full circle. Our ancestors, the hunter-gatherers, had an intimate and connected relationship with the natural world. Today, we strive to rediscover that to find a kinship with the nature. Back then the impulse was intuitive, automatic, a simple expression of the desire to survive. Today, it's different. This time we consciously choose to rejoin nature – electing to shed our old, destructive ways to develop an intentional relationship with nature.

In the past chapters, a historic summary was provided to give a context for each one chapter. In this chapter we look forward towards the dawn of the Age of Integration. We see a glimmer of hope in an otherwise frightening future. Humanity has already begun to find ways to harness self-interest and short-term thinking to the benefit of long-term, empathetic societies. We come together through the common goal of preserving our shared future. All of the case studies and examples shown throughout the book provide the evidence we need to see that a new course has already been charted. It was seeded in the 1960s, the great empathic awakening and the beginning of the slow movement towards social equity and the beginning of a recalibrated relationship with the planet Earth. Today, there are literally hundreds of thousands of people, if not millions, working on behalf of the new worldview, challenging business as usual with the collaborative economy based on trust, accountability, and transparency.

15.2 Regenerating nature

In Chapter 7, scenario planning was explained as the process of imagining possible scenarios and then charting courses of action to address threats or capitalize on opportunities. In the near future human-induced climate change will continue to alter the planet in increasingly profound ways. Even if we were to unilaterally agree to stop emitting greenhouse gases, the arc of global warming and associated climate impacts will continue. The approaches of mitigation, regeneration, and adaption, as covered in Chapter 7, will become more and more the focus of design. Even the more recent focus on health, well-being, and quality of life will give way to more basic considerations for much of the world's population – survivability. Therefore, sustainability remains an elusive goal, a utopian ideal, still suitable to capture the imagination and to organize human behavior around the goals of restoring the ecological health of the planet. Don't worry though, nature will bounce back. It always does.

15.3 Humanity

So here we are, the latest iteration of the human race, now armed with a coherent and holistic set of sustainable values and a powerful set of sustainable design principles. We know what is happening to the planet, and we have the technologies needed to reverse course, but, will we have the courage? Will self-interest continue to widen an ever-growing gap between rich and poor? Will the natural world finally surrender to the great demands and technologies of humans or will it hurl maelstroms of violence upon the earth, ending the reign of humans? While overly dramatic, the situation requires a giant empathic leap, a blossoming of empathy across time, space, and difference. Sustainable design needs to take its rightful place at the center of all human activity, helping to create a possible or even probable sustainable future.

You are here on earth at this time, reading this book for a reason. Now is the time to consider your role in the ongoing drama of human evolution. Will you begin a new path? In Figure 15.0a (preceding page), a series of positions on sustainable design are shown. Where do you fall? Are you a "Laggard" incapable or unwilling to change? Are you a code compliant designer, happily taking a paycheck while slowly contributing to the destruction of the planet? Are you a green designer, making inroads, and seeing some early results…or will you jump over the *razor's edge* to become a sustainable designer in the broadest sense? Hopefully, this book has helped to move you along. The rest of this chapter will show some pathways forward to begin to understand how to advocate for change and fight for the future you want to see.

Expanded Maslow's hierarchy

Maslow intended his hierarchy of needs (Chapter 4) to relate to individuals and their needs. Given that we are heading to an age with more collaboration, Richard Barrett has developed a second set of hierarchies that define the needs of a group. This is extremely helpful in supporting the pursuit of sustainability in an organization. The upper levels of the Pyramid shown in Figure 15.3a (below) are discussed below:

- Level 4: **Transformation**
  - Those who have reached the Transformation stage in development are more likely to empower others to do the same thereby creating a more equitable workplace.
- Level 5: **Shared Values**
  - The development of Shared Values is critical to successful team dynamics. A shared purpose helps to define useful activities and eliminate those that are contrary to the shared vision.

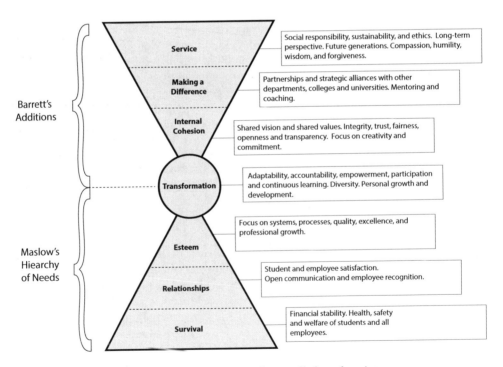

Figure 15.3a Barrett's Full-Spectrum Consciousness® as applied to education.
Source: Barrett Culture Center.

- Level 6: **Making a Difference**
  - Partnerships across vertical and horizontal value chains are critical to achieving sustainability. Suppliers, manufacturers, retailers, and customers constitute a vertical value chain. Partnering with competing companies to solve common environmental problems is an example of working across a horizontal value chain.
- Level 7: **Service**
  - If all of those conditions are in place, the pursuit of SERVICE in the form of sustainability is more possible, thereby meeting the needs of society as a whole today and into the future.

In Table 15.3a (below), there are three scales of focus: Individual, Organization, and Community/Society and the seven levels are represented and also notice the terms "Self Interest" and "Common Good." As a sustainable designer, the values in all of the cells need to be reached across all scales in order to achieve a truly holistic and lasting level of sustainability.

The process of change itself is not trivial. The emergence of sustainability has generated the need to transform organizations seeking either to comply with new regulations or take advantage of new business opportunities. Organizations change and evolve from outside pressures/opportunities, but also from changes in the values and beliefs of the individuals in the company. If a leader of a company changes the way they think, their actions and behaviors will eventually follow and will have an impact on the entire organization, creating a change in culture. Then, the company will begin to change its collective actions in the form of new policies, procedures, production, and marketing.

Starting small

For many, the enormity of the issues included in the sustainability challenge seems overwhelming, often leading to rejection, apathy, and inaction. It is critical that we begin to

Table 15.3a Barrett's full spectrum consciousness across scale

Barrett's hierarchy	Scale of activity			
	Personal	*Organizational*	*Community*	
Service	Selfless service	Service to humanity	Global sustainability	Common good
Making a difference	Making a positive difference in the world	Internal/external collaboration	Strategic alliances	
Internal cohesion	Finding meaning	Sense of progress	Strong cohesive culture	
Transformation	Letting go of fears	Ongoing improvement	Democratic processes	Self-interest
Self-esteem	Feeling a sense of self-worth	High-performance systems and processes	Institutional effectiveness	
Relationship	Feeling protected and loved	Positive relationships	Sense of belonging	
Survival	Satisfying basic needs	Financial viability	Economic stability	

Source: Barrett Cultural Center.

transform the concept of sustainability from an *"issue"* into an *"opportunity"* – an opportunity to generate wealth and to serve the common good. In this way, entrepreneurship becomes a vehicle for positive change and for profit.

At the same time, we are beginning to realize the limits of what one person can achieve. By teaming with other like-minded people who share the same values, we can begin to attack different aspects of the sustainability opportunity. If the organizations we create have alignment between stated values and actions, the chances for success will be greater. Ultimately, if enough groups chip away at enough parts of the sustainability opportunity it will eventually disappear and become business as usual. And that is exactly what is happening, albeit very slowly.

Finally, there has been great progress in the last 20 years towards a sustainable future, and there is a good reason to believe that more progress is coming, even if we have to take major steps backward once in a while to go forward.

You will absolutely be surrounded by those who either don't care, don't know, or who want to stop your efforts. It's sobering to imagine that the world has so much potential but that many people have not yet opened their eyes, or worse yet, choose not to see, lest they are jolted from the comfortable confines of their current worldview.

We have provided some tools for your use within your organizations and communities. Remember, it's a marathon, not a sprint. Results will vary and will be measured in increments of years, not months or minutes.

15.4 Sustainable values and the sustainable future

The development of the sustainable values of economic prosperity, social equity, ecological regeneration, and beauty will continue to find their way into hearts and minds across populations. New and powerful communication technologies powered by solar and wind power will not only move society towards a low-carbon existence, but will actually begin to usher in a new consciousness – a new worldview. Even when things seem bleak, evidence of a new worldview based on integration is all around us, you just have to know where to look.

For example, even the most profitable corporations are making inroads towards a sustainable future. The Global 100 released by the Corporate Knights evaluates companies based on 17 sustainability metrics and lists the top 100 in the world – which hail form 22 different countries. Here are some remarkable facts about the Global 100, according to the Corporate Knights program:

- They paid an average of 27% more taxes
- They had three times as many top female executives
- They generated six times more clean revenue than their global peers.

The Global 100 companies since 2005 have outperformed benchmarks by "close to a third" with a cumulative return on investment of 163% by 2017 (Cotterman 2018). The upshot is simple: discovering and developing new ways of doing business that incorporate sustainable values will often win out, providing not only financial return but the all-important expression of empathy at large scales and across time.

15.5 Integral Sustainable Design

Applying the sustainability values through Integral Sustainable Design on all levels, on all projects, at all the times, and at all scales is a game changer. This is a holistic way of thinking and doing that is necessary for reaching higher levels of project performance, social equity, ecological restoration, and aesthetic excellence. Integral Sustainable Design offers an important "mental map" that is easy to understand and supports a holistic approach to solving complex sustainability opportunities.

The consideration of performance, systems, culture, and experience combats the impulse of design team members to focus solely on their own areas of expertise or concern. For example, the Experience Perspective allows designers to engage with engineers to become partners in pursuing beauty and meeting the deeper psychological needs of the design process. The Culture Perspective includes occupants and the community to expand designers, understanding of the role the project plays in supporting the social structure of the organization or neighborhood.

The Systems Perspective allows engineers and scientists to collaboratively engage designers in the evaluation of critical ecological and infrastructure elements. The Performance Perspective keeps the entire team focused on the efficient use of energy, environmental and financial resources, while still meeting the needs for light, air, and views. All of these collaborations lead to the possibility of reaching a higher, more holistic level of sustainability, engaging all stakeholders, and helping those resistant, to change and to understand the advantages.

15.6 Bio-inspired design

Throughout the book, the case has been made for the use of bioinspired design as an integral aspect of effective sustainable design. The integration of biophilic principles is becoming more and more common because of the deeper psychological and physical benefits along with the pure aesthetic desire for more plants and gardens.

Biomimicry requires that we truly analyze nature and natural systems and then apply the base concepts to our complex problems. The solutions that result are innovative and allow us to break the shackles of the industrial revolution and approach meeting our needs by using local, abundant resources that don't require the brute force of high temperatures or toxic chemicals. We are learning ways to work with nature instead of against it.

15.7 Resilience

From mitigation to resilience and adaption

This transition, a primary focus for sustainable design, is not something to be taken lightly. It is a critical shift in mind-set, technological applications, and design approach. Thinking of the built environment as a viable asset for 100–200 years or more is something very new in the modern architectural profession. Resilient buildings and infrastructure will still need to exhibit the energy and resource efficiency typically associated with sustainable design but will need to go further. Design directives should now be considering the power outages, extreme weather, natural disasters, and other social and economic upheavals that are becoming more a part of our world. This may not be pleasant but is necessary to achieve a truly resilient project.

The addition of battery backup, on-site water purification and food production, or passive and living systems that function without generated power sources makes resilient design more challenging.

15.8 Health and well-being

From design as an act of creation to design as an act of healing

The rise of health and well-being as a major driver of sustainable design is a welcome addition. Corporations, organizations, and especially health-care facilities are making conscious efforts to include features that improve human health. We are finding that with some slight changes and minor additions to the typical design directives, long-term human health, productivity, quality of life, engagement, and social structure can all be improved. In Chapters 6 and 8, we looked at how the design community can move from doing "less bad" to doing "more good." Landscapes, buildings, and interiors are opportunities for phys-ical and physiological healing. In this way, design becomes a *restorative practice* focused on a broader and deeper set of design considerations.

15.9 Cocreativity with nature and coevolution

The need for transdisciplinary collaboration should be obvious by now. Sustainable design is a tremendously complex and difficult process, but it does get easier over time. Building the right project team with aligned goals is critical for success. Alignment of mind-set early in the process along with reinforcing loops means that sustainable designers are as much educators and motivational speakers as they are creative professionals.

The need to bring all the stakeholders to the table, including nature, is even more crucial. Without a healthy natural world, all of our ecosystem goods and services are threatened, making it difficult for survival let alone the economic and social progress that many would like to see. In other words, a healthy environment is a prerequisite to any human activity.

Design as service

Bringing everyone to the table as an equal and valued team member means checking egos at the door and focusing on solving the problem holistically. This means that all professions, community members, employees, maintenance people, contractors, and designers are all included early in the process.

15.10 Global sustainable design

The United Nations, though the creation of the Sustainable Development Goals (SDGs), has already built a powerful foundation to achieve long-term environmental health, while enabling the initiatives to improve human health and build a more equitable society. It is the hope that the SDGs will continue to build momentum and be adopted by more and more government agencies. As government agencies require compliance from the private sector, the SDGs will influence larger sections of society.

This is one of the classic ways by which innovation occurs. New requirements, stand-ards, and regulations force design projects to evolve. In the U.S., the government is leading by example and requiring all new construction to meet the United States Green Building Council's (USGBC) Leadership Energy and Environmental Design (LEED) standards. The result is that a large number of professionals are designing to meet these standards in gov-ernment projects. Experience with the new rating systems paves the way for the principles to be applied in their other projects as well. There are still design firms that do not want to have members of their team educated in sustainable design practices, but the design disciplines are slowly changing towards sustainable practices.

15.11 Sustainable cities

Cities will continue to be placed under stress as climate change continues to affect established weather patterns and sea levels. Migration to urban centers is predicted to continue as millions of people look for work, financial stability, and culture. The relationship between weather pattern change, migration, urban population rise, and civil unrest is an emerging and a significant trend taking place in the world. Urban areas will need to not only cope with the stress of increasing populations but to discover ways for people thrive in dense populations without losing their connections to nature while promoting culture and social equity. This is a key aspect of Integral Sustainable Design.

Smart cities where technology is used to improve the flow of traffic, monitor air quality, and optimize energy will be a critical part of the sustainable city. Biophilic cities will help urban areas be more beautiful, more ecologically integrated, and more socially equitable. Resilient cities will allow extreme weather events to be less damaging and respect the embodied energy and resources used to create them. These are tall challenges and will require the kind of societal shifts in a mind-set consistent with the new worldview of integration. Examples of the sustainable city can be found from the soon to be carbon neutral Copenhagen, to the technology-enhanced Masdar City, and to the resilient redesign of New York City's Manhattan Island.

15.12 Districts and sites

Perhaps the most impactful scale of sustainable design can be found at the district and site scales. The difficult prospect of designing net zero energy buildings becomes much easier when the entire site and neighborhood is considered. Access to more land, more rooftops, and more infrastructure offers the best chance to pursue ambitious sustainability projects. Consideration of the district scale allows for shared collection of water and sun and shared food production. These are powerful strategies for sustainable design, but they require great leaps of thinking, cooperation, and technical innovation. Since ecodistricts are made up of diverse communities with varying goals and worldviews, the challenges are very real. Ecovillages offer a hopeful pathway as these are intentional communities that come with all the benefits of ecodistricts but more importantly offer alignment in worldview, probably the most important asset in building a sustainable community. From a resilience point of view, ecovillages can function as autonomous communities off-grid which means they can continue to function in difficult times.

15.13 Building and interiors scale

What does a brick need to be?

Louis I. Kahn dramatically and poetically asked the question, "What does a brick want to be?" This statement underscores everything that is wrong and right about architecture. For the 21st century we change the question, "What does a brick want, *and need*, to be?" The brick wants to be beautiful, connecting to the earth in the most permanent ways, and reaching to the sky igniting the imagination and fueling spirit. But the brick also *needs* to be locally sourced, made from a zero carbon processes, built by workers making a fair wage, and reused at end of life. In Integral Sustainable Design, the wants of the client *and* needs of society, together, create beautiful *and* meaningful projects. For too long the questions: How much more will it cost? "What do we get besides a plaque with LEED?" "Isn't sustainable design just part of good design?" have dominated discussions about sustainable design.

"Eyes which do not see"...... again?

(Braham 1995)

All of these questions reflect a certain kind of blindness, mostly caused by the eternal pull of self-interest, resistance to change, and the fear of the unknown. Le Corbusier lamented his frustrations regarding the lack of progress towards the widespread adoption of modernist principles and expressions by mainstream society. Today, the same frustration exists with the movement towards sustainable design. Just as Le Corbusier saw the obvious next step of modernism, so too does the next step of sustainable design stand right in front of us, daring us to move forward to an eye opening new design consciousness. And as each person makes the leap, they come together to form groups of people who share the same visions and values. If this book has illuminated anything, it showed the massive groundswell of activity and effort around sustainability. In design schools, students wait for their professors to assign a sustainable project, allowing the rest of the work to find its way into the usual business of "wants" based personal and idiosyncratic expressions of built from. So too, professional architects wait by the phone, hoping for the call from a client wanting sustainability, when it was right there all along beckoning from the corner underneath the long buried green spec that would achieve LEED silver levels of sustainability without the client ever knowing. And finally, when will architects, like doctors, assume their Hippocratic Oath to "do no harm" and better yet, to "regenerate" the planet's resources through ambitious sustainable design practices? At the printing of this book, 21 Living Building Challenge projects have been completed with hundreds more on the boards. Much more is needed.

Technological advances within the architectural profession are changing daily, moving us closer towards truly sustainable buildings. Adopting the long view of the structures we design and materials we use allows us to expand our view of their long-term effects and possibilities. John Ruskin (2016), a writer, social thinker, and professor at Oxford University in the late 1800s says it best:

> When we build, let us think that we build forever.
>   Let it not be for present delight nor
>   for present use alone.
>   Let it be such work that our descendants will thank us for...
>   and that men [or women] will say,
>   as they look upon the labor and wrought substance of them, "See! This our fathers did for us."

We need to view the buildings and the structures we are designing today as a gift to future generations and to the world at large. Our current society will be judged by the structures we leave for those who follow us. The choice is ours to carefully and holistically analyze the multiple options for every aspect of the buildings that we design to create a regenerative, truly sustainable structure with the potential to last hundreds of years – beautifully enhancing the community it serves.

15.15 The last word: tiny revolutions

Finally, we have reached the end of this book and the beginning of your journey, the first small steps towards a different future. There is no doubt that the prospect of transforming your mindset to pursue sustainable design is daunting, but there is no turning back now. Once you've seen it, it's hard to unsee. Bill Mollison, creator of permaculture, stated:

> The greatest change we need to make is from consumption to production, even if only 10% of us do this, there is enough for everyone. Hence the futility of revolutionaries who have no gardens, who depend on the very system they attack, and who produce words and bullets, not food and shelter.

Ultimately, you can't do it all at once. The problems are too large and many partnerships are needed to see the kind of change the planet needs. Instead of becoming paralyzed by the thought of the massive change, try and make a small change each day. Maybe it's a change to a design project, or a good deed, or the confrontation of a person exhibiting bigotry. Maybe it's going to a meeting to talk about protecting a local watershed, or maybe you take another course on sustainable design. All of these, and many more types of actions, are "Tiny Revolutions," small acts of bravery, that, in themselves, don't add up to much, but when multiplied by millions of other tiny revolutions, over time, will lead to overwhelming positive change. The potential for massive positive change then comes into focus. You are a microcosm of the world and your individual actions matter, and perhaps more importantly, your inactions matter even more. So, as you close this book, it's time to get started. What will be your first tiny revolution?

References

Braham, W., 1995. *Eyes That Do Not See? The Practice of Sustainable Architecture*. University of Pennsylvania, http://repository.upenn.edu/arch_papers/27 edn.

Cotterman, R., 2018-last update, The Sustainability Paradox: Why Business Leaders Need to Evolve Their Approach [Homepage of Sealed Air], [Online]. Available: https://sealedair.com/insights/sustainability-paradox?gclid=CjwKCAjwma3ZBRBwEiwA-CsblHD4n5K08OcCmKHuZxn-6RqE5A8XxICcnjdJVsomu6HiZFpGT-hYzbhoCexgQAvD_BwE [July 1, 2018].

Ruskin, J., 2016. *The Seven Lamps of Architecture*. Norwich, CT: CreateSpace Independent Publishing Platform.

Index

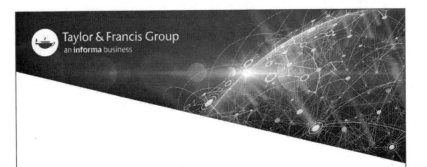